# An Incomplete Theory

## The Search for Quantum Gravity

(a story)

MEGAN HENRY

Second Edition, First Printing, April 2023

Copyright 2021 by Megan Henry

Published in 2023 by Megan Henry

Edited by Lucia Gordon
Original book design by Jessi Nassif
Illustrations by Emily Henry


ISBN: 979-8-9881084-0-5

# CONTENTS

# Book III: Frontiers of Physics

# Book IV: Mach's Principle

# Book V: Dead Ends and Loose Ends

## Book VI: The Search for a Conclusion

# PREFACE

Megan Henry gives us a delightful reading of philosophical ideas behind physics — Einstein's theory of general relativity mainly among them. The unifying theme of the book is "Mach's principle", a principle that is natural to most physicists. As Dr. Henry describes it: "theory should be based in the measurement of observables, efficiently expressing relationships among phenomena." It is well known that such a theory puts all observers on the same footing, nothing is privileged. This in principle suggests the theory incorporates Mach's ideas. But does it? This book is the history of this "Incomplete Theory".

Written with excellent teaching skills, the author starts giving us the history of various concepts in physics starting from the ancient Greeks and its passage through the Arab world into the Middle Ages, reaching Copernicus, Kepler, Galileo and Newton, Faraday, Maxwell, Einstein (of course) and key figures of quantum mechanics. She includes many interesting historical notes that make a lively prose easy to read, even when touching on deep subjects. The chapters on frontiers of physics, with the history of the Japanese school of nuclear physics of Yukawa and Nishina, the issue of infinities in quantum field theory, the quest for a theory of everything, cosmology, the Big Bang, dark matter, dark energy and the detection of gravitational waves bring us to modern times, showing how it's built on shoulders of giants.

Megan tells us how she started her own physics reading passion and curiosity, and tying up dead ends and loose ends, poses a challenge for physicists and philosophers: is general relativity compatible with Mach's principle? She says no, and this could be a clue for advancing the theory — but the theory is still incomplete, as we learn when we finish the book.

Jorge Pullin and Gabriela González, Baton Rouge,

— June 14, 2021

# Introduction

Ok, I'm starting over. I've made a few attempts at writing a short "professional paper," but that kind of writing is just not me. And what's the point anyway? Galileo did not write an academic paper readable to only a few, so why should I? Galileo wrote to the people, to convince them that the Earth moves. This seems like such a small thing when I write it, but it was not. Galileo looked through a telescope and showed us that there is imperfection in the heavens. He imagined away resistance and found perfection on the Earth, in the movement of a falling body. Galileo destroyed Aristotle's dichotomy between the perfect, unchanging, everlasting heavens and the imperfect, ever-changing Earth. In his quest to convince the world that the Earth moves, he showed us that the universe is one, that it is knowable.

In this way, Galileo set the stage for Newton, who put the pieces together to develop universal laws of motion and gravity that are still taught today and good enough to navigate a rocket ship to the Moon. And with his principle of relativity, simply stated that we can't detect constant motion, Galileo developed one of the pillars of physics that has survived the many revolutions in modern science. And it all began with his attempt to convince the world that the Earth moves, which wasn't even his idea.

I do not claim to have anything as revolutionary as that, but I too have a quest. I believe that all motion is relational, that one cannot define motion with respect to space but only with respect to other objects. This is not my idea but rather began as a critique to Newton's absolute space practically as soon as he introduced it. The idea gained greater articulation with Ernst Mach in the late 1800s and became coined as "Mach's Principle" in the early 1900s by Einstein, who used this principle as inspiration for the development of general relativity.

Although Einstein believed on philosophical and observational grounds, like many physicists, that Mach's principle was correct, he found that general relativity was not consistent with it. His efforts to modify the theory to make it compatible did not work. Similar attempts by other physicists have likewise failed. I believe, and will try to convince you, the reader, that any such attempt must fail. And I believe that this inherent incompatibility with Mach's principle is fatal to general relativity regarding its application to cosmological questions such as the beginning and end of the universe. Furthermore, I believe that if we don't abandon general relativity as a starting point for a theory of gravity, we will never make any progress toward a theory of gravity compatible with quantum mechanics.

Is this important? Well, maybe not if all we care about is getting to the Moon. But it certainly is interesting...

# BOOK I

---

# The
# Story
# of
# Gravity

---

# 1

# Inspiration From a Book

It is a pleasure to stand upon the shore,
and to see ships tossed upon the sea;
a pleasure to stand in the window of a castle,
and to see a battle and the adventures thereof below:
but no pleasure is comparable to the standing upon
the vantage ground of Truth...
and to see the errors, and wanderings,
and mists, and tempests, in the vale below.

— Francis Bacon

Years ago, I bought a book, *Principles of Cosmology and Gravitation* by Michael Berry. It had been recommended to me by an astronomy professor after I explained my theory that matter and antimatter must repel each other. It's just the opposite of electrical charges: like masses attract and opposites repel — why not? Since light from an antimatter star is identical to light from a matter star, we don't actually know what we're looking at, and my idea avoids clumsy explanations of why we haven't discovered substantial amounts of antimatter. After the universe settled down a bit after the big bang, matter and antimatter each formed their own little pockets of stability based on which was dominant and repelled each other. Simple! At least at the time that I researched this problem, no one had experimentally shown that matter and antimatter attract. And every once in a while, we observe some unexplained gamma ray burst. Couldn't that be a matter star colliding with an antimatter star?

I eventually debunked my own theory. Since photons are their own antiparticle and they bend in the presence of a large mass, they must also bend in the presence of a large antimatter mass. If photons attract both matter and antimatter, matter and antimatter must attract each other. So much for my idea. But I still had the book.

In Berry's book, I was first introduced to Mach's principle. Mach's principle has a complex history, but essentially it is the belief that all motion is relational and can only be determined by and described with respect to other objects. To see how this differs from classical physics, imagine that you are outside on a clear night. When you look up, you see many stars. Now imagine that while looking at the stars you turn around in a circle several times. You will see the stars moving and probably feel a little dizzy. According to classical physics these two things are a coincidence; we feel dizzy when we spin because we're accelerating with respect to "fixed space." On average the stars happen to be more or less unaccelerated with respect to fixed space, which is why we feel dizzy when we see the stars move in circles. Mach's principle, rather, implies that acceleration with respect to the stars and all the other matter in the universe *causes* the dizziness.

Biologically, dizziness is quite complicated. However, the physical basis for the dizziness in this example is a fairly straightforward application of what is called an inertial force. Inertial forces arise in classical mechanics when considering motion from the perspective of an accelerating observer. When we accelerate in one direction, we feel pushed in the opposite direction. For example, when we're on a train that stops quickly, we lunge forward. If we stand on a scale in an elevator accelerating upward, it will read a value greater than our weight as if we were being pushed down. And when we spin in a circle, our inward acceleration causes the fluid in our inner ear to be pushed outward, which then results in a swishing that our brain interprets as dizziness.

Classical mechanics prefers to view motion from unaccelerated reference frames, called inertial frames. There are many ways to define an inertial frame, but my favorite is any reference frame in which we don't feel motion. If you're on a plane with your eyes closed, most of the time it feels exactly like you are at rest. This is to a very good approximation a classical inertial frame. When on a train, we can always feel some small vibrations even when moving at constant velocity, but this can still be considered an inertial frame for most purposes if you can hold a mug filled with coffee without the coffee spilling. When we're standing still on the surface of the Earth it truly feels like we're not moving. And yet if a pendulum is suspended in such a way that it can rotate freely, famously called Foucault's pendulum, the plane of its swing remains constant with respect to the stars, *not* the Earth's surface. As the Earth rotates with respect to the stars, the plane of the pendulum's swing precesses with respect to the Earth. Therefore, the Earth is not actually an inertial reference frame, but we usually treat it as one because we can't feel its motion. The Earth orbits around the Sun, the Sun orbits around the center of the Milky Way and so on, so any real reference frame we deal with is only an inertial frame in some approximation.

When considering motion in an inertial frame, the net force on an object equals its mass times acceleration. Here, acceleration is defined with respect to the inertial frame. When motion is considered in the object's own reference frame, the acceleration term disappears and a new force is introduced called the inertial force that is in the opposite direction as the acceleration. The inertial force balances out the other forces; the net force on an object in its own frame must be zero. In classical mechanics, the inertial force is a "fictitious" force that appears when considering motion in a non-inertial reference frame. From the perspective of Mach's principle, the inertial force is a real force caused by an object's acceleration with respect to other masses.

If the inertial force is a real force, we need an equation to describe it. The equation must be consistent with the classical expression for the inertial force after adding up contributions from all the masses in the universe. In addition, the inertial force between any two objects must be symmetric in their masses. This is a direct result of Newton's 3rd Law, which states that the force of one object acting on a second object is equal and opposite to the force of the second object acting on the first. It is also common sense, for looking at the situation from either object's perspective should yield the same result. Furthermore, the equation must be consistent with the observation that inertial effects act in reference to distant stars.

An equation for the inertial force between two masses that satisfies the criteria outlined above was first suggested by Dennis Sciama in the early 1950s. As required by Mach's principle, the equation only depends on relative quantities; there is no reference at all to space or any coordinate system. With this equation, he estimated the density of the universe, obtaining a value that is about twenty-five times larger than the density of observable matter. This is at least in the ballpark of what we would expect. The fact is, we don't know the density of the universe.

Many aspects of this discussion intrigued me, first, that Mach's principle embodied in a mathematical theory allows us to calculate the density of the universe. This seemed and still seems hugely important because of all the evidence that most of the matter in our universe is non-radiating "dark" matter. Furthermore, the fate of our universe depends on its density. Will the universe expand and cool forever, creating what is known as the "Big Freeze"? Or will it eventually collapse on itself, creating the "Big Crunch"? Or will the universe reach a stable equilibrium? Current theory tells us that the answer to these questions depends on whether the density of the universe is smaller, greater, or equal to some critical density that is about ten times larger than the density of the observable universe. And current theory has no good way of performing this calculation.

I was also very interested in the equation Sciama used to approximate the inertial force, called the Law of Inertial Induction. He took this equation straight out of electromagnetism, substituting charge for mass and making appropriate changes in constants. This suggests a deep connection between gravity and electromagnetism that Sciama discussed in his paper, "On the Origin of Inertia." If we can write the laws of gravity as a set of equations similar to those in electromagnetism, as Sciama suggested, this leads to the very specific prediction that gravity must be mediated by detectable gravitational particle-waves consistent with quantum mechanics. General relativity in its present form is known to be incompatible with quantum theory. Couldn't a theory of gravity modeled after electromagnetism have the side effect of resolving this incompatibility? It is interesting to note that electromagnetism is consistent with Mach's principle and was also the first force successfully incorporated into quantum theory.

The calculation of the density of the universe in Michael Berry's book, which he introduced almost apologetically as "not a proper general-relativistic treatment," spurred a flurry of reading into the subject. I read about Mach's principle, theories of inertia, the relationship between gravity and electromagnetism, experimental evidence for Mach's principle, experimental evidence for general relativity, the many failed attempts to make general relativity compatible with Mach's principle, and the incompatibility between general relativity and quantum theory.

It is easy at any point in history to believe that all the great discoveries have already been made, that all the great theories have been formed. But here I had before me one of the greatest problems: to figure out gravity and how it fits in with everything else. So I bought myself the book *Gravitation* by Misner, Thorne, and Wheeler and began teaching myself general relativity, for in order to challenge a theory, I must first understand it.

# 2

# Aristotle

For after he had already crossed into Asia, and when he learned that certain treatises on these recondite matters had been published books by Aristotle, he wrote him a letter on behalf of philosophy, and put it in plain language.
And this is a copy of the letter. "Alexander, to Aristotle, greeting. Thou hast not done well to publish thy acroamatic doctrines; for in what shall I surpass other men if those doctrines wherein I have been trained are to be all men's common property? But I had rather excel in my acquaintance with the best things than in my power. Farewell."

— Plutarch

Aristotle was born in 384 BCE in a small Greek city called Stagira. His father died when he was young, leaving him with some wealth. After running through this money, he left to become a soldier. When this did not work out, he fell to selling drugs. At the age of seventeen or eighteen, Aristotle entered Plato's academy in Athens. It is not known why he chose this school among the other schools in Athens. It may be because his guardian was a friend of Plato. Another possible reason is that he had read some of Plato's dialogues and was attracted to his philosophy.

Here at Plato's academy, Aristotle found a home. He sat in on lectures and proved to be a good student. He spent large amounts of time by himself reading, often preferring this quiet study to public discussions. He studied and taught courses in dialectic methods. He conducted research in astronomy and the natural sciences. And he collected proverbs, justifying this pursuit by claiming that although no one reaches the whole truth, no one entirely misses it. This collection constituted part of his research in science as well as other areas; he searched for grains of truth within commonly held beliefs.

Aristotle remained at the academy for about twenty years, leaving shortly after Plato's death in 347 BCE. He spent the next twelve years in various places throughout the Mediterranean, first at the court of the tyrant Hermias in Asia Minor and later in Macedonia where he tutored the king's son, Alexander. After the king was murdered and Alexander became king, Aristotle remained in Macedonia for some years as a citizen. Not much else is known about this period of his life except that by the time he returned to Athens in 335 BCE, he was ready to open his own school.

Aristotle's school, at least primarily, was outdoors. Aristotle was not a citizen of Athens and could not, therefore, own property there. Ancient writers refer to this school as the Lyceum school, indicating that classes were held in or near the Lyceum, one of the public gardens in Athens. Alternatively, it was called the Peripatetic school, indicating that classes may have been held outside on a promenade or while strolling.

There are no external accounts about the structure and content of Aristotle's school. He wrote no personal accounts, and ancient writers, who enjoyed poking fun at Plato and other popular philosophers, remained silent about Aristotle and his school. All of the ancient writings about Aristotle are "fictionalized biographies" in which most of the details appear to be made up. However, we know a great deal about Aristotle's school through his own writings, which were, with a few exceptions, lecture notes for the many courses he taught, revised throughout the years based on discussions at his school.

Through these writings, we learn of the purpose of Aristotle's school — to create a place where people could come together to share in a life of philosophy. It was a life chosen not for money or power but rather as a way of realizing one's unique potential as a human being, the best route toward happiness. In these respects, it was similar to the ideals of Plato's academy. At Plato's academy, however, the discussion of ideas was centered on face to face interaction. Arguments were held in public and won or lost, often based on emotional appeal. A goal of teaching was to educate rulers and those seeking political power. For these students it was most important, whether right or wrong, to be convincing.

Aristotle, rather, valued the written word. He was one of the first to collect books and create a library, and he believed that it was possible to set up an argument in a book without ever having a face to face conversation. He also valued rational argument over emotional persuasion. Moreover, he believed that it was important to have times of quiet reflection in order to absorb and compare different ideas. The goal of this kind of contemplative life was to find truth rather than to win an argument.

However, Aristotle was not proposing a solitary life, for he also valued friendship. He envisioned a life shared with friends collaborating, studying together, and rejoicing together. He coined a new term for this kind of life, *symphilosophein*, which means "to do philosophy together." This was the purpose of his school, to create a setting for such a life.

Aristotle taught courses on practically every subject imaginable. He founded the subject of logic and established the validity of scientific inquiry with the presupposition that in nature, nothing acts on or is acted upon by any other thing at random. Everything occurs because of a cause, and it is the goal of scientific inquiry to determine that cause. Even events that appear to be random only appear so because of the innumerable causes that are linked to the event. In this way, he claimed that events appear random solely because of the fact that it is impossible to keep track of all the causes in these situations.

Furthermore, Aristotle established the scope of scientific inquiry. Specifically, he included astronomy along with plants, animals, and simple bodies under the heading of physics, the study of the physical world. He contrasted all of these with mathematics, which can exist even apart from any physical phenomenon. This constituted a great and important deviation from the philosophy that he had been taught at Plato's academy.

Plato believed that the world of the mind was separate and superior to the physical world. He suggested, therefore, that the most important academic subjects are the most abstract, those pertaining to the things that can exist without physical representation. He first identified arithmetic and geometry as such subjects; it is possible to conceive of numbers and shapes even if these objects don't exist in the physical world. He also included astronomy, claiming that the beauty and perfection of the heavens draws a person into the world of the mind. Moreover, he suggested that the best way to observe the heavens is *through* the mind, writing that we should "let the heavens alone" if we are to approach the subject in the right way.

Aristotle disagreed. He grouped astronomy with the other physical sciences and separated these from mathematics. And he specifically affirmed the senses as a way of learning; information gained by the senses combined with memory yields experience, and those with experience succeed better than those who have theory alone.

In fact, throughout his writings Aristotle was critical of those who ignored observation as an important way of assessing a physical theory. For example, he spoke of thinkers who claimed that the heavens are moving so fast that they must make a great sound. So why don't we hear it? Aristotle's answer: we don't hear it because it is not there. Similarly, he found fault with Pythagorean's fire-centered model of the universe, which assumed a counter-Earth that is always 180° out of phase with the Earth and therefore never visible. Aristotle objected, claiming that they are not looking for causes to account for observed facts but rather forcing observations to fit into their own theories.

Furthermore, Aristotle rejected the idea implicit in Plato's writing that what is unchanging is superior to what changes. Nature, according to Aristotle, is "a principle of motion and change" and worthy of studying. Nevertheless, Aristotle agreed with Plato on a very important point, that reason is a more reliable way of knowing than direct observation or experimentation. Although we can gain knowledge of the world through the senses, he believed that sense perception is "common to all, and therefore easy and no mark of Wisdom." Furthermore, he believed that the most exact sciences are those involving first principles, and first principles are most readily grasped with the mind.

Aristotle applied his philosophy of nature to the universe and everything in it and sought connectedness in all he considered. For example, he suggested that a theory of elements should not just propose a set of elements randomly as other ancient thinkers had done, but it should explain *causes*. Although he agreed with many that earth, water, air, and fire combine to create all of the matter found on Earth, he believed that these cannot be elementary since they turn into each other. He, therefore, suggested that they are made out of elementary qualities that themselves are not matter but become matter when joined together.

Aristotle also agreed with other ancient thinkers that the first principles must be contraries. He suggested that there are two pairs of contrary qualities: hot and cold, and dry and moist, which couple together in four different ways to create earth, water, air, and fire. For example, cold and dry combine to create earth. Fire is made up of hot and dry and can be generated from earth by changing the cold into hot. So if earth is heated up through friction, it becomes fire. In this way, Aristotle not only gave an answer to the question of *what* regarding the elements, but he also attempted to explain causes, why and how certain substances transform into other substances.

Furthermore, Aristotle considered the motion of elements, first dividing all motion into natural or violent. Natural motion is motion whose cause is internal to the object, while violent motion is the result of a push or a pull. The natural motion of each element depends on its heaviness. Earth is absolutely heavy, water is relatively heavy, air is relatively light, and fire is absolutely light. Heavy things naturally fall toward the center until they come to rest below, and light things fall away from the center until they come to rest above. The "center" refers both to the center of the Earth and the center of the universe, as these two centers coincide.

An object moving in circular motion has a different simple motion than any of the four earthly elements; hence, it must be made of a different element. This led Aristotle to the conclusion that the heavens must be made of a fifth element. Moreover, since circular motion has no contrary, unlike upward and downward motion, the fifth element is not made of contraries. It is entirely different from earth, water, air, and fire in every way.

Using his theory of elements and natural motion, Aristotle began to develop a theory of the universe. At its center was the center of the Earth. Heavy matter falls toward the center of the universe until it comes to rest. The Earth must be, therefore, at rest at the center. This conclusion was also well grounded in sense experience. If the Earth were not at the center, objects would not everywhere fall directly down. And if the Earth moved, we would feel it. They didn't feel it, and so it must be stationary.

The Earth must not only be stationary, but also spherical, since matter falling uniformly from all sides would naturally form a sphere. That the Earth is spherical is corroborated by observation, for we see different stars at different locations, showing that its surface must be curved. Furthermore, the Earth's shadow during a lunar eclipse is circular, indicating that the Earth is spherical.

The heavens are also spherical, claimed Aristotle, because they are perfect and should be created in the most perfect shape, which is a sphere. Sense experience confirms this. We can see that the Moon is spherical. We can see this directly and more clearly during a solar eclipse when we observe its circular shadow. If one of the bodies in the heavens is spherical, the rest should be too. They are made of the same substance and have the same motion; it is reasonable to assume that they have the same shape. If the objects that make up the heavens are spherical, we can assume that the heavens themselves are also spherical.

But is there anything beyond the heavens? To answer this, Aristotle appealed to his theory of motion. The speed of a falling body, he suggested, is inversely related to the resistance of the medium through which it falls. For example, if water has ten times the resistance of air, a rock moves ten times faster in air than in water. If a void existed, a body moving through that space would have infinite speed. Since this is impossible, Aristotle concluded, it is impossible for a void to exist. Therefore, outside the heavens there is nothing, not even emptiness or a void; the universe must be finite.

Aristotle could also explain why the heavenly bodies move in predictable ways, while the things on Earth are ever-changing. The four earthly elements are made of elemental contrary qualities. They are generated from each other by changing a quality into its opposite: hot into cold, dry into moist, and so on. With its change in composition, the body also changes its natural motion. For example, when earth heats up and turns into fire, its natural motion changes from down to up.

14

However, circular motion has no contrary. Therefore, a body that moves with circular motion cannot be generated from or into any other body. It must be immutable and its motion continuous and everlasting. Therefore, the motion of the heavens must be continuous and everlasting. Aristotle appealed to observation to confirm his conclusion: in all of recorded time there has never been a change in any part of the heavens. And so, he concluded, "Our theory seems to confirm experience and be confirmed by it."

In this way, Aristotle was able to answer the age-old question, was there a beginning of time? No, he answered. Time is infinite. Since circular motion is continuous and everlasting, without a beginning or end, motion has always existed. Since motion defines time, time has always existed. The universe and all the matter in it have always existed and will always exist, the four earthly elements continuously turning into each other, while the fifth element remains perfect and unchanged, circling forever.

Aristotle taught at the Lyceum for about twelve years, until 323 BCE when he was chased out of Athens with charges of impiety. He died the following year. Sometime between leaving Athens and his death, he wrote a will. In it, he detailed the care of his family, namely his two children and Pytheas, his companion and the mother of his son. He provided generously for the people who had served him. He specified that statues be erected to loved ones who predeceased him, including his guardian, brother, mother, and wife. He made requests pertaining to his burial. Throughout the document, no mention was made of Athens or his school.

Even so, Aristotle's school continued, for in Athens he left his books. He left his library, which he had collected throughout the years from various parts of the Mediterranean. And he left his own writings, which were like an encyclopedia of what was known or believed at that time in ancient Greece, synthesized and augmented with his own analysis and research. These books were passed down, replicated, and dispersed so that they survived even after his school, which had by that time gained structural permanence, was destroyed in 86 BCE.

# 3

# Transition

We ought not to be ashamed of appreciating the truth
and of acquiring it wherever it comes from,
even if it comes from races
distant and nations different from us.
For the seeker of truth nothing
takes precedence over the truth,
and there is no disparagement of the truth,
nor belittling either of him who speaks it
or of him who conveys it.
No one is diminished by the truth;
rather does the truth ennoble all.

— al-Kindi

Alexander, whom Aristotle had tutored in Macedonia until he became king, spent most of his reign on a massive and hugely successful military campaign, first within Greece and then throughout the Persian Empire. In 331 BCE, he founded a city in Egypt where there had been a small port town. This city, which became known as Alexandria, grew into one of the most important intellectual, cultural, and economic centers in the ancient world.

Alexandria was home to the Museum, the "Shrine of the Muses" — a school created in the tradition of Plato's Academy and Aristotle's Lyceum. Very few details about it are known except that it was for "men of learning" and was led by an appointed priest. We don't even know if it included any teachers. There is no question, however, that it had books.

The collection of books at the Museum was its treasure: the Great Library of Alexandria. Scholars at the school added to this collection through their own writings. In addition, they preserved, corrected, and explained old works. Here in Alexandria began the western tradition of textual criticism, with margin notes and introductions inserted into existing works. Scholars at the Museum enriched the books in their collection in this way, and these versions were spread throughout the ancient world.

The library at the Museum grew and grew. They collected works of history, poetry, literature, philosophy, mathematics, and science. In addition to Greek books, they collected books from the Romans, Persians, Indians, Babylonians, and Hebrews, translating "barbarian" texts into Greek. They bought books and they stole books, searching ships that entered the Alexandrian harbor, making copies of all the books that were found. They then returned the copies to the owners and kept the originals. The goal of the Museum was to create an archive of *all of the books in the world.*

The size and scope of the library grew to such an extent that, as a practical necessity, they developed a way to organize the collection, grouping the books together by genre and arranging each group in alphabetical order by author. A new position evolved, the Librarian, who acted as the guardian of the books, editing and cataloging them. The Library of Alexandria became not only the treasure of the Museum but also a monument to all the ancient civilizations accessible to Alexandria by land or water.

And then in 46/47 CE, forty thousand books, or perhaps sev-

17

en hundred thousand books, were consumed by flames in Caesar's Alexandrian war. But this was not the end of the library. Scholarship at the Museum continued, both textual criticism of old works and the addition of new works to the library. It was during this time that Ptolemy wrote his great astronomical work, a treatise detailing an Earth-centered model of the universe capable of accurately predicting the positions of the stars, Moon, Sun, and planets. However, at the end of the fourth century, we hear additional tales of destruction — the intentional destruction of the library by Christians in order to eliminate "pagan" writings. The Great Library of Alexandria was gone, and Alexandria, which had long since lost its youthful intensity, faded into the background of the ancient world.

In 762 CE, the Abbasid Caliph al-Mansur commissioned a city to be built on the Tigris River as the new capital of the Islamic Empire. It was designed so that it could be easily defended, with a tall, brick, circular wall enclosing the main part of the city and a deep moat surrounding the wall. The original name of the city was *dar ul-Islam*, "The Realm of Peace." It soon became known as Baghdad.

Baghdad was a cosmopolitan center, built at the crossroads of several civilizations. Inhabitants of the city included Syrian Jews and Christians, Arab Christians and Muslims, and Persians. It was a diverse group that had helped the Abbasids overthrow the former dynasty, united in their opposition to the authoritarian rule of the previous caliph. Now the Abbasids had to justify their own rule.

Already, they had been subject to revolts, mainly by Persian Zoroastrian revivalists. They would have to find support, especially with this group. To this end, al-Mansur initiated a project to have Zoroastrian and other Persian texts translated into Arabic. More and more Persians spoke Arabic as their primary language, and so these translations served to preserve Persian culture and history, which may have otherwise been lost. Furthermore, al-Mansur had Greek texts translated into Arabic. According to legend, Alexander the Great had taken the Greek corpus from a Persian king whom he had conquered. Now, these writings would be restored to their proper place.

With this, al-Mansur began a movement to acquire and translate books. He used these translations mainly to gain support among the Persians. Therefore, the majority of the early translations, both Persian and Greek, were writings related to astrology, a subject valued by the Persians. Al-Mansur also used the astrological predictions of these translated texts to legitimize his rule.

For later caliphs, the translations were used to gain an upper hand in religious discussions. They had philosophical works translated, especially Aristotle's, to help form logical arguments defending Islam to Jews and Christians. They also had his scientific works translated to gain knowledge, especially regarding cosmology, to aid in these debates. The Muslim caliphs embraced Greek science and used it to claim superiority over Christian leaders who rejected such learning.

Although the translation movement began with the caliphs, it soon spread throughout the empire. Engineers wanted scientific and mathematical works translated. Medical persons wanted medical texts translated. Government secretaries and scholars wanted books related to their work translated. As the demand for translations rose, the quality of the translations improved. Translators mastered Greek so that they could translate from the original texts rather than through intermediary Persian or Aramaic texts. They also translated non-Greek works, especially from India and China. Many added clarifying comments and explanations to their translations so that they would be more useful to the sponsor. Copies of these texts were made on cheap paper and spread throughout the realm in public libraries, accessible to most of the population, the first of such libraries in the world.

Around the turn of the millennium, the translation movement slowed down and then came to a stop. Practically every scientific, mathematical, and philosophical work of value within reach had already been translated into Arabic. At this time, original Islamic scholarship flourished, especially in the regions outside of Baghdad, where important advances were made in algebra, optics, and the experimental scientific method.

Baghdad, by this time, had already begun to experience political upheaval. The city had lost and regained the capital twice. Successive invasions battered the city. And then in 1258 CE, a Mongolian invasion completely destroyed it. All the people were killed. Palaces and mosques were burned. The Mongols set fire to the books, and the books that would not burn were thrown into the Tigris River.

In 1085 CE, when King Alfonso VI conquered Toledo, the city already had a history of more than a thousand years. It had witnessed the rise of Christianity from its disorganized roots and the rise of the

Holy Roman Empire. As the Roman Empire weakened, it had welcomed the nomadic Visigoths from the north, who later made the city its capital. It had seen the influx of Jews and attempts by Christian leaders to control them with their laws. It had suffered invasion by Arabs and Berbers, which gave Muslims control of the city for over three hundred years.

When King Alfonso and his army took Toledo away from the Muslims, he did not destroy their history. He did not destroy their palace. He did not destroy their mosques. He did not destroy their books.

At this time, in the part of the world dominated by the Catholic Church, any creative scholarship that existed was used primarily in the service of the Christian faith. "Pagan learning" was acceptable only to avoid the greater sin of ignorance. The available Greek works were, with few exceptions, simplified translations; most of the original writings by now had been lost or destroyed. To some, the opening up of the Arab libraries seemed like the discovery of buried treasure.

Around 1140 CE, the Italian scholar Gerard of Cremona traveled to Toledo to learn Arabic so that he could translate Ptolemy's *Almagest* into Latin. He translated this and other texts into Latin with the help of Muslim and Jewish scholars who remained in Toledo. He translated several works by Aristotle. He translated Euclid's *Elements*, al-Khwarizmi's *The Book of Addition and Subtraction*, and several books on medicine. He remained in Toledo for the rest of his life translating; about eighty translations are attributed to him in all.

Gerard was the most prolific of about a dozen scholars who made the journey to Toledo to translate Arabic works into Latin. This was the beginning of a massive translation movement throughout the region as other Muslim cities were conquered and their libraries opened up. The sponsors of these translations, at least at the beginning, were all private citizens.

This early translation movement coincided very closely with the birth of the university system in Europe. The University of Bologna was established in 1088 CE, and many others soon followed. The curriculum throughout these universities was uniform, dominated by the *quadrivium*: arithmetic, geometry, astronomy, and music. The *quadrivium* had been part of the core curriculum in Europe's educational system since the sixth century; it was natural that the universities would adopt it as the foundation for their programs.

The universities quickly adapted their curriculum to the new knowledge. Aristotle had enjoyed an elevated status within the Islamic libraries, and this status was transferred along with the books to the universities. They taught Aristotle's philosophy as part of the core curriculum and recast some aspects of the quadrivium to reflect Aristotelian thought. Furthermore, Plato's cosmology was replaced with the cosmological model developed by Aristotle and extended by Ptolemy. Euclid's writings were used to supplement the mathematics curriculum.

The texts used as the base of the university educational system were not the original translated texts. Rather, consistent with the educational traditions at the time, they were simplified "scholastic" summaries. Ptolemy's quantitative theory was replaced by a qualitative version. Theorems and proofs were removed from Euclid's writings. The mathematical sciences that had developed in Alexandria were reinterpreted in terms of Aristotelian terminology. Natural science was portrayed as a coherent whole, wrapped in Aristotelian doctrine and taught with the help of a uniform system of questions and comments.

The Church, too, embraced Aristotelian thought, despite many apparent contradictions with its own teachings. This was due largely to a 13th century Italian priest, Thomas Aquinas, who wrote commentaries on many of Aristotle's writings reconciling various aspects of Aristotelian doctrine with Catholic ideology. In his writings, he made an important distinction between God's absolute power and his ordained power. God has absolute power to do what he pleases, Aquinas suggested, but he used his freedom to create the universe with natural laws that guide its behavior. In this way, Aquinas validated the study of the physical universe for Christians; by trying to understand the laws of the universe, they are trying to understand God's own handiwork. Following Aquinas's lead, the Church accepted Aristotle as their own, the authority on everything secular.

By the beginning of the 15th century, Aristotelian doctrine had gained almost uncontested supremacy within both academic and religious institutions throughout Europe. Even those challenging Aristotle's views did so within the framework of his theories. But then another flood of ancient writings entered Europe in the 1450s, when Constantinople fell to the Ottomans. In the aftermath, many Greeks emigrated, bringing with them their ancient books. They brought numerous dialogues by Plato. They brought original Greek versions of Aristotle's texts and Ptolemy's astronomical treatise. They brought texts reflecting the Pythagoreans, Stoics, and atomists. With the help of the recently invented printing press, these texts were copied and distributed throughout Europe, stimulating a new translation movement that was accompanied by excited debates. It was in this period, which we now call the Renaissance, that Nicolaus Copernicus was born.

# 4

# A Change in Perspective

I know that I am mortal and living but a day.
But when I search for the numerous
turning spirals of the stars,
I no longer have my feet on the Earth
but am beside Zeus himself,
filling myself with god-nurturing ambrosia.

— Ptolemy

Niklas Koppernigk was born in 1473 in Toruń, an old Prussian city in the Kingdom of Poland. He had been groomed from an early age to become a priest and rise through the hierarchy of the Catholic Church by his uncle, who had taken over his care upon the death of his father. Koppernigk completed his general studies in Krakow and then left for Italy to study canon law in Bologna, later earning his doctoral degree. He studied medicine in Padua in preparation for becoming a "healing physician" for those in the religious order. After completing his studies, he worked for his uncle, who at that time was the bishop of Varmia. He eventually settled in Frauenburg, a town in Varmia, where he served as a canon for the rest of his life. At some point, he Latinized his name to Nicolaus Copernicus.

Copernicus had a productive career in the Church. He was a physician, secretary, and accountant for his order. He oversaw its mill, bakery, and brewery and was active in local economic matters. He traveled around the province of Varmia administering the chapter's vast land holdings, and during a military conflict, he aided in the defense of one of the chapter's castles. Copernicus was well respected, but he never advanced through the hierarchy of the Church. He had other ambitions.

Copernicus first became interested in astronomy as a university student in Krakow. He purchased astronomical tables to calculate the positions of the planets, attaching them to blank pages where he added miscellaneous astronomical notes. He continued his astronomical observations while working for his uncle, recording important events such as the alignment of planets and a lunar eclipse. He studied Ptolemy's Earth-centered model of the universe through an abridged version of his treatise. Copernicus referred to Ptolemy as a most outstanding astronomer, and indeed he was. Ptolemy had named his work of astronomy *Mathēmatikē Syntaxis,* "Mathematical Treatise." By the time it was translated into Latin, it was called *Almagestum,* "The Greatest."

Ptolemy had collected, synthesized, and expanded on all the known astronomical knowledge available to him. He offered a detailed introduction to the advanced mathematics required for spherical astronomy, including rigorous proofs and relevant mathematical tables. He also included tables for the rising times of the zodiac constellations and the day length according to time of year and latitude. He gave instructions on how to construct and use an astrolabe, a parallactic instrument, and an early version of a quadrant. He created a star catalog of more than a thousand stars, organized by constellations. While it is certain that Ptolemy built on others' work, it is not known in many cases what was borrowed and what was new. His treatise exceeded all past astronomical works to such an extent that the others were soon neglected and then eventually lost. With regards to Ptolemy's personal life, we only have what he left in his writings. We know that he lived in Alexandria during the second century because of his astronomical records.

In his treatise, Ptolemy presented a mathematical theory describing the motions of the stars, Moon, Sun, and planets. At the center of his universe, following Aristotle, is the Earth, stationary and immovable. Surrounding the Earth are fixed stars, forming a spherical globe. When Ptolemy used the word "fixed" in reference to the stars, he did not mean that they are stationary, only that they are stationary with respect to each other. The entire sphere of stars rotates about an axis once a day. In addition, there is a small precession of the celestial globe such that each star moves about one degree in longitude every century, using the position of the Sun at the equinox as a reference. This precession had been observed by the Greek astronomer Hipparchus in the second century BCE when comparing his measurements to those of earlier astronomers, and Ptolemy confirmed this precession with his own measurements.

Ptolemy also confirmed the ancient observation that the motion of the fixed stars about the Earth is uniform and circular. The motion of the Moon, Sun, and planets is also uniform and circular, suggested Ptolemy; however, from the perspective of the Earth, they do not appear so. Relative to the fixed stars, they change speed and, in the case of the planets, sometimes even reverse their direction.

Ptolemy proposed two hypotheses utilizing uniform circular motion to explain these apparent irregularities. In the eccentric hypothesis, the uniform motion takes place on a circle that is not concentric with the center of the Earth. In the epicyclic hypothesis, the uniform motion takes place on a circle, called an epicycle, whose center moves along the circumference of another circle, called a deferent circle, which is concentric with the center of the Earth.

However, Ptolemy showed that even with these two hypotheses, it was not possible to explain the motions of the planets. He, therefore, introduced a third device to be used with the epicyclic hypothesis: the equant. The equant is a point separate from the geometric center of the epicycle from which uniform speed is observed. So from one point, circular motion is observed, and from another, uniform speed.

Copernicus disagreed with Ptolemy. He objected specifically to his use of the equant, claiming that it violated the assumptions of uniform circular motion. He believed it would be possible to rid the theory of the equant by placing the Sun rather than the Earth at the center of the universe. Ptolemy had already considered this; he rejected it, claiming that if the Earth moved, the resulting violent motion would be so great that it would throw all of the objects on the Earth in the opposite direction of its motion. Furthermore, since heavy objects fall the fastest and the Earth is made up of the heaviest element of all, the Earth would fall down faster than all the other objects, leaving them behind and falling out of the heavens. Ptolemy concluded that such things are "utterly ridiculous even to think of."

Still, Copernicus did think of such things. Around 1511, he wrote a short essay describing a Sun-centered model of the universe. He sent it out to at least one person outside Varmia, and it was circulated somewhat. His reputation as an astronomer reached Rome, and he was called upon to obtain a more accurate estimate for the length of a year to help with calendar reform. Copernicus worked on this and was able to determine the length of a year to within seconds, far surpassing the precision of clocks at that time. But Copernicus wanted more. He wanted to create a new cosmology. And it couldn't just be a suggestion that might be ridiculed by mathematical astronomers. He needed to counter Ptolemy's objections with even stronger arguments and support it rigorously with mathematics. Copernicus wanted to create something as great as Ptolemy's *Almagest*.

In March of 1513, Copernicus set to work building an observation tower, an elevated platform near his residence that gave him an unobstructed view of the sky and a level surface for his astronomical instruments. He began recording measurements of the stars, compiling data for his own star catalog. He recorded the positions of the Moon, Sun, and planets. He used the most modern astronomical instruments, which were updated versions of the same instruments that Ptolemy had used and written about fifteen hundred years before.

In 1515, a full text of Ptolemy's *Almagest* became available in Latin. This was the version translated by Gerard of Cremona from several Arabic translations, dated back to 1175, but had only just been

26

published. Copernicus obtained a copy and read it, filling the margins with notes and diagrams. He continued to record astronomical observations, form conceptual arguments supporting his Sun-centered model, and work out the mathematics of his theory. By 1535, he had pretty much finished writing a book that he titled *De Revolutionibus Orbium Coelestium,* "On the Revolutions of Heavenly Spheres."

Copernicus modeled his book after Ptolemy's, with similar structure, tables, diagrams, and conceptual explanations where their theories overlapped. He included the same mathematical introduction to spherical astronomy, replacing the Greek number system with the Hindu decimal system for ease in calculation. He discussed astronomical instruments and included a star catalog. Like Ptolemy, he not only offered a mathematical model describing the motion of the stars, Moon, Sun, and planets, but practically everything an astronomer would need to understand, verify, and apply the theory.

The essential elements of Copernicus's theory as presented in *De Revolutionibus* were the same ones he had sketched out decades earlier. The fixed stars are stationary and immovable, forming a spherical globe that encases the universe. The Sun is stationary near its center. The Moon orbits the Earth, and the Earth and the planets all orbit around the Sun. The apparent motions of the stars, Moon, Sun, and planets can all be explained without the equant, using only eccentrics and epicycles that assume uniform circular motion.

In the decades between the two writings, Copernicus had worked out the mathematical details of his theory and made his own observations, which largely supported the work of earlier astronomers. He also discovered a new motion. By combing through almost two thousand years of astronomical data, he found that the precession of the equinoxes, which Ptolemy and Hipparchus had believed to be a constant one degree per century, actually increases and decreases in a periodic manner. Copernicus incorporated this into his theory, again using only variations of uniform circular motion.

However, this was not enough. Ptolemy's model, too, could have incorporated this new motion. There was enough flexibility in both theories to allow for almost perfect correspondence with the actual positions of the celestial bodies as viewed from the Earth. Copernicus's model could not compete with Ptolemy's by merely offering another mathematical way to calculate celestial positions. He needed something compelling, something beyond the mathematics in order for his theory to be viewed as more than just an interesting alternative.

Towards this end, Copernicus demonstrated that his Sun-centered model could account for some astronomical observations better than Ptolemy's. For example, it could explain why Mercury and Venus are always seen close to the Sun, only visible around sunrise and sunset. This is because they are the innermost planets, closest to the Sun. Similarly, Copernicus could explain why Mars, Jupiter, and Saturn appear larger when they are seen rising in the evening; it is at these times that they are on the same side of the Sun as the Earth and therefore closer to it. It could also explain why the apparent backwards motion of the planets is more pronounced in the planets that pass closest to the Earth. In Ptolemy's theory, all of these observations are simply a result of carefully chosen parameters.

However, the strength of Ptolemy's theory was its consistency with Aristotelian doctrine. Heavy objects fall, light objects rise, and objects that are neither heavy nor light travel forever in constant, circular motion. All of these motions are in reference to the center of the Earth, which is at the center of the universe. In order to displace the Earth from the center of the universe, Copernicus had to crack the very foundation of Aristotle's physical theory. And he did this, even while holding fast to one of its most cherished assumptions, that of the uniform, circular motion of heavenly bodies.

According to Aristotelian theory, a simple body has a simple motion. Copernicus argued against this by considering the motion of the planets. Since they are sometimes closer and other times farther from the Earth, they must have a compound motion consisting of falling, rising, and circling. Therefore, either each simple body cannot have a simple motion, or the Earth cannot be at the center of their motion. He also objected to the Aristotelian assumption that a falling object has a fixed speed until it comes to rest at the surface of the Earth; it can be determined through observation that an object speeds up as it falls.

Aristotelian theory distinguishes between natural motion, the motion an object has due to its internal composition, and violent motion, motion that is caused by something external. One of the strongest arguments against a Sun-centered model was that if the Earth moved, it would move violently and break into pieces, throwing all the objects on the Earth in the opposite direction of its motion. Copernicus claimed, rather, that if the Earth rotates, the rotation would be natural since circular motion is natural.

On the other hand, Copernicus argued, if rotation is unnatural, we should be much more concerned about the rotation of the universe — it would have to go extremely fast on account of its size to complete one full rotation a day. If the heavens were thrown out, they would have to move faster still to complete a rotation in twenty-four hours, and so they would be thrown out even more, eventually becoming infinite in size! It was much more reasonable, Copernicus suggested, to allow the Earth to move.

He then supported this idea that the motion of the Earth is natural by appealing to the observation that we can't feel constant motion. He used a ship at sea as an example, writing that "when a ship floats over a tranquil sea, all the things outside seem to the voyagers to be moving in a movement that is the image of their own, and they think on the contrary that they themselves and all the things with them are at rest." Therefore, he claimed, it is possible that the Earth is moving without us feeling it.

Finally, Copernicus addressed the question, if the Earth is not at the center of the universe, then why do heavy objects fall toward the center of the Earth? He suggested that there may be multiple centers of motion, writing, "I myself think that gravity or heaviness is nothing except a certain natural appetency implanted in the parts by the divine providence of the universal Artisan, in order that they should unite with one another in their oneness and wholeness and come together in the form of a globe." In this way, he not only responded to objections raised against a Sun-centered model but also anticipated a new theory of motion that could accommodate such a universe.

By about 1535, Copernicus had done all of this. He had built his mathematical model, completed tables, diagrams, a star catalog, mathematical proofs, and conceptual explanations; he had written his great work. Copernicus had a compelling theory, but still he was reluctant to publish it. To obtain satisfactory agreement with astronomical data, his theory required over fifty circles. He worried that it still was not perfect. He feared the public's reaction, especially from other mathematical astronomers. Friends tried to persuade him to publish. A bishop urged him to share his writings and offered to pay for them to be copied, writing to Copernicus that he was zealous for his reputation and "eager to do justice to so fine a talent." Still, he did not publish.

# AN INCOMPLETE THEORY

In 1539, the young mathematician Georg Rheticus traveled hundreds of miles to visit Copernicus, hoping to learn the new astronomy. He came unannounced and brought with him as gifts three bound volumes containing five important astronomical texts. He convinced Copernicus to share his work and stayed on for two years helping to prepare *De Revolutionibus* for publication. On May 24, 1543, just moments before Copernicus died, the final pages of the manuscript, returned from the printer, were laid in his hands.

# 5

# Kepler's Laws

Let all philosophers,
new as well as ancient, be silent!
Let the very theologians,
interpreters of the divine mysteries, be silent!
Let the mathematicians,
describers of the heavenly bodies, be silent!

— Tycho Brahe

Tyge Brahe was born in December of 1546 in Knutstorp Castle. Two years later, his aunt and uncle took him away, without his parents' knowledge or permission, to their castle in Tostrup. There, they raised him as their own. As far as records show, his parents never protested this abduction or tried to get him back. When his uncle died, Tyge, who by this time went by Tycho, inherited nothing since his uncle had not completed the formal process to make him a legal heir. When his father died, he shared his father's estate and fortune with his brothers and sisters. Yet he grew up separately, raised as an only child in the castle of his aunt and uncle.

If he had grown up with his parents, Tycho probably would have taken the same path as his four younger brothers, the path that most sons of Danish nobles took. After studying for a few years in Copenhagen, he would have gained experience in Denmark and abroad, serving in other noble households. He would have then returned to Copenhagen as a courier or knight in the king's court. After this, he would have been ready to govern and defend a royal fief. His uncle, a Brahe, may have chosen this path for him too. But his aunt was an Oxe, and the Oxe family valued education. Their men were not sent to foreign courts; they were sent to foreign universities.

Accordingly, at the age of fifteen, Tycho was sent to Saxony to attend the University of Leipzig, one of the largest and most important universities in Europe. Here, he began to study astronomy, at first in secret with the help of books that he had purchased. He studied maps of constellations. He studied astronomical tables. And he began to observe, often staying awake all night peering through a skylight. He tracked the motion of the planets; he determined their positions with the help of a string held taut between two stars and recorded their positions on a small globe. He soon discovered that neither the ancient Alfonsine Tables, based on Ptolemy's model, nor the Prutenic Tables, based on Copernicus's model, determined the positions of the planets with an acceptable degree of accuracy. Someone should produce better tables. Tycho decided that *he* would be that someone.

He continued to observe, using the most modern astronomical instruments. When he found that these weren't good enough, he began designing his own. He designed a new pair of compasses. He created the "Great Quadrant" twenty feet high, which required forty men to set it in place. He continued his university studies, moving on to universities in Wittenberg, Rostock, and finally Freiburg, until he had learned everything he could learn from both teachers and books. In order for Tycho to take his work to a new level, he needed an observatory.

The uncle that had raised Tycho had died, but he had another uncle, his mother's brother, who was sympathetic to his pursuits and had recently acquired a large estate that had formerly been a Cistercian monastery. He moved in with this uncle and together they transformed the estate, building an astronomical observatory, a paper mill, an alchemical laboratory, and an instrument factory. Here, Tycho invented a new astronomical instrument called a sextant. And it was here, on November 11, 1572, that he discovered a new star.

The star was brighter than any other star in the sky. It was even brighter than Venus. Tycho knew that it was not a planet because of its location in the sky, north of the zodiac constellations. It could not be a comet because it had no tail. He tracked it over many nights and found that it was stationary relative to the fixed stars. He also tracked it at various times in a single night and found that its position against the stars remained constant even as his observation point relative to these stars changed. The Moon *does* change its position against the stars with different observation points, which astronomers call parallax. This change in position was used to estimate its distance from the Earth. Since this new star exhibited no parallax, it was farther away than the Moon. This was a clear violation of Aristotelian cosmology, which assumed that everything in the heavens is perfect, eternal, and unchanging.

Tycho was alone in arriving at this conclusion, although others had also seen the star. Some claimed that it was a comet even though it had no tail. Others believed it was below the Moon even though it exhibited no parallax. Still others insisted that it was not new, and they maintained that it did not constitute a change in the heavens even though it visibly dimmed during the months following its initial appearance. Tycho stood firm. He wrote up his observations and conclusions in a small book, *De Stella Nova*. He published it too, although those of his rank did not publish books.

The nobility did not publish books and they did not teach. Since the Reformation, the Danish nobility did not partake in the leadership of either church or university, which were both reserved for the educated middle-class. However, with the publication of *De Stella Nova*, there arose great interest among scholars in Copenhagen to have Tycho teach at the university. Several noble students wrote up and signed a petition asking him to give lectures to them, who were in his social class. Others would also be welcome to attend. The king signed the petition as well, and with passion and knowledge Tycho began to teach, sharing his unorthodox views on the workings and influence of the heavens.

Tycho began teaching in the fall of 1574 and stopped abruptly the following spring after his father's estate had been settled. He was now twenty-eight years old, financially independent, and had started his own family. He sought a permanent residence to live and pursue his work on a larger scale. He took a trip abroad to explore his options with the blessing of King Frederick. In fact, he went in part for the king, to help him find the best architects, engineers, sculptors, and other experts to transform a medieval fortress. But after his return, King Frederick became worried that he might lose Tycho to a foreign nation and offered him generous enticements to remain in Denmark: an island, a guaranteed income, and a large sum of money to build an observation tower.

Tycho traveled to the proposed island, the island of Hven. He surveyed the land. He considered the advantages and disadvantages of its isolation. He took note of the peasant population, since they would be his primary workforce; as their lord, each farm would owe him two days per week of unpaid labor. He observed the skies, well aware of the limitations that the northern latitude would impose. After three months, he decided to accept.

Tycho chose the center of the island as the building site for his observatory and named it Uraniborg after Urania, the muse of astronomy. It took many years for the estate to be completed. When it was finished, it had generous living quarters for his family and guests as well as space for students and assistants. It had a three thousand book library with a large brass-plated celestial globe. It included two observatories, one on top of the palace and another away from the main building and mostly underground, designed to securely hold his large astronomical instruments. It had gardens, fishponds, running water, and an intricate communication system made with bells and strings. It also had a paper mill, an alchemical laboratory, and an instrument factory. Tycho worked here for over two decades with students and assistance from all over Scandinavia and Europe. Continually innovating new and better observational techniques, he produced astronomical data estimated to be about five times more accurate than the best data that had ever been produced.

Johannes Kepler was born in December of 1571 in Württemberg, a Lutheran duchy within the Holy Roman Empire. He was born sickly and two months premature into an overcrowded house that his family shared with his grandfather, aunts, uncles, and cousins. His family had financial problems, despite their noble lineage. His father was often away fighting as a mercenary and eventually abandoned the family completely. Kepler suffered multiple illnesses throughout his childhood, leaving him with nearsightedness, double vision, abdominal problems, and crippled fingers.

Württemberg had established a free school system, and it was here that young Kepler began his education. He started out at a German school where he learned to read and write. His teachers, noticing his exceptional ability, transferred him to a Latin school. He moved on from there to seminary schools, where he studied the Bible and classical texts in both Latin and Greek. Kepler loved learning and was good at it. He also delighted in riddles, games, puzzles, and poetry.

At seventeen, Kepler set out on foot to the nearby Tübingen University, where he was accepted on scholarship to study theology. He excelled there, completing two years of general studies and nearly three years of theological studies. His intention throughout was to become a Lutheran minister. However, right before finishing his studies, a seminary school in Graz contacted Tübingen seeking a mathematics teacher, and the University called upon Kepler to take the position. With the provision that he could return to the ministry, Kepler reluctantly agreed and began the long journey to southern Austria.

Kepler's teaching responsibilities in Graz included advanced mathematics and astronomy. Furthermore, as the district mathematician he was to create annual astrological calendars, which made predictions about such things as weather, war, and the best times for physicians to perform surgery. Although he did not have a strong belief in the validity of astrology to predict the future, he had experience with astrological predictions from his days in Tübingen and continued here in Graz with a surprising amount of success. Through his astronomy classes and horoscopes, Kepler began to more carefully consider astronomical models.

His former teacher at Tübingen was one of the few who took the Copernican model seriously as a physical model, rather than simply a mathematical tool that could be used to make predictions and explain observations. Kepler shared this view and began to probe the model more deeply. He began asking questions that none of his contemporaries were asking, questions that neither Ptolemy nor Copernicus had asked. He began asking questions that even Aristotle, who grouped astronomy together with physics and sought the causes for all physical phenomena, didn't ask. Kepler began to ask, what are the *causes* of the heavenly motions? Why did God create the universe the way he did? In this way, Kepler found purpose in his new calling; he wanted to look into the mind of God and discover the reasons behind his creation.

Kepler began to play around with the Copernican model, looking for patterns. He paid particular attention to the ratios of the orbital separation of the planets. He noticed that the five Platonic solids inscribed within spheres that coincided with the planets' orbits gave the correct distance between their orbits. He believed that he had discovered a relationship between the size of each orbit and the length of the planet's year. He also looked for patterns in the speeding up and slowing down of the planets as they approached and receded from the Sun. This could be explained within the Copernican model with the idea that the Sun somehow *causes* the planets' speed. Kepler wrote up his ideas in a book called *Mysterium Cosmographicum*, the first published defense of the Copernican system. He sent a copy of it along with a letter to Tycho Brahe, praising him as the "prince of mathematicians," not only of the present but of all times.

By the time Tycho received Kepler's book and letter in the spring of 1598, he had left Denmark. His generous sponsor, King Frederick, had died ten years earlier. In 1596, Frederick's son Christian had come of age and assumed the throne. Shortly after his coronation, King Christian began transferring fiefs from one noble to another, purposely creating discord among the nobles to weaken their power and strengthen his own. He took away Tycho's Nordfjord fief, which had been a significant part of his income, and let him know that his support of Uraniborg would not last forever. As it became clear that departure from Uraniborg was inevitable, Tycho rushed the completion of his star catalogue, omitting the usual cross-checking between the two separate observatories. He then moved his family and a few remaining assistants to his mansion in Copenhagen. When Tycho set up a new observatory in one of the towers, King Christian complained that the instruments ruined his view and ordered that they be removed.

And so, Tycho and his household began another journey, with wagons and carriages loaded with astronomical instruments, a printing press, laboratory equipment, his giant brass globe, and thousands of books — everything Tycho would need to carry on with his life and his work. This time he traveled out of Denmark into Germany, where he sought out the very wealthy and highly respected scholar Viceroy Heinrich Rantzau. Although Rantzau was not in a position to provide Tycho with an appropriate patronage, he did offer some advice and the use of one of his castles.

Tycho settled into his new lodgings, began to rebuild staff, and resumed observations. He put the finishing touches on a book he had been working on, a book describing his numerous astronomical instruments, beautifully illustrated with thirty-one woodcuts and engravings. He also completed his star catalog. Tycho dedicated both of these books to Rudolph II, the emperor of the Holy Roman Empire. He sent elegantly bound copies to princes, bishops, and archbishops throughout the lands neighboring Prague, where Rudolph held court. Tycho hoped to present specially bound copies of these books, along with a third book of solar and lunar positions, to the emperor in person.

Tycho was still in Germany when he received Kepler's book and letter. He disagreed with Kepler's espousal of the Copernican model. If the Earth orbits around the Sun, the stars should exhibit parallax as the Earth's position with respect to the stars changes throughout the year. Through all of his careful observations of the heavens, Tycho had never detected even the slightest change in the appearance of the stars; he concluded that the Earth must be stationary. However, he agreed with Copernicus that the *planets* must be orbiting around the Sun. He, therefore, adopted a hybrid model in which the Sun and Moon orbit around the Earth while the planets orbit around the Sun.

Although Tycho disagreed with Kepler's conclusion regarding the Copernican model, he recognized in his work a brilliant mind. He wrote back, suggesting that Kepler's analysis might benefit from his own more accurate measurements. Kepler did not like Tycho's hybrid model. Although it was mathematically equivalent to the Copernican model, it was inconsistent with his own idea that the Sun causes the motions of the planets. But he very much wanted Tycho's data.

By the time Tycho's response reached Kepler, Kepler's situation in Graz was becoming precarious. The Counter-Reformation had reached the city; Protestant patients were passed by in hospitals without care and Protestant sacraments were forbidden. Finally, the archbishop expelled all Protestant teachers and ministers upon penalty of death. Kepler was allowed to stay because of his role as district mathematician, but his school was closed. And the situation only seemed to be getting worse. Kepler heard about Tycho's grand procession into Prague, and when an invitation came offering him a ride to Prague and an introduction to Tycho, he accepted. He did not know that an invitation from Tycho, for himself and his family, was already on its way.

In the summer of 1599, Tycho and his entourage arrived in Prague. Within days, Tycho was introduced to Emperor Rudolph, who had been eagerly awaiting his visit since he heard from abroad about the beautifully illustrated astronomical books that Tycho had given as gifts. The emperor received his books enthusiastically and gave Tycho a position as the Imperial Astronomer, offering him a choice of castles and generous support. Tycho chose Benátky, a castle on a high hilltop with an indoor water system, and began planning its transformation.

Kepler arrived in Prague early in 1600. As soon as Tycho received this news, he sent his eldest son to greet Kepler and escort him to Benátky. Tycho welcomed him warmly and offered to reimburse his travel expenses.

Everything went downhill from there. Tycho was distracted with the massive renovation of the castle. He was also occupied with financial concerns when Rudolph's promised money was not forthcoming and costs kept mounting. Practically the only time Kepler could get his attention was at the dining table, but with all of Tycho's family, assistants, and visitors eating together, most of them speaking Danish, mealtime was overwhelming and uncomfortable. Kepler disliked being treated as a guest, or worse, as a hired hand with no salary or security. Furthermore, suspicious after an assistant had plagiarized his work, Tycho wouldn't let Kepler see any of his data, leaving him with nothing to do.

Soon, however, Tycho gave Kepler an assignment: to analyze Mars's orbit. He first had Kepler work with one of his trusted assistants. Later, pleased with his work, he allowed Kepler to work on his own, first having him sign a pledge promising not to reveal any of his secrets.

With Tycho's permission, Kepler used the Sun as the center of Mars's orbit rather than the Earth, as Copernicus had done. Analyzing the data from this perspective, he provided support for the idea that the Sun is the cause of Mars's motion; Mars speeds up as it approaches the Sun and slows down as it recedes. With a clever manipulation of the data, he was able to repeat this analysis for the Earth's orbit and found the same result. The Earth is one of the planets, like Mercury, Venus, Mars, Jupiter, and Saturn! It did not take long for Kepler to begin speculating about the existence of life on other planets. Why would God create these planets and not put life on them?

This success only made Kepler more eager to obtain the rest of the data and have time to do independent research. He also became impatient to solidify his employment. Although Tycho generously took care of all of his needs, Kepler could not ask his family to join him under these circumstances, where they would all be treated as guests. He started to put pressure on Tycho to give him, or find him, a salary. Contract negotiations began through a variety of intermediaries, but the process was taking too long. Kepler decided to write up his own contract: he wanted a salary from both Tycho and the emperor, he wanted every afternoon free so that he could pursue his own research, he wanted to be excused from all observation on account of his poor eyesight, and he wanted to live with his family in a separate house. He also asked for Sundays and holidays off, which insulted Tycho since he gave that to all of his assistants. The negotiations dragged on, reaching their climax when Kepler exploded, right in the middle of dinner, and Tycho responded in kind. Kepler left Benátky the next day.

Tycho could have let Kepler leave forever. Kepler was not particularly pleasant to be around and he had his own ideas of the universe, believing in the Copernican model rather than Tycho's system. But Kepler was *very* good at mathematical analysis. There were few people in the world capable of such work. Tycho was eventually able to diffuse the situation and Kepler decided, at least tentatively, to remain working with him and left for Graz to bring his family back with him.

This was in the spring of 1600. For the next year and a half, both Kepler and Tycho remained in limbo. Just as Tycho was almost settled in Benátky, Rudolph summoned him to Prague so that he would be available for daily astrological counseling. Soon after, the emperor suffered a temporary mental breakdown and leaned more and more on Tycho not only for astronomical advice, but also political and psychological counsel. Tycho had neither time nor a suitable place to conduct his work.

Kepler, upon returning with his family to join Tycho in Prague, became sick with a recurring fever. When he was well enough to work, Tycho wanted him to help prosecute the former assistant who had plagiarized his model, which Kepler considered a waste of time. Tycho persisted even after the prosecuted assistant had died. Again frustrated, Kepler made an effort to find other employment, but nothing came through. It seemed to Kepler that destiny was conspiring for him to remain with Tycho.


Tycho was now in his fifties and feeling his age. He had spent practically his entire life making the most accurate observations ever, and nothing had come out of it. He had an idea for a new model of the universe, but it remained only an idea with no analytical support. All of his other mathematically-able assistants had returned, homesick, to Denmark. Only Kepler remained; Tycho finally decided to put his entire trust in him.

In October of 1601, Tycho introduced Kepler to Emperor Rudolph with the following proposal. Together, he and Kepler would create a new set of astronomical tables, more accurate than any that had ever been produced. They would be called the Rudolphine Tables. Tycho asked nothing for himself but requested that Kepler be given a salary. The emperor was delighted with the idea.

A few days after this meeting, Tycho became severely ill during a dinner party and never recovered. As he lay delirious on his deathbed, he repeated over and over, "Let me not seem to have lived in vain."


After Tycho's death, Emperor Rudolph appointed Kepler his successor, giving him the title of Imperial Mathematician. As such, Kepler was expected to create annual astrological calendars, give astrological advice, and complete any of Tycho's unfinished work, especially the Rudolphine Tables. Other than that, he was given the freedom to pursue whatever projects he wished, all of which would give credit to Rudolph and his empire.

In return, Kepler was given a separate home for himself and his family and a modest salary. Although the salary was neither consistently paid nor great enough to hire an assistant, it was enough to live a comfortable life. He found friendship within the court and among the common people of Prague. He gained respect well beyond Prague through his writings. He engaged in lively correspondences with intellectuals and acquaintances throughout Europe, and important people visiting Prague sought out his company. Kepler remained in Prague until Rudolph died in 1612. After that, he moved to Linz in Upper Austria. This was within the Holy Roman Empire and he was able to retain his position as Imperial Mathematician. At this point, he gave up Tycho's instruments and books. However, he was allowed to keep all of Tycho's observations.

Kepler's writings were many and varied. Tycho had told him about a strange observation that the Sun when viewed through a pinhole and projected on a screen appeared *larger* than the Moon, which would make a total eclipse of the Sun impossible. Yet many total eclipses had been observed. This sparked Kepler's interest in optics and now that he had time, he delved into the subject more deeply. He discovered an inverse-square law guiding the relationship between the brightness of light and the distance to its source: since the rays from a source spread out spherically, the brightness decreases as the area of the sphere increases, according to the square of the distance from the source. Later, after the invention of the telescope, Kepler used rigorous mathematics to explain the operation of two lens magnification and came up with a new and improved design for the telescope.

Kepler wrote up his optical work in two treatises. After a new star appeared in 1604, Kepler studied it and then wrote a book demonstrating that the star was far beyond the Moon and planets, as far away as the fixed stars. He wrote a seven-volume treatise expounding the Copernican system. Eventually, many years after Rudolph's death, he completed the Rudolphine Tables.

Kepler also continued his work on planetary orbits. With Tycho's data, he had the confidence to move forward knowing that if his model disagreed with the data, there was something wrong with his model. In this way, Kepler discovered how the motion of Mars varies as its distance from the Sun changes; the distance it travels in one day is inversely proportional to its distance from the Sun. He also discovered that Mars's orbit is not circular or made up of circles, but rather it is in the shape of an ellipse with the Sun at one focal point. He discovered all of this with the Mars data and then found that these laws also applied to the Earth and all the other planets.

Kepler spent many years focusing on the Mars data and many more years working with the data for all the planets. He wanted to know why the planets moved with different speeds, why they were at different distances from the Sun, and why their orbits deviated from circular orbits by different amounts. This time, he drew inspiration from music, relating various quantities to musical intervals, as if the entire universe were singing in harmony. He summarized his discoveries in a book he titled *Harmonices Mundi*, "The Harmony of the World."

In the spring of 1618, just as Kepler was preparing to publish this work, he discovered another law. He found that the square of a planet's year is proportional to the cube of its distance to the Sun. The precision of Tycho's data made him confident in the exactness of this relationship. He was ecstatic. He added this law to the book and then shared his excitement:

Now because 18 months ago the first dawn, 3 months ago broad daylight but a very few days ago the full sun of the most highly remarkable spectacle has risen — nothing holds me back. I can give myself up to the sacred frenzy, I can have the insolence to make a full confession to mortal men that I have stolen the golden vessel of the Egyptians to make from them a tabernacle for my God far from the confines of the land of Egypt. If you forgive me I shall rejoice; if you are angry, I shall bear it; I am indeed casting the die and writing the book, either for my contemporaries or for posterity to read, it matters not which: let the book await its reader for a hundred years; God himself has waited six thousand years for his work to be seen.

Kepler had arrived at great truths and had written them down, but he knew well that the mathematics involved to understand them was beyond all but the best minds. Kepler was excited; he wanted to share his discoveries with the world. But he was patient.

# 6

# The New Science

It is proof of a base and low mind for one
to wish to think with the masses or majority,
merely because the majority is the majority.
Truth does not change because it is, or is not,
believed by a majority of the people.

— Giordano Bruno

In 1597, Kepler received a letter from Galileo Galilei, an Italian mathematics teacher working at the University of Padua. Galileo had gotten hold of a copy of Kepler's book *Mysterium Cosmographicum* and had read the preface. He wrote to Kepler that he looked forward to reading the book, that he also was a Copernican. In fact, Galileo continued, he had written at length in support of Copernicus but did not plan to publish while these views were so generally scorned. Kepler wrote back, encouraging Galileo to support Copernicus's model openly. Galileo did not reply.

Galileo had reason to be cautious. It was in Venice, where Galileo spent much of his leisure time, that Giordano Bruno had been arrested by the Venetian Inquisition and then sent to Rome to be tried for heresy. Bruno had espoused ideas even more radical than Copernicus, that the universe is infinite with many suns that appear to us as stars, and these suns are surrounded by innumerable inhabited planets. He had been found guilty of heresy and, when Galileo responded to Kepler, Bruno was still imprisoned in Rome. Three years later he would be burned at the stake. But Galileo had other reasons to be cautious. He had worked hard to gain his position and did not want to jeopardize it.

Galileo was born in 1564, the oldest child in a large family that never had enough money. His mother was well-born with high aspirations for her children. His father was employed in Florence as a court musician and wrote books on music theory, challenging established teachings. He worked in the wool trade to supplement what he earned as a musician; even so, their family always struggled to make ends meet.

Galileo's father wanted something more profitable for his eldest son. Finding that he was an exceptional student, he pushed Galileo toward medicine, enrolling him at the University of Pisa. While there, Galileo became fascinated with mathematics. He made the acquaintance of a court mathematician who, impressed with his outspokenness and deep interest in mathematics, began to give him private instruction in secret and against his father's wishes. His father desired that he stay with medicine. Advanced mathematics was hardly more practical than music — its only real use was in astrological calculations. When Galileo failed to get a medical scholarship that would allow him to continue, he left the University without a degree.

This was in 1585, and Galileo spent the next several years trying to gain credibility as a mathematician while doing what he could to earn an income. He traveled throughout Florence and the surrounding area giving private instruction in mathematics. He invented a little balance and wrote a short scientific paper explaining how its precision in calculating the density of solids and liquids surpassed the ancient method devised by Archimedes. He wrote up a paper calculating the center of gravity of variously shaped objects and shared it with leading mathematicians. Galileo wanted a university mathematics teaching position, and finally he succeeded. In the fall of 1589, he returned to the University of Pisa as a professor.

Back at his old school, Galileo attacked his superiors without fear or respect. He dressed carelessly, ridiculing the long, professorial robes worn by his colleagues. He told his students that if you wear a toga, you have to follow certain rules. *He* was above any rules. Galileo's three year contract in Pisa was not renewed. Now, in Padua, he had another chance. Perhaps his reluctance to take a radical position alongside Kepler had less to do with his fear of losing his life than losing his job.

Galileo wanted and *needed* to keep his job. His father died soon after one of his sisters got married, and he was left responsible for paying her dowry. Additional obligations on behalf of other siblings soon followed. Galileo supplemented his income with the help of another invention, a military compass, a complicated device designed to facilitate calculations needed for warfare. He sold it to noblemen and instructed them in its use. He didn't take any radical positions or poke fun at his peers. When his contract came up in 1599, it was renewed for seven years.

Galileo was still teaching at the University of Padua in autumn of 1604 when a new star was sighted in the Milky Way in the constellation of Sagittarius. Tycho's star had disappeared years ago and was mostly forgotten, but now this other star reminded some of the earlier one. It started out as a faint reddish speck in the sky. It grew quickly over a few days until it was almost as bright as Venus. Then it grew fainter and fainter over many months, eventually disappearing. Kepler tracked it for about a year and wrote up his findings, claiming that it was a transitory star deep in the Milky Way. The star became known as Kepler's star.

Galileo began to observe this star about a month after it was first sighted. He knew well how to determine distance using parallax; he had taught many noblemen how to do this with the help of his military compass. He applied these same principles to determine the distance to the new star. Unable to measure any parallax, he concluded that the star must be far beyond the Moon. He scheduled three lectures at the University of Padua to talk about the star.

In this series of lectures, Galileo made his first public appearance as an astronomer. He filled the Great Hall and spoke eloquently, with intelligence and humor. He indirectly and carefully attacked Aristotelian doctrine, which claimed that the heavens are perfect, unchanging, and everlasting. These lectures brought him some amount of fame within Padua and his normal lectures, overcrowded, had to be moved outside.

As the star faded, so too did Galileo's interest in the sky. He busied himself with pulleys, magnets, and falling bodies. However, his interest was quickly revived during the summer of 1609 when he heard that a Dutch spectacle maker had invented a spyglass. Galileo began experimenting with lenses and quickly, without ever having been told how it worked, reinvented it. He continued tinkering with the design and by autumn of 1609, he had created an instrument that could magnify an object twenty times its original size — more than three times the magnification of the best Dutch instruments. He now had a spyglass capable of observing the sky. Admirers of the instrument soon named it the "telescope."

With his hand-crafted telescope, Galileo saw the heavens more clearly than anyone had ever seen them. He pointed it toward the Moon. He could see the details of the dark patches on its surface. Observing them night after night, he saw that the size and the shape of the dark spots changed as the Sun shone on the Moon's surface from different angles. They are shadows! Like the Earth, the Moon's surface is jagged and uneven.

Galileo pointed his telescope toward the stars. Even magnified twenty times, they still appeared to be points. But he could now see many, many stars that had never been seen before. He looked at the Milky Way, a cloudy band that stretches across the sky. Through the telescope, he could see that this band is actually made up of tiny, individual stars — there must be hundreds of thousands of them!

Galileo pointed his telescope toward Jupiter, visible in the winter sky. He saw three tiny dots arranged in a line, two to the east of

Jupiter and one to the west. The next night, they were all to the west of Jupiter. He continued to observe them and discovered a fourth dot in line with the others. The dots changed positions from night to night, sometimes disappearing as if behind or in front of Jupiter. The dots were orbiting Jupiter! Jupiter has moons!

Galileo saw Jupiter's moons in January of 1610, but already he had started working on a little book to share his telescope discoveries. He was forty-five years old, almost forty-six, and although he had begun to write several books, the only work that he had finished and published was a manual to accompany his military compass. With this new book, he sped toward completion. He continued to observe and write, even while parts of the book were being printed. The book came out on March 13th. All 550 printed copies sold within a week.

Galileo titled his book *Sidereus Nuncius*, "The Starry Messenger." He boldly introduced the work: "THE HERALD OF THE STARS unfolding GREAT, and HIGHLY ADMIRABLE Sights, and presenting to the gaze of everyone, but especially PHILOSOPHERS, and ASTRONOMERS." He composed the book not just for the intellectual elite, but for *everyone*. It was mainly a picture book, a collection of detailed sketches with words used to clarify and explain the drawings, a guided tour through the heavens — the stars themselves were speaking!

And the heavens spoke clearly. The Moon is not perfectly spherical, as Aristotle had claimed. Aristotle's concept of a celestial sphere, where all the stars are at an equal distance from the center of the Earth, is unlikely given the densely populated band of stars across the sky. The existence of moons around Jupiter demonstrates that not everything orbits the Earth. Aristotle was wrong.

One matter that the heavens did not take a stand on was whether the Earth moves. Nothing Galileo saw could decide this one way or the other. He had looked for parallax of the stars as proof that the Earth moves. Like all of those who had searched before him, he found none. The heavens did not take a stand, but Galileo did. He made this clear not at the beginning or end of the book, but right in the middle, claiming that the Earth should not be excluded from the "dance of the stars." He promised to demonstrate this in another book with many arguments and experiments.

Kepler received two copies of *Sidereus Nuncius*. The first was one passed on to him by Emperor Rudolph. Shortly after, he received a second one from Galileo himself accompanied by a letter asking for his opinion of the book. Kepler wrote back within a week, making it clear that he believed all that Galileo had written even before he could confirm the observations with his own eyes. He also offered some historical context for the work and found some points of support for Copernicus's model in Galileo's observations. He later published his reply in Prague as a thirty-five page book that was reprinted in Florence and Frankfurt. It took Galileo four months to write back to Kepler, but when he did he was appreciative: "I thank you because you were the first one, and practically the only one, to have complete faith in my assertions."

Galileo had for some time been growing restless. He longed to have the freedom, like Kepler, to pursue his work without the distractions of teaching responsibilities. He had started writing books on motion, mechanics, and the structure of the universe. He contemplated writing others, books on military science, light, sound, and the movement of animals. He wanted time to complete these books and time for experiments and astronomical observations. And he wanted to return to Florence. With his new instrument, Galileo saw an opportunity for all of this. He directed his attention toward Cosimo II of Medici, the Grand Duke of Tuscany.

Galileo had known Cosimo as a youth. He had instructed him in the use of the military compass and then returned to the court over summer holidays as his personal tutor. Cosimo had just recently reached legal age and assumed his current title. Galileo now honored him by naming the four moons of Jupiter after him and his three brothers; the satellites of Jupiter would be called the "Medicean stars." He then sent Cosimo a copy of the book along with his best telescope. In July of 1610, Galileo officially became the Grand Duke of Tuscany's chief mathematician and philosopher.

With this new freedom, Galileo continued to observe the heavens. He pointed his telescope towards Saturn. It had a strange appearance with bulges on either side. He planned to include this in a second edition of *Sidereus Nuncius* but didn't want to risk someone else claiming the discovery for themselves; telescopes, and some very good ones, were appearing throughout Europe. He sent a message to Kepler, first in code then later in plain language: "Saturn is not a single star, but three together, which touch each other." Kepler published this discovery on Galileo's behalf, as he had wished. Galileo announced his observations of Venus's phases in the same way, through Kepler. In both cases, he communicated with Kepler through a Medicean ambassador. He was in Florence now within the domain of the Roman Inquisition and could no longer communicate directly with a Protestant.

Galileo's observations of Venus's phases were particularly significant. At its smallest size, Venus appeared full. But at its largest size, Venus appeared as a crescent. This implied, firstly, that Venus must shine with reflected light from the Sun, like the Moon. This was important because the absence of emitted light from the Earth was used to argue against its movement. Galileo had used the Moon to invalidate this argument; the Moon does not have its own light and yet, according to all theories, moves. Now Venus provided additional support for this position. Furthermore, the observation of Venus's phases clearly demonstrated that Venus orbits the Sun, sounding the death knell for Ptolemy's theory for anyone who would trust their senses.

In early January of 1611, Galileo began planning a trip to Rome to share his discoveries. This would be his first official trip as the philosopher and mathematician to Grand Duke Cosimo. He was excited to bring the new science to the Church! The religious leaders greeted him cordially and with great interest. The Jesuits at the Roman College, who had been observing Jupiter's moons for over two months, embraced him as a colleague and an inspiration. He met with Pope Paul V, who praised and blessed him.

While Galileo was in Rome, Cardinal Robert Bellarmine, an intellectual leader and the pope's chief advisor, looked through the telescope. He had been one of the Inquisition judges who had condemned Giordano Bruno of heresy sixteen years earlier. Galileo heard with some relief that he was pleased with what he saw. Bellarmine then asked the Jesuit scientists to clarify five points: Was the Milky Way really made up of countless tiny stars? Yes it was, they responded. Was Saturn actually three stars joined together? They could not say. Saturn appeared to be oval in shape, but they could not see any separation between the parts. Did the Moon have an uneven surface? It seemed to, but they couldn't be sure. Does Jupiter have moons? Does Venus have phases? Yes, they could confirm both of these observations.

While in Rome seeking approval from religious leaders, Galileo was embraced by another group, the Academy of Lynxes. It was a secret society dedicated to the pursuit of knowledge, especially scientific. The idea of the organization was that each member of the society would instruct the others in an area of science to the mutual benefit of all. It had been established in 1603 by a young, rebellious nobleman, Prince Cesi. Eight years later, they only had four members — all nonscientists less than thirty years old. Kepler had written to Cesi about Galileo's book. When Cesi heard that Galileo was in Rome, he sought him out and invited him to join as their fifth member. With Galileo, the society quickly grew in size and stature.

Over the next several years, Galileo's attention alternated between floating bodies and sunspots. He was back in Florence surrounded by old friends and former students. He was also in the company of rivals, especially professors from the University of Pisa who resented his fame and the influence he had gained at their institution from his new position as chief mathematician and philosopher to the Grand Duke. The subject of floating bodies arose in this second group, when Galileo engaged in a dispute with a couple of professors of philosophy about the properties of ice. They argued that ice is compressed water, denser than water, and that it floats because of its shape. Galileo, on the other hand, argued that a body's density determines whether it sinks or floats; since ice floats, it must be *less* dense than water. He insisted that a piece of ice with any shape when submerged in water will rise to the surface. Likewise, a body that sinks will sink whatever its shape.

The dispute occurred in July of 1611, and soon after, the contestants settled upon a date for a public duel. The weapons of choice would be their experiments. For several reasons, not the least of which was the Grand Duke's reluctance to have his chief mathematician and philosopher engage in such a battle, the duel was postponed. In the meantime, the opposing party performed a public demonstration that supported their side, with balls and chips of ebony. The spheres sank and the chips floated on the surface. They loudly proclaimed victory and ridiculed Galileo.

Galileo responded with an essay addressed to the Grand Duke in which he reviewed the situation and defended his own position. Regarding his opponents' demonstration, he claimed that the chips on the surface were not completely solid but rather had pockets of air within them. The composite object, therefore, had a density less than that of water. If a chip were held underwater so that water filled those spaces, it would sink just like the sphere. He ended the essay with a rant against his opponents, who must have used Galileo's credentials as a mathematician to argue against his ability to consider matters of physics:

Here I expect a terrible rebuke from one of my adversaries, and I can almost hear him shouting in my ears that it is one thing to deal with matters physically, and quite another to do so mathematically, and that geometers should stick to their fantasies and not get entangled in philosophical matters — as if truth could ever be more than one; as if geometry up to our time had prejudiced the acquisition of true philosophy; as if it were impossible to be a geometer as well as a philosopher — and we must infer as a necessary consequence that anyone who knows geometry cannot know physics, and cannot reason about and deal with physical matters physically!

Galileo was a mathematician *and* a philosopher. He had insisted that this last designation, "philosopher," be added to his title when he accepted his position. Galileo had found imperfection in the heavens, which had broken the Aristotelian dichotomy between the perfect, unchanging, everlasting heavens and the imperfect, impermanent Earth. The universe was one. If the motions of the heavens could be described mathematically, so too could the things on Earth!

Galileo never published this essay, but he did eventually meet his adversaries. He and the professors were invited to the Grand Duke's dining table to battle in front of visiting cardinals who enjoyed intellectual debates. With simple experiments, Galileo clearly proved his point and left the duel triumphant, the uncontested winner. The Grand Duke requested that he write down his arguments in a treatise, which he did. In this, he described experiments that his readers could replicate at home to verify his conclusions. Along with his experiments and conceptual arguments, he added mathematically rigorous analysis.

The treatise was published in the spring of 1612 and gained immediate popularity due to the widespread publicity of the debate. Through this work, Galileo gained new supporters as well as new adversaries. Within a few months four books were published that vehemently defended the opposing position. These authors banded together with others, forming a unified force against Galileo. There was so much interest in Galileo's work that he was able to publish a second edition the same year with additional material.

Meanwhile, Galileo had been brought into another debate, one about sunspots: dark spots that appear as blemishes on the Sun's surface. A wealthy German banker, Marc Welser, had received three letters concerning sunspots from an anonymous stargazer writing under the name of "Apelles." Apelles suggested that the dark spots could be due to little stars orbiting the Sun. Welser sent copies of these letters for Galileo to read and asked for his opinion: Were the sunspots made of starry matter or not? Where are they located? How do they move?

Although Galileo had been observing the sunspots for over a year, he could not yet answer these questions with any degree of certainty. He had been looking directly at the Sun, at dawn or dusk, through the telescope. This would temporarily blind him, making it difficult to record anything that he had seen. Father Castelli, a Benedictine monk who was one of Galileo's former students, suggested that he project the Sun on paper. With this method, Galileo began to systematically track the sunspots, drawing sketches to indicate their positions, shapes, and sizes.

By studying these sketches, Galileo soon noticed that as the spots moved toward the edge of the circular image of the Sun, they became narrower and closer together and appeared to move more slowly. He concluded that these must be visual effects due to viewing them from the side rather than head-on. He then used geometric analysis and determined that the spots must be on the surface or right above it. He also showed that the spots, taking perspective into consideration, moved in uniform motion along the surface of the Sun.

Welser then became an intermediary in a debate between Galileo and Apelles, who was later revealed to be the Jesuit priest Father Christopher Scheiner. In his letters, Galileo attacked both the accuracy and interpretation of Scheiner's observations and abused him for clinging to Aristotelian notions. He also aggressively claimed the discovery of sunspots for himself. Welser did not publish Galileo's letters, although he had published Scheiner's, excusing himself by suggesting that it would be too expensive to print all of Galileo's large sketches. The Academy of Lynxes gladly took the letters from Welser, whom they had recently initiated into their society, and published them at the society's expense. They were always eager to spread radical ideas! They printed and sold well over a thousand copies, half with both sets of letters, the other half with only Galileo's side of the debate.

Although Galileo had shown the sunspots to religious leaders in Rome during his visit, he had not asked for their official position on them. Neither had he discussed the possibility of the movement of the Earth. Shortly after his first letter to Welser, Galileo brought up these ideas to a prefect of the Inquisition in Rome. He wanted to know whether the Bible insisted on perfect, unchanging heavenly bodies, or was that just Aristotle? Also, what was the Church's position on the motion of the Earth? The prefect replied that it was Aristotle who insisted on the immutability of the heavens, not the Church, and that the Bible actually indicates the contrary. As to the motion of the Earth, the Bible doesn't address this matter clearly, but Galileo should be careful not to assert that it moves until he can offer compelling proof.

Prince Cesi worked closely with the Inquisition censors to produce a final manuscript. The censors were satisfied with Galileo's observations and analysis of the sunspots but objected to his arguments that the new astronomy, with mutable heavens, was more consistent with the Bible than the old astronomy. The censors deleted this as well as other attempts to interpret the scriptures. Kepler, too, had been advised to remove such theological interpretations from his *Mysterium Cosmographicum*. It was acceptable for scientists, as scientists, to put forth radical ideas. But they should not attempt to be theologians.

Galileo took the scriptural interpretations out of his publication but included some in a letter to Father Castelli. Castelli was a devoted Copernican and trusted friend. Upon Galileo's recommendation, he had recently obtained Galileo's old position as the chair of mathematics at the University of Pisa. While having breakfast at the Grand Duke's palace, Castelli defended Copernicanism to several other university professors with the help of scriptural passages. He had written to Galileo about the discussion; Galileo responded with forceful arguments that might help his case.

With Galileo's permission, this letter was widely circulated. It was received by, among others, a couple of Dominican priests. A year later, one of them delivered an impassioned sermon denouncing Galileo. The other sent the letter to the Inquisition in Rome, formally charging Galileo with heresy. The Inquisition judges spent a year collecting evidence and then finally decided that Galileo was not guilty. But Galileo was not content merely to be let off the charges. He was a crusader on a mission to convert the Church to the new science! He had visited Rome to help make a case for himself against the accusers, and now he remained in Rome campaigning on behalf of Copernicanism, urging the Church to reconsider their views.

And so, the Church reconsidered their views. A panel of theologians chosen by Pope Paul V, none of whom were mathematicians or scientists, came to the conclusion that Copernicanism was both absurd and heretical. Cardinal Bellarmine, on behalf of the pope, delivered the message that Galileo was to neither teach, defend, nor discuss Copernicanism either orally or in writing. The formal decree, which was pronounced in March of 1616, was more ambivalent: the works of Copernicus were to be amended and all books teaching Copernicanism as true were to be banned.

After this, Galileo no longer dared to fight openly for Copernicanism, but he could still observe the sky. He attempted once more to measure the parallax of the stars. This would have been convincing physical evidence that the Earth moves. If he could measure the parallax of stars, the Church would *have* to reconsider its position. Again, he found none.

Galileo continued to study Jupiter's moons. He had been observing them now for many years and could accurately predict when each moon would eclipse Jupiter. Since the eclipses would all occur at the same time throughout the world but at different positions in the sky, they could be used by sailors to determine their position at sea. Specifically, Jupiter's moons could help determine longitude, which would otherwise be impossible because of the continual movement of the stars across the sky during the night. Galileo designed a hat with several mounted telescopes protruding outward so that sailors could easily observe both the sky and the sea. He also invented a small device that could be used to translate the observations of Jupiter's eclipses into useful information.

During the summer and fall of 1618, three comets appeared in the sky. Galileo heard about the comets but was too sick to observe them. His friends and supporters were eager to know his views, whether the comets could be used to support the Copernican system. Galileo initially kept silent but another stepped forward, the Jesuit Orazio Grassi, who claimed that the comet's lack of parallax showed that it was beyond the Moon, in contradiction to Aristotle's belief in the immutability and permanence of the heavens. Grassi then used this to give support to the Tychonic system in which the planets orbit the Sun, which orbits the stationary Earth. Galileo, hiding behind a thin shield of one of his students, attacked Grassi's paper, defending Aristotle's view of comets! His arguments were seemingly designed only to weaken support for Tycho's model and humiliate his opponent.

At any point, Galileo could have gained freedom by traveling abroad. Kepler had recently published *Epitome of Copernican Astronomy*, which gave strong mathematical and physical support to the Copernican system based on his laws of planetary motion. The book was immediately put on the Index of Banned Books, but Kepler remained untouched. Kepler wanted to convert the world to Copernicanism; Galileo wanted to convert Rome. Eventually political changes there made it seem possible.

First, the two leading figures involved in Galileo's Inquisition trial, Pope Paul V and Cardinal Bellarmine, died. Then in August of 1623, Cardinal Maffeo Barberini was elected pope. Cardinal Barberini had been one of the guests at Grand Duke Cosimo's during the duel with the Aristotelian professors. He had cheered for Galileo during this debate and had been an enthusiastic supporter ever since. Barberini was often among the first to be sent Galileo's new discoveries and writings, and he was always appreciative. Barberini had sent Galileo a poem of praise with a letter signed, "your brother." With confidence that it would be enjoyed, Galileo dedicated his latest work to Barberini, now Pope Urban VIII. The book was read to the pope at dinner, and he roared with laughter.

With a greater sense of freedom than he had felt in many years, Galileo returned to the work he had begun decades earlier: his defense of the Copernican system. He wrote slowly, easily distracted. He tinkered around with magnets, wrote about miracles, and calculated gambling probabilities. He stopped writing altogether for about three years. After recovering from a serious illness, he renewed his resolve to finish the book. Finally, in May of 1630, Galileo arrived in Rome with a completed manuscript.

The book was written as a dialogue, a popular form in Galileo's day. The culmination of the dialogue was a theory of tides presented as physical proof of the motion of the Earth. The book was meant to be enjoyed, with humor mixed in among the intellectual arguments. Like all of his works with the exception of *Siderius Nuncius,* it was written in vernacular Italian. He made this choice to broaden his audience, believing that there were many people without a formal education who had the intelligence to follow his arguments. This choice of language also had the effect of making the dialogue even more vibrant.

The dialogue takes place between three characters: Salviati, Sagredo, and Simplicio. Galileo based these characters on people in his own life. Salviati was a brilliant, skeptical, young nobleman, a friend of Galileo from Florence who had been Galileo's companion while studying floating bodies and sunspots. Salviati would be the hero of the dialogue, Galileo's mouthpiece. The supporting role would be played by Sagredo, a Venetian noble and former student from Padua. He and Galileo had exchanged letters, sharing with each other their thoughts, humorous anecdotes, and life stories for almost a decade, sometimes as frequently as twice a week. Simplicio was the incarnation of all his opponents and enemies throughout his life. His friends Salviati and Sagredo had died many years before. His most important enemies were still very much alive.

The book was good entertainment. Galileo polarized the debate by making Simplicio an Aristotelian even though by this time most people had abandoned Ptolemy's Earth-centered model in favor of Tycho's system. Salviati led the attack, with Sagredo quickly won over. The two of them then ganged up on Simplicio, who offered straw arguments that were easily defeated. Nowhere in the dialogue was there any reference to Kepler's laws, which would have provided the most compelling arguments in favor of a Sun-centered model. Galileo had never studied Kepler's works; he found the mathematics too difficult. The only reference to Kepler was in the final section of the book, where Galileo ridiculed his theory of tides. Kepler thought that the tides were caused by the Moon. Galileo, rather, believed that the tides were caused by the motion of the Earth despite the fact that this theory led to the erroneous prediction that there should only be one tide a day.

Galileo was welcomed by the pope in Rome and spoke with the censors about his book, who recommended small changes here and there. Since his theory of tides was used to prove the motion of the Earth, Galileo had named his work *Dialogue on the Ebb and Flow of the Sea*. The censors objected to this title, believing that it presented the dialogue as being too one-sided. Instead, they suggested *Dialogue Concerning the Two Chief World Systems*, which gave equal weight to both positions. Galileo left Rome with the book nearly approved. The censors added a preface and conclusion that made clear the hypothetical nature of the arguments for a moving Earth, and the manuscript was ready for publication.

The Lyncean Academy had dissolved after Prince Cesi's death and there remained no one to oversee the publication in Rome. Taking the advice of his friend Father Castelli, now a mathematician at the Roman College, Galileo published the book in Florence. In May of 1632, the first copies arrived in Rome.

Galileo's *Dialogue* was read by Father Scheiner with particular interest. Scheiner had not forgotten how Galileo had ridiculed him for his early work on sunspots and again more recently under the protection of the new pope. Here in Galileo's *Dialogue* was the final straw. Galileo had taken Scheiner's model of the sunspots from a recent publication, replicating it so closely that he must have had the book in front of him while writing. He not only did not give Scheiner credit, but he also claimed that he had come up with the model years earlier, before Scheiner had written the book. Galileo then used it to argue for the Copernican model.

Scheiner joined together with his fellow Jesuit, Father Grassi, who had been attacked by Galileo over his work with comets. These Jesuits were intelligent, courageous, open-minded thinkers who had been inspired by Galileo's work. Now, they joined together to destroy him. They enlisted the help of their Jesuit brothers to put pressure on the censors: Didn't they notice how *weak* the arguments were that Galileo made against Copernicanism? And the introduction and conclusion written by the censors were italicized, clearly separating them from the rest! In response to the Jesuits' involvement, the censors began reviewing the book and quickly started confiscating the copies that had been released.

The Jesuits then convinced Pope Urban V that the simple-minded character Simplicio was himself – that Galileo had dared to poke fun at him! All of the trust, respect, and affection toward Galileo that had built up over their twenty-year relationship crumbled. With this, Galileo lost his most important protector.

At first, there was some hope that the removal and correction of the book would resolve the matter. But then the Inquisition forces dug up evidence from the 1616 trial revealing that Galileo had been commanded not to teach, defend, or discuss Copernicanism. It was Pope Urban V, then Cardinal Barberini, whose influence had prevented the use of "heresy" to describe Copernicanism in the earlier trial. Now, he ordered Galileo to recant his belief in Copernicanism, thereby proclaiming it heretical and Galileo a heretic. The Inquisition forces put Galileo's *Dialogue* on the Index of Banned Books and burned all available copies.

After the trial, Galileo was detained in Rome for over a year. He was then permitted to leave under the supervision of a sympathetic archbishop for the sake of his health; the stress of the situation had taken a toll on his mind and body. He was filled with bitterness at the betrayal by the Jesuits, which despite great efforts by others on his behalf, could not be undone. At the same time, he was filled with remorse and self-reproach, realizing that his pride and ambition had caused his downfall.

Galileo reached the bottom soon after the death of his eldest daughter, Sister Maria Celeste. Her letters had been his greatest source of comfort during his trial and imprisonment. When she became sick, he was allowed to move back home, to within walking distance of the convent where she lived. Galileo never left her side during the months leading up to her death, and once she was gone, he felt that he had lost everything.

Slowly, Galileo began to revive. He occupied himself with gardening and playing the lute. He was cheered by the many letters that streamed in from around Europe. A long-time correspondent in Paris sent him good news about the translation of his *Dialogue* into Latin. He also arranged that a portrait be made of Galileo. The painter encouraged Galileo to talk while he sat; when he talked, his face came back to life.

And Galileo began to work again. He had been strictly forbidden to speak, write, or in any way discuss Copernicanism. He was not even allowed to argue against it. But if the universe is one, the new science applies to the Earth as it does to the heavens. He was a mathematician *and* a philosopher, a geometer of celestial bodies *and* bodies on Earth! Now, at the age of seventy, he returned to the work that he had begun many years before while still a professor in Padua and began to write it all down.

Galileo wrote this new book, like his earlier one, as a dialogue. The characters were the same: Salviati, Sagredo, and Simplicio. However, Simplicio was no longer the simple-minded character of his earlier work but rather a younger version of Galileo himself, asking the same kinds of questions that he had once asked. The work also took a more serious tone, replacing derisive humor with definitions, theorems, and propositions in the tradition of the great geometers and astronomers.

In this book, Galileo began by discussing how the strength of an object depends on its size and shape. He established the importance of scale, that strength and size do not increase proportionally. Animals and plants, for example, are proportionally stronger when they are small. And smaller animals can fall from a greater height than larger animals and not be harmed.

Furthermore, Galileo considered how shape affects strength. He proposed that, given a fixed amount of material, a beam bent in the shape of a parabola is stronger than in any other shape. A hollow cylinder is stronger than a solid one made out of the same amount of material. Although most of his mechanical analysis was conceptual, he suggested a mathematical law guiding the balancing of unequal weights: it is possible to create a balance between two unequal weights if their distance from the balance point is inversely proportional to their weight.

Galileo then considered motion, focusing on the motion of a falling body. If a marble egg and a hen's egg fall in water, the marble egg will reach the bottom well ahead of the hen's egg. If these same two eggs fall in air, they will fall at nearly the same rate. Therefore, he concluded, in a medium completely devoid of resistance, two bodies should fall at the same rate. He could not test this directly as he had no way to observe a body falling in a vacuum. But he could *imagine* it.

Galileo next demonstrated that a falling body accelerates: if a block falls onto a stake from a greater height, it will push the stake in deeper when compared to one falling from a lesser height. Therefore, a falling body must speed up as it falls. He hypothesized that the acceleration of a falling body is uniform; if it falls from rest, it will be moving twice as fast after two seconds as it was moving after one second. He then proposed an experiment to test this hypothesis by rolling a marble down a ramp. He argued that the ramp should slow down the marble's motion without altering the character of the motion. With a carefully constructed ramp, he tested his hypothesis, verifying that a falling body accelerates at a constant rate.

Galileo could not create a surface without resistance, but he could imagine it. A marble rolling down a ramp speeds up. A marble rolling up a ramp slows down. A marble rolling on a perfectly smooth horizontal surface should move with uniform, eternal motion. And if a projectile is thrown upward at an angle so that it moves both horizontally and vertically, its horizontal motion should be constant while its vertical motion should be uniformly accelerated. In this case, in the absence of resistance, the path of the projectile should be a perfect parabola. Galileo had found imperfections in the heavens, and now he found perfection on Earth.

Even before finishing the book, Galileo began to make inquiries regarding its publication. He was still under house arrest, as he would be for the rest of his life, with his movements and communications restricted. Rome had banned all of his books, including any that he might write in the future. The Inquisition sent notices throughout Italy, making it clear that nothing at all written by Galileo was to be published anywhere ever.

Galileo saw an opportunity when he heard that Count de Noailles, a French ambassador who had been one of his students in Padua, was returning home to Paris after a visit to Rome. He obtained permission to meet the ambassador. In secret, he gave Noailles the manuscript to bring to Paris. There, it was translated into Latin and given a title, *Discourses and Mathematical Demonstrations Relating to Two New Sciences*. Galileo added an introduction to the manuscript that made it clear that he did not intend for the book to be published. He only wanted to make sure it would not be lost in case someone, somewhere, someday wanted to read it.

But Galileo *did* want it published. While still struggling to figure out how he would do this, he was informed that the court in Amsterdam was offering a prize to anyone who could devise a method to determine longitude at sea. He had already done this – with Jupiter's moons! He gathered together his writings on the subject, including descriptions of his telescope helmet, and sent them to Amsterdam. The Dutch, impressed with his work, offered him gold and a distinguished chair at the Athenaeum of Amsterdam. Hearing about the offer, the church in Rome forbade Galileo from accepting the gold. Even better than gold, the Dutch offered to publish his book. And so, from Holland, Galileo's new science was sent out into the world.

# 7

# Newton's Universe

A Frenchman who arrives in London finds a great alteration
in philosophy, as in other things. He left the world full;
he finds it empty. At Paris you see the universe composed
of vortices of subtle matter; at London we see nothing of the kind.
With you it is the pressure of the moon which causes the tides of the sea;
in England it is the sea which gravitates toward the moon...
Among you Cartesians all is done by impulsion;
with the Newtonians it is done by an attraction
of which we know the cause no better.

— Voltaire

Isaac Newton was born on Christmas Day in 1642 in a small manor house in Woolsthorpe in the county of Lincolnshire. He was born prematurely; his mother said that he was so small he could fit into a quart pot. He was given the same name as his father, an illiterate farmer who died three months before his son was born. With neither a father nor siblings, Newton enjoyed his early years with his mother without rival. But just after he turned three, his mother remarried and moved away, leaving him in the care of her mother. When he was eleven, his mother's husband died and she returned, bringing three new children. She then sent him away to a nearby market town, Grantham, where he went to grammar school and boarded with the apothecary. At sixteen years, she called him home to become a farmer so that he could take over his father's estate.

As a youth, Newton spent much of his free time building. He built a water clock with a dial plate at the top indicating the hour. He created a miniature windmill, which he sometimes powered with a mouse. He enjoyed designing and constructing kites, carefully choosing their dimensions and the perfect point to attach the string; he would occasionally fly a kite with a paper lantern hanging from the string on a dark night so that it looked like a comet, frightening the neighbors. He created an elaborate sundial, improving it over the years until it could accurately tell time throughout the seasons. Newton had one friend, the daughter of one of his mother's friends. For this girl and her friends, he would build little tables, cupboards, and other playthings.

When he wasn't building, Newton was usually reading or writing. He wrote melancholic phrases in the pages of his Latin exercise book: "A little fellow; My poore help; Hee is paile... I will make an end. I cannot but weepe. I know not what to doe." With money his mother had given him, he bought a small notebook. Inside the cover he wrote, *Isacus Newton hunc librum possidet*, "Isaac Newton owns this book." He wrote in tiny script, often less than one-sixteenth of an inch high. He copied parts out of the book *The Mysteries of Nature and Art* by John Bate to aid with his building projects. He copied drawing instructions and recipes for making inks, salves, and powders. He copied astronomical tables to use with his sundial, which he supplemented with computations from his own observations.

After being called home from school to become a farmer, Newton quietly rebelled. When he went to the marketplace accompanied by a servant to do business, he would abandon the servant to do the work while he ran off to borrow books from the apothecary. At home he would lie reading in the meadows when he should have been tending the sheep. He left the fences in disrepair and let the swine run loose; for these negligences he was fined in the manor court. Eventually, his old schoolmaster and his mother's brother interceded on his behalf, and Newton was sent back to school.

In June of 1661, Newton entered Trinity College at Cambridge University, where his uncle had gone to study for the clergy. His mother could have afforded to pay tuition; she now had income from the estates of her two deceased husbands. She chose instead to enroll him as a subsizar, the lowest category of students who earned their keep through service to other students and ate their leftovers. She did give him a modest allowance, which he managed carefully.

Although Trinity was only about a hundred years old at this time, its curriculum was two hundred years older than that, inherited from the medieval colleges that merged to form it. Consistent with the college's ancestry, Aristotle was the authority on everything secular, including natural philosophy. Nature is a principle of motion, and motion is change. If we understand *causes*, we can understand the workings of the universe. The things on the Earth are made of four earthly elements, which rise or fall naturally depending on whether they are light or heavy, or move violently if subject to a push or a pull. The things in the heavens are made of a completely different element, which is, like the heavens themselves, perfect, eternal, and unchanging. Within Trinity's curriculum, Ptolemy's fifteen-century-old Earth-centered model of the universe still reigned supreme.

But outside Trinity, and even within the large holdings of Trinity's library, Aristotelian doctrine was crumbling. In England and throughout Europe, astronomers were looking at the heavens through telescopes. They saw the jagged surface of the Moon, the phases of Venus, the moons of Jupiter, and the hundreds of thousands of stars in the Milky Way. They knew that Aristotle's conception of the heavens was incorrect. Many had discarded Ptolemy's model in favor of Copernicus's, in which Earth is a planet orbiting around the Sun along with the rest of the planets. Speculative works began to appear, both fiction and nonfiction, considering the possibility of intelligent extraterrestrial life, and people began dreaming up flying ships that could reach these other lands.

Galileo Galilei, an Italian astronomer and philosopher, had suggested that the natural motion of an object is uniform and *all* matter is heavy, accelerating toward the center of the Earth at the same rate in the absence of resistance. With these assumptions, it was possible to mathematically determine the path of a projectile as it moves through the air. Natural philosophy was merging with practical innovation as gunners used Galileo's method to calculate the trajectory of cannons. Even the ancient prejudice against manual work was fading. Books on mechanical inventions were being published, and it was becoming fashionable for nobles to try their hand at the new science.

Through Trinity's curriculum, Newton was fully immersed in the Aristotelian tradition. He read Aristotle in different translations with commentary and debate. He respected and admired Aristotle, the collector of books who believed that the universe was knowable and sought to understand it in its entirety. However, through Trinity's library and his own book purchases, Newton read a lot that blatantly and decisively contradicted Aristotle. Aristotle had disagreed with his teacher, writing "Plato is my friend, but truth is my greater friend." Newton copied this expression into his notebook, inserting Aristotle into the sequence: "Plato is my friend, Aristotle is my friend, but truth is my greater friend." He then went forward, in the tradition of Aristotle, to seek the truth.

Newton's new notebook was one hundred and forty pages. He filled the front and back of it with Aristotle, writing in esoteric shorthand to conserve paper and encrypt his work. In his second year, he began a new section in the middle of the notebook, which he titled *Questiones Quœdam Philosophicœ* — "Some Philosophical Questions." In this, he organized his knowledge of the natural world, separating each part with a label: matter & atoms, quantity & place, time & eternity, motion, celestial matter & orbs, gravity & levity, heat & cold, magnetic attraction, colors, sounds, memory, and so on — forty-five topics in all. He sometimes filled in these topics based on his reading and other times based on his own reasoning or speculation. Occasionally, he left a topic blank altogether. This was a beginning.

During his third year at Trinity, Newton's studies took a turn. He purchased an astrology textbook at a fair and was puzzled by a diagram of the heavens that assumed knowledge of trigonometry, which he hadn't studied. He bought a book on trigonometry but couldn't understand the proofs. He then began to study Euclid. He searched for more books and soon had at hand the most advanced mathematical texts from all over Europe. It was around the same time that Cambridge, for the first time in its history, had a chair of mathematics, Isaac Burrows. Newton attended his lectures while continuing his independent work.

Newton's mathematical studies were interrupted in the middle of his fourth year, in December of 1664, when a comet appeared. He stayed up tracking its position against the background of the stars all night until it disappeared in the morning light. He did this night after night until he finally collapsed from exhaustion.

To many, a comet was a fearful sight, an omen of bad fortune. In England, the sighting of this comet coincided with another bad omen, news from abroad of a deadly plague spreading throughout Europe. A scattering of suspicious deaths appeared in London. Soon, it was obvious that the threat was real, and city officials began taking action to contain the disease. Fear spread throughout England, and in June of 1665, all of Cambridge University was shut down.

Newton returned to Woolsthorpe and settled back into the manor house of his birth. He created a study in one of the upstairs rooms, building a bookshelf for his many books. He had a new notebook, a thousand-page book that he had inherited from his stepfather. He had written a few things in it already, but it was mostly empty. He labeled it "Waste Book." This was to be his reading notebook.

Newton had studied Euclid, but now he read it more carefully. He worked through the geometric proofs as mathematicians had since ancient times, using only a straightedge and compass. This was the way that geometry was supposed to be done, without the aid of a marked ruler or protractor, tools used by common craftsmen.

But the old philosophy was crumbling, and everything that had been assumed was now being questioned. The French scholar René Descartes had proposed a new philosophy, a new method to discover truth. As an example of his philosophy he suggested a new way to do geometry: points, lines, and shapes could be described with algebraic equations — with numbers, letters, and symbols. He sketched out these ideas in an appendix, La Géométrie, which was soon translated from French into Latin. By the time Newton read it, a second Latin edition had come out with additional explanations and commentary.

Newton worked through this book as well, filling in steps, expanding, asking questions and then answering them. The union between geometry and algebra was rich with possibilities. Descartes had found equations for shapes in two dimensions. Newton now extended these to three dimensions. Descartes described shapes on one set of axes. Newton figured out how to transform the equations from one set of axes to any other. He drew a hyperbola and attempted to calculate its area by setting up a series and summing its terms. In his notebook, he neatly recorded the calculation of the series to fifty-five decimal places.

Newton also began considering what none of his predecessors had dared to consider: infinity. Descartes had used "x" and "y" to denote the positions of points on a plane. Lines, curves, and shapes could be written in terms of these quantities through equations. Newton realized that for every equation there is an infinite number of $x$'s and $y$'s separated by an infinitely small amount — the infinite and infinitesimal can work together to create finite values. He had calculated the area under a hyperbola to fifty-five decimal places. With an *infinite* sum, its area could be determined exactly.

Newton continued to innovate and extend. He had worked out the mathematics of the infinite and the infinitesimal, calculating maximums, minimums, slopes, and areas. Next he sought a physical connection. A curve could represent a shape, but it could also represent the position of an object as it moves in time. In this case, its slope is the object's velocity, how quickly and in what direction its position changes. The slope of its velocity is its acceleration.

Newton considered simple motions in addition to compound ones. What if an object moving in a circular path were also moving perpendicular to the plane of the circle at a constant velocity? One could simply add the algebraic expressions. It would yield a spiral! What would be the path of a point moving around a circle that was also rolling, like a point on the rim of a wheel? A cycloid! The motions could be described as curves, which could then be added to yield more complicated motions. Similarly, if one could determine the motion of two objects separately, it was possible to determine their relative motion by subtracting the two. With his new mathematics, Newton could determine velocities from positions and accelerations from velocities. He could even do the reverse. The possibilities seemed endless.

Meanwhile, the plague raged on in London killing hundreds and at its worst thousands per day. Before it was over it would kill almost a quarter of the city. It spread to other areas of the country, but Newton was safe in the seclusion of his home. And the quiet of this seclusion gave him time to absorb new ideas and allow each new idea to give birth to others. He soon turned his attention to the *cause* of motion. Why does a stone fall? What causes the Moon to orbit the Earth? What determines the motion of a comet?

Galileo believed that the natural motion of an object is constant, that in the absence of any resistance an object moving on a horizontal surface would continue in a straight line at the same speed forever. He also claimed that the acceleration of a falling body in the absence of resistance is constant. Newton felt that this last statement must be incorrect; rather, the acceleration of an object must be smaller when it is farther from the Earth's center. Is it possible that whatever it is that makes an object accelerate toward the Earth is the same thing that causes the Moon to deviate from a straight path? Could this influence centered on the Earth extend outward infinitely, just getting weaker and weaker?

Newton began to play around with numbers. He had Galileo's value for the acceleration of a falling body in a new translation, about one hundred cubits per five seconds, but he needed more accuracy. He tried to determine his own value using a ball tied to a string moving in a horizontal circle, a conical pendulum. In his notebook, he recorded that it went through 1,512 "ticks" in an hour. This corresponded to a value for the acceleration that was about twice as large as the one based on Galileo's calculations. He needed other values too, the radius of the Earth and the distance between the Earth and Moon, numbers recorded in various sources with an assortment of units: miles, passūs, braci, and pedes. He needed a precise value for the motion of the Moon against the stars.

With his best estimates of all of the required data, Newton calculated that the acceleration of a falling body at the Earth's surface was about 4000 times the acceleration of the Moon about the Earth. Could the influence of the Earth decrease according to an inverse-square law, as brightness decreases as the distance from the source increases? He arrived at the same inverse-square relation when considering Kepler's laws. Everything seemed to point to an inverse-square law, but the numbers didn't quite work out. The distance between the Earth and the Moon is about 60 times the Earth's radius.

According to an inverse-square law, its acceleration should be smaller by a factor of $60^2 = 3600$, not the 4000 he had calculated.

Descartes had an entirely different idea. Rather than assuming some kind of invisible influence that extends through space, he proposed that all of the large-scale motion in the universe can be explained mechanically through small-scale collisions among particles of matter. The universe is full. At its creation, it was set in motion — every movement of one particle leads to the displacement of another. This formed stable "vortexes" throughout the universe. The larger a body is, the easier it is to transmit its motion; therefore, it was natural for the largest body in a region to form a center of motion — the Sun is the center of the planetary system, planets are the center of the orbits of moons, and so on. An object appears to have "weight" because it is being pushed toward the center of a vortex by the pressure of the matter surrounding it.

Descartes had created a complete theory to rival Aristotle's. He had new laws of motion and a new theory of elements in which everything in the universe is made up of different forms of light. It was a compelling theory, and Newton considered it. According to Descartes, light is a pressure that we "see" when the particles of light hit our eyes. This was something Newton could test, so he slid a needle into his eye socket and pressed it against his eyeball. When he rubbed his eye with the point of the needle, he could see clear circles, which faded when he kept the needle still. He stared with one eye at the Sun. When he then turned to look at a dark wall, he saw circles of color. He repeated this many times until he began to fear that he would permanently ruin his vision. He then hid himself in a dark room for several days until his eyes recovered.

The plague was abating. Newton had obtained a new book from London, *Micrographia*, written by the English natural philosopher Robert Hooke. Hooke was the Curator of Experiments for the Royal Society of London, a newly founded society dedicated to promoting and communicating the "New Philosophy," which valued experimentation as a way of arriving at scientific truth. As curator, he performed experiments on the nature of air, respiration, barometric pressure, color, magnets, falling bodies, and pendulums. In *Micrographia*, the first major publication of the Royal Society, Hooke described and drew detailed pictures of the miniature world as viewed through a microscope: the tip of a needle, linen cloth, flakes of ice, charcoal, petrified wood, leaves, mold, insects, and various other living and nonliving entities.

In addition to these pictures and descriptions, Hooke recorded observations of light through his microscope. He observed glowing objects and viewed light through thin plates and prisms. Based
on his observations and experiments, he concluded that light is a
quick, short, vibrating motion. Furthermore, blue light is a vibration
in which the weakest part precedes the strongest part when hitting
the retina of the eye. Red is the opposite; the strongest part precedes
and the weakest follows. He claimed that with this hypothesis, it is
possible to explain *all of the observations of color in the world.*

This was another interesting idea. Newton didn't have a microscope, but he did have a prism that he had bought at a fair. With
this prism, he began to perform his own experiments. He recorded
his observations in his Waste Book along with his infinities, infinitesimals, and the inverse-square law. The plague was now over; Newton closed his notebook and returned to Trinity.


Back at the university, Newton continued his mathematical investigations and shared some of his work with Burrow, the mathematics
chair. Burrow then showed him a new book that demonstrated how
to calculate logarithm values using an infinite series. This was just
a special case of Newton's own theory! With encouragement, Newton
agreed to share some of his work with the mathematician and member of London's Royal Society, John Collins. Collins received it enthusiastically and begged for more. Could he calculate the interest on an
annuity? Would he share the general theory? Newton replied with
the formula to calculate interest. He did not share his general theory.

Newton quickly climbed through the ranks at Trinity. In
1665, just before Cambridge closed, Newton had earned his Bachelor
of Arts degree. Upon returning in 1667, he was elected as a Minor
Fellow. The next year, he became a Major Fellow and earned his Master of Arts degree. The year after that, in 1669, Burrow resigned as
mathematics chair, and Newton was chosen to take his place. In this
position, Newton had freedom and security. His only academic obli-
gation was to read a lecture on mathematics once a week and deliver
it to the university library.

By this time, Newton had effectively abandoned his mathematical studies. He studied old alchemy texts, taking extensive notes
on his readings. He built a laboratory in a shed outside of his room

where he performed his own alchemical experiments. He also studied the Bible. He compared the new English translation of the Bible with older versions in Latin, Greek, Hebrew, and French. He attempted to apply systematic analysis and logic to the interpretation of scripture; he came up with a list of fifteen rules as a guide.

Newton continued to do optical research. It was on this subject that he lectured, when he did give a lecture, sometimes reading his notes to a completely empty room. He had discovered that different colors of light bend by different amounts when entering another material, which results in the distortion of an image magnified with a lens. He designed a telescope that avoided this problem altogether by using mirrors rather than lenses to magnify the object. He built it in his laboratory, carefully hand grinding his own mirrors. It was short and wide, able to gather a lot of light. He kept it for two years, observing Jupiter and its moons, the phases of Venus, and other celestial objects. He then lent it to Burrow who brought it to London to share with his friends at the Royal Society. They received it enthusiastically and invited Newton to join the Society as a member.

The Secretary of London's Royal Society was a German who had immigrated to London during the Thirty Years' War on a political mission. He was born Heinrich Oldenburg. He later Latinized his name to Henricus and then finally assumed the English first name "Henry." He was a master of languages and with his international connections made a perfect focal point for the Society's communications. He received scientific news from a variety of sources and published them under the title, *Philosophical Transactions.* Newton had been reading these for years, taking careful notes. After Oldenburg received the telescope from Newton, he reached out to him, inviting him to claim credit for his invention in writing.

In a long letter addressed to Oldenburg, Newton narrated the journey that had led to his invention. It was a personal account, a story of discovery. He shared the details of his experiments along with the pleasure, confusion, and excitement he felt as the pieces began to fit together. Oldenburg read the letter aloud at a meeting, as Newton had intended. Robert Hooke, who was in the audience, quickly dismissed Newton's work, contradicting both his theory and the validity of his observations.

Hooke's response to Newton's work initiated a series of letters written by the two men, Hooke attacking and Newton defending. They were all addressed to Oldenburg and published in his *Philo-*

*sophical Transactions.* Newton, who labored over every response, became exhausted. Hooke had found a soft spot, a weakness in Newton's work. Newton had no "Hypothesis" for the nature of light; he only had the certainty of his experiments and mathematical analysis. He withdrew from the Society and from all communications for two years. He then returned with his own hypothesis, which he explained in a long letter to Oldenburg. He did not pretend to have certainty here but rather presented it as if he believed it to be true without proof. The letter was to be read aloud to the Society but not published. Furthermore, he did not wish to answer any objections.

Newton had found a hypothesis, and he had found a weakness in Hooke. Hooke had claimed the discovery of an optical phenomenon called diffraction. However, as Newton pointed out in his letter, the French Jesuit Honoré Fabri had already written about it before Hooke, and Fabri had learned about it through the Italian mathematician Francesco Grimaldi. Hooke then told members of the Society that Newton had taken his theory of light from *Micrographia*, merely putting the finishing touches on what he had started. Oldenburg shared this with Newton and then read his response in front of the Society — yes, Newton had built upon Hooke's work, but to such an extent to make it almost unrecognizable. Anyway, Hooke had only slightly modified what Descartes had said first. After four years, Newton and Hooke had reached a stalemate; they agreed that all future debates should be private.


Newton and Hooke interacted civilly over the next several years. Hooke had been converted to Newton's theory of colors and in this area sided with him against others. Newton congratulated Hooke when, after Oldenburg died, he was chosen to be the new secretary of the Royal Society. Hooke assured Newton that he would keep him informed of what was going on at the Society as Oldenburg had done.

In November of 1679, Hooke wrote a letter to Newton inquiring about his activities. He asked if Newton had any objections to his idea, published five years earlier, concerning the motion of the planets — that a uniform, straight-line motion combined with an attraction toward the Sun could produce a planetary orbit. Hooke also referred to a recent paper in which he had attempted to prove the motion of the Earth. Newton's answer was evasive. He had recently returned from Lincolnshire where he had been for many months tending to his sick mother and, after she died, taking care of her affairs. He referred to this and wrote that he did not recall hearing of Hooke's work. Anyway, he had practically given up on experimental philosophy. Even so, he offered Hooke an idea to prove the motion of the Earth: drop an object from a great height and it should spiral, landing to the east of where it was dropped.

This was an error and Hooke pounced. He read Newton's letter aloud to the Royal Society, although he had promised days before that he would keep their communications private. He pointed out Newton's mistake, explaining that, ignoring resistance, the path should be an eccentric elliptoid. With resistance it should be an eccentric ellipti-spiral that should make the object land not east, but south-east. Newton was furious. He responded once more, correcting his own mistake and suggesting that the path would be even more complicated than Hooke claimed. He also hinted that he had begun a mathematical solution. He sarcastically retracted in advance any mistakes he might make.

Hooke wrote another letter to Newton, clarifying the problem at hand: the attractive force should be an inverse-square relationship to the distance. With this and Newton's mathematical method it should be easy to determine the specific curve and suggest a reason for the inverse-square law. But Newton was done with Hooke. He never replied.

In June of 1682 during a meeting of the Royal Society, a new value for the distance of a degree of the meridian was announced. Newton took note and used it to recalculate the diameter of the Earth. He then returned to the calculations he had performed fifteen years earlier while at his home in Woolsthorpe during the plague. Now the calculations worked out! The same force that causes an object to accelerate near the Earth's surface can account for the acceleration of the Moon! He then composed a series of propositions related to the motion of the planets around the Sun. He communicated these to the Society, offering up the problem for others to work on, and returned to his alchemy and scriptural studies.

A couple years later, Hooke was at a coffee house in London discussing planetary motion with Christopher Wren, one of the founding members of the Royal Society, and a young astronomer, Edmond Halley. All three men had independently arrived at an inverse-square law for the attraction of a planet toward the Sun. Hooke boasted that he had already worked out the mathematics of planetary motion but didn't want to share his calculations until more people had tried and failed. In this way, others would have a greater appreciation for his work.

Halley doubted that Hooke's claim was true and took a trip to Cambridge to visit Newton, checking on any progress he might have made on the problem. In August of 1684, Halley asked Newton directly: What kind of curve would an inverse-square law of attraction toward the Sun produce? Newton answered that it would be an ellipse. He had calculated it long ago. He would redo it and send it to him. In November, Newton sent him a nine-page paper in which he not only showed that the path of a planet subject to an inverse-square law of attraction would be an ellipse, but also that the planet would speed up and slow down as it approached and receded from the Sun, consistent with Kepler's laws. Halley wanted to publish it, but Newton told him to wait. He was not finished.

Newton was writing in Latin now, the language of the universities, the language that would be understood by scholars throughout Europe. This was not a journal entry written in esoteric shorthand or a personal narrative of discovery. It was an authoritative treatise in the tradition of the ancient geometers and astronomers. He was not beginning a discussion to be continued but rather ending a conversation, establishing the final word on motion with definitions, laws, correllaries, scholiums, lemmas, propositions, theorems, examples, and problems.

Newton began with definitions. The current vocabulary was insufficient for what he was attempting to do. He needed a word to quantify the amount of matter in an object. He called this *mass*. The *quantity of motion* is a body's mass times its velocity. *Inertia* is the innate force of matter to maintain uniform motion. An *impressed force* is something that, when exerted on a body, can change its motion. A *centripetal force* is a force that acts toward the center, like gravity or the force acting on a ball on a string when it is whirled around in a circle. *Weight* is not a constant for an object but rather depends on its mass and gravity.

Newton was writing a treatise on motion; he would have to consider space and time. He declared both absolute. Time flows evenly, without regard to anything external. Space is rigid and fixed. With absolute time and absolute space, we have absolute motion. He then established the three Axioms, or Laws of motion: 1) every body retains a constant state of motion unless it is compelled to change by an impressed force; 2) a change in motion is proportional to and in the same direction as the impressed force; 3) for every action there is an equal and opposite reaction. In the context of his definitions and laws, Newton wrote up his work on the inverse-square law of attraction and its connection to elliptical orbits. He generalized this further, showing that the path of an object subject to such a force could be any conic section, so it could also be a parabola or hyperbola.

This was a start. He rummaged through his old notebooks; he had thousands of pages of his writings lying about in Cambridge and at his home in Woolsthorpe. He presented a geometric version of his theory of infinities and infinitesimals, just enough that a reader might understand the mathematics of his theory of motion. Then he applied this theory to practically every mechanical situation that he or anyone else had ever studied. He applied it to simple machines — levers, planes, wheels, pulleys, screws, and wedges as well as combinations of them. He applied it to simple motions, such as the motion of a falling body, a pendulum, and a cycloid. He applied it to compound motions, such as cycloids rolling inside and outside of a globe and the motion of an oscillating body in a cycloid that is rolling around a globe. In all of these examples, he first considered idealized situations in which any resistance acting on the objects was ignored.

Newton next began to include resistance in his analysis, offering various mathematical models for the force of resistance. He then described experiments that he had performed to determine the resistance of different fluids and their effect on the motion of objects. He shared his experimental method and ways that he revised experiments to improve their precision. He included error analysis, indicating in what situations theory and experiment did not agree satisfactorily and suggesting reasons for any observed discrepancies. He also considered the motion of fluids, sound, and light. And he considered the properties of vortices, demonstrating that a pressure model of gravity, like Descartes', was inconsistent with Kepler's law of periods.

Newton wrote to London asking for the most recent astronomical data and a table of the tides. The structure of his treatise was taking shape — it was organized in three books, the last of which he titled "Systems of the World." He had originally planned to write this section in a popular style to widen his audience. However, he felt that his conclusions lost their strength without mathematical support. Therefore, to avoid unnecessary argument, he wrote this last book like the previous two, justifying conclusions with rigorous proofs backed by observations. To allow interested readers to understand this last book without reading the entire first two books, he indicated which sections were needed and promised to point to additional propositions as necessary. After this preamble, Newton began to apply his theory to the universe.

First, he reviewed observational evidence. The moons of Jupiter and Saturn obey Kepler's laws of planetary motion. The orbit of the Moon also obeys these laws. Furthermore, the specific acceleration of the Moon can be explained with the same inverse-square law that accounts for the motions of falling bodies and projectiles near the Earth's surface. All of these observations can be accounted for by a universal law of gravitation, where every body attracts every other body with a force proportional to the product of their masses and inversely related to their squared separation. With this assumption and astronomical data regarding various distances and sizes, he estimated the mass and density of the Sun, Earth, Jupiter, and Saturn. Furthermore, he showed how the weight of a body changes according to the gravity on a planet.

Newton agreed with others that the center of the universe is stationary and immovable. However, according to a universal law of gravitation, all celestial bodies accelerate toward all other celestial bodies. Therefore, neither the Earth nor the Sun is stationary, as had been assumed. Rather, the center of mass of *all* the bodies in the universe should be stationary. The stars, as evidenced by their lack of observed parallax, are too far away to exert a sensible force. Using the masses of the known celestial bodies, he showed that the center of the universe should coincide very nearly with the center of the Sun, even in the event that all of the planets happened to be on the same side of the Sun.

Newton then proceeded to explain many diverse observations that had previously been unexplained. He explained why some planets are thicker around the equatorial plane, that it was a result of their spinning. He accounted for the observed speeding up and slowing down of clocks at different locations, attributing this to differences in gravity due to a change in altitude. He explained irregularities in the Moon's orbit by taking into consideration the influence of the Sun. Similarly, he explained the observed irregularities in the orbits of Jupiter's moons. He created a theory of tides that included the gravitational effects of both the Moon and Sun. He explained the slow and irregular precession of the equinoxes as a result of the gravitational pull of the Sun and Moon on the Earth's equatorial bulge. Finally, he developed a theory of comets.

In April of 1686, Newton was nearly finished. Although the Royal Society had agreed to print his book, they had run out of money publishing a book on the history of fish. Halley stepped forward and not only agreed to pay for the expenses of the printing but also to oversee the publication process. At the end of the month, the treatise, which Newton titled *Philosophiæ Naturalis Principia Mathematica*, was presented in summary to the Royal Society. Newton was in Cambridge at the time but received a report from Halley afterwards. Halley informed Newton that Hooke claimed priority for the inverse-square law of gravity, although he conceded that the mathematical demonstration of the curves was entirely Newton's. Hooke felt that Newton should mention him in the preface. Newton received a more candid report through another contact in London: Hooke publically and directly accused Newton of plagiarism.

Newton was sent into a rage. He had *labored* over this treatise. He didn't just have the "idea" of gravity, he had created an entire structure of laws, definitions, and proofs. He had cross-checked *everything* with examples upon examples corroborated with experiments and observations so that the entire structure was virtually indestructible. And now Hooke wanted to share credit for the theory, with Hooke the inventor and Newton merely the mathematician! Newton went back to his manuscript and struck out all mentions of Hooke and threatened to destroy the treatise, or at least parts of it, unless it was accepted as it was.

Newton softened somewhat after Halley shared with him more fully the context of the accusations. Hooke was in the audience when the presentation was made before the Royal Society. John Hoskins, the vice president of the Society and one of Hooke's closest friends, praised both the novelty and dignity of the subject. Another member suggested that it was even more remarkable because it was invented and perfected by the same man. No one made any mention of Hooke, not even Hoskins, although Hooke had shared with him his ideas about gravity. After the presentation, the members continued talking at a coffee house. Hooke tried to convince the others that he had given Newton his first ideas about gravity, but no one took him seriously. They told him that if he had, he should have done a better job staking his claim to the discovery.

Newton reviewed the letters he and Hooke had exchanged and admitted that Hooke had corrected his suggestion of the path of an object dropped from a great height. Hooke had also renewed his interest in gravity and goaded him on, talking as if he had already determined the elliptical paths of the planets. In the end, Newton gave him credit, along with Wren and Halley, for independently deducing the inverse-square law of gravity. He also mentioned Hooke in reference to his observations of the recent comet and his barometric experiments. But he would not admit Hooke as his "master." Newton was both the inventor and the perfector. This was *his* theory of gravity and his alone.

Hooke persisted in his accusations without success. He became bitter and couldn't stand to be in the same room as Newton. Although he retained his position within the Royal Society, he had lost his standing. And as Hooke fell, Newton rose. Newton began to attend Royal Society meetings and social outings. He connected with other learned men who came to London, conversing and corresponding on a large range of topics. He became active in Cambridge politics and in 1689 was elected to be their representative in the English Parliament. He was chosen to help England standardize their coinage; he was given the official title Warden of the Mint in 1696 and three years later, Master of the Mint. After Hooke died in 1703, Newton was elected the new president of the Royal Society, and in 1705, he was knighted by the Queen.

Newton published a second edition of *Principia* in 1713. The Dutch mathematician and natural philosopher Christiaan Huygens, one of the few who could understand it, worked with Newton to compile errata for the new edition. Huygens promoted Newton's theory for its mathematical correctness, although he had some reservations about the idea that gravity could mysteriously act by one body on another without anything in between. Huygens accepted the theory anyway, but others were more critical. Many preferred Descartes' universe where ethereal matter acts as a medium among all bodies, causing stones to be pushed down and planets to be whirled around in vortices. Newton had no way to explain how gravity works; he had no hypothesis. He only had the certainty of his mathematical theory and its ability to explain and predict natural phenomena.

Instead of inventing a hypothesis to explain gravity, Newton created a new philosophy. He laid down its final form in a third edition of *Principia* published in 1726 with a set of rules. The first three rules, which were present in some form in the earlier editions, established the ideal of simple, universal laws to explain natural phenomena. The fourth and final rule stated that we should consider propositions arrived at through experiment and observation as true, or nearly true, until other observations or experiments help improve or give exception to these propositions. Experimental conclusions should never be evaded by a contrary hypothesis. With this, Newton ushered in a new era of science.

# 8

# General Relativity

*Oh leave the Wise our measures to collate*
*One thing at least is certain, LIGHT has WEIGHT,*
*One thing is certain, and the rest debate –*
*Light-rays, when near the Sun, DO NOT GO STRAIGHT.*

— Arthur Eddington

On April 22, 1715, the Royal Society of London gathered together to observe a solar eclipse. The midday sky was bright, and then, just at the predicted time, the sky darkened showing the silhouette of the Moon surrounded by a faint glow. Edmond Halley, secretary of the Royal Society, had published a map of the path that the Moon's shadow would take as it crossed England so that others could observe it. He asked that people note the time and the duration of the eclipse using a pendulum clock. This event, which in former times would have been considered auspicious, was now splendid confirmation of Newton's theory.

Other confirmation followed. Newton had calculated that the Earth should bulge at the equator as a result of its spinning. Descartes' model predicted the opposite, that the Earth should be narrower at the equator. In 1733, the French Academy of Sciences set out to resolve this question, sending expeditions northward to Lapland and southward to Peru. A decade later, the expeditions returned with data supporting Newton's theory.

The sighting of a comet in 1758 gave additional support to Newton's theory. Many years before, not long after the publication of *Principia*, Halley had combed through old data for comets that went back hundreds of years. He believed that he had discovered a comet that repeatedly orbited the Sun, sighted in 1531, 1607, and 1682. He obtained the most accurate data on its recent occurrence and, using Newton's laws, determined that its path was elliptical with the Sun at one focal point. With this, he was able to calculate when and where it should next appear. In December of 1758, about fifteen years after Halley died, it was sighted just as he had predicted.

Advances in telescopes led to the discovery of many more moons, all of which seemed to orbit according to Newton's theory. And in 1781, a seventh planet was discovered, which was named Uranus. In 1845, when irregularities were discovered in its orbit, Newton's theory wasn't questioned. Rather, scientists postulated the existence of another planet beyond Uranus. The next year, Neptune was found, right where calculations showed it should be!

In 1859, another irregularity was discovered, this time in Mercury's orbit. It was observed that the axis of its elliptical path slowly changed with respect to the fixed stars, about 5,600 arcseconds per century. Newton's theory could account for 5,557 arcseconds by taking into consideration the perturbations to its orbit by the other planets.

Everyone expected that the discrepancy of 43 arcseconds would be explained by the existence of another planet or a group of small orbiting objects. Astronomers searched and searched for this planet and even gave it a name, Vulcan. But no planet was ever found.

Albert Einstein was born on March 14, 1879 in Ulm, a medium-sized city in Bavaria where his father, Hermann Einstein, had a thriving business in featherbeds. A year after Albert was born, his father abandoned his featherbed business to join with his brother, Jakob, to establish an electrochemical factory in Munich. With recent technological advances, the future lay in electricity! Jakob, a trained engineer, would be the technical director, and Hermann would be in charge of the business side. They all moved together into a cottage in the suburbs of Munich. There, the brothers set up their workshop, stockroom, and store.

As a consequence of his father and uncle's business, Albert Einstein grew up surrounded by technological production and innovation — dynamos powered by steam-driven engines, electrical lamps, telephones, telephone switchboards, and microphones for broadcasting music. These were a part of his daily life. Through this exposure and the kind employees who indulged the young boy's curiosity, Einstein received an extensive, hands-on education in electrical engineering.

He also grew up with music. His mother played the piano, especially Beethoven sonatas. When Einstein was six, she arranged for him to take violin lessons. Later, he fell in love with Mozart's sonatas and began playing in earnest, patiently practicing intricate passages. And thanks to a poor Jewish student, Max Talmud, who joined the family every Thursday in their midday meal, he grew up with books.

Talmud introduced Einstein to Aaron Bernstein's *Popular Books on Natural Science*, a set of books covering a wide range of scientific topics. The books were written in an accessible way, often interweaving history and practical application into theory. Bernstein shared the story of Neptune's discovery to highlight the power of Newtonian mechanics. He used street lamps to illustrate the inverse-square law applied to light. He challenged the reader by posing questions such as "What would we see if we could ride on a beam of light?"

Talmud obtained this collection for Einstein and many others. He introduced Einstein to Alexander von Humboldt's *Cosmos*, a five-volume set that included the development of science from its ancient roots. He introduced Einstein to philosophical works such as Immanual Kant's *Critique of Pure Reason* and Ludwig Büchner's *Force and Nature* that dealt with fundamental questions on the nature of science. Einstein eagerly absorbed all of these, but he wasn't just content with reading them. He insisted that Talmud discuss each book in detail with him.

Through books and discussions with Talmud, Einstein gained a thoughtful and broad-based scientific education. And through his uncle, Einstein began his studies of mathematics. The uncle playfully introduced the young boy to algebra: "Algebra is the calculus of indolence. If you do not know a certain quantity, you call it x and treat it as if you do know it, then you put down the relationship given, and determine this x later." His uncle also taught him some geometry, including the Pythagorean Theorem. Einstein took it upon himself to tackle the proof, and after three weeks of patient effort, he succeeded.

With so many wonderful opportunities to learn at home, it should not be surprising that Einstein was dissatisfied with much of his formal education. In particular, he disliked the gymnasium where he attended secondary school with its curriculum focused on Latin and Greek grammar and its drill sergeant-like teachers who emphasized rote memorization. He attempted to learn despite the less than ideal learning environment. When he was twelve, he obtained a geometry book for school and studied it on his own before the teachers could take the joy out of the experience. Similarly, he studied more advanced topics on his own. By sixteen, he had mastered the entire mathematics curriculum offered at the school, including differential and integral calculus.

Meanwhile, because of financial difficulties, Einstein's father and uncle decided to liquidate the factory and relocate the family and shop to Milan. Einstein's parents wanted their son to earn his degree at the gymnasium so that he could continue on to university to study electrical engineering. They left him and his sister with relatives in Munich and departed for Italy.

It was not long before Einstein, not able to tolerate the situation, devised a plan to escape. He obtained a doctor's note certifying that he had a nervous breakdown and needed to return to his family. He obtained a note from his mathematics professor saying that he had mastered the mathematics curriculum, hoping this might serve in the absence of a degree. With everything in place, Einstein packed his bags and surprised his parents when he arrived, unannounced, in Milan.

With his newly found freedom, Einstein officially renounced his German citizenship and began enjoying the sights and sounds of Italy. He took delight in its art and music and hiked in its mountains. Soon, his father put an end to this holiday. The business was going badly, and he was no longer able to support his son. Young Einstein would have to get a job. He, therefore, began to think of his future. He was drawn toward physics, especially theoretical physics. It would be difficult to gain admittance to a university without a high school degree, but he might be able to get into a technical school where he could study to become a teacher. In the end, he decided to travel to Switzerland to apply to the Polytechnic School in Zürich, the most prestigious technical school in central Europe outside Germany.

Einstein began preparing for the entrance exam by studying Jules Violle's three-volume *Textbook of Physics*, translated from the original French into German. This prepared him well for the physics portion of the exam. He also had no problem with the mathematics, but he failed the entrance exam on account of his poor performance on the French and botany sections.

Einstein was encouraged to complete his degree at a Swiss secondary school, which he did. He was impressed with their more liberal philosophy in which students were encouraged to think, teachers were available to discuss, and laboratory experiments accompanied science courses. He recovered from his aversion to school and completed the coursework within a year.

With a degree in hand, Einstein entered the Zürich Polytechnic in the fall of 1896. He approached his studies there with an independent spirit, skipping classes that he didn't like, especially mathematics. He delved deeply into the subjects that he did like, such as electromagnetism and the kinetic theory of heat, reading the "greats" in these areas — James Clerk Maxwell, Heinrich Hertz, Ludwig Boltzmann, and Hermann von Helmholtz. And he found companionship in a quiet, free-thinking Hungarian woman, Mileva Marić, who shared his passion for the great physicists.

In 1900, Einstein finished up at the Polytechnic with a mathematics and teaching degree and began looking for a job. He applied for assistantships at the Polytechnic and at various institutions throughout

Europe. In most cases, he didn't even receive a response. He found tutoring jobs and a substitute teaching position, but nothing permanent. He wrote to friends asking for help. His father, without his knowledge, wrote a letter to a professor appealing to him on his son's behalf for a position. Meanwhile, Einstein applied for Swiss citizenship. After a lengthy process it was granted, which opened up additional job opportunities. Finally, he received good news. A position would soon be available at the Swiss patent office in Bern that his friend assured him he was almost guaranteed to get. Hopeful, Einstein packed up his bags and left for Bern.

In Bern, Einstein immediately posted an advertisement offering tutoring. Through this, he met a couple of young men, Conrad Habicht and Maurice Solovine, who shared his interest in physics and philosophy. The three formed a little reading group that they referred to with humor as the "Olympia Academy." Einstein, the youngest of the three, was their Knight and president. They met regularly, usually at one of their homes, often staying up late into the night eating, drinking tea, and talking. They discussed books by Ernst Mach, David Hume, Baruch Spinoza, and Henri Poincaré. They debated beliefs on causality, the validity of hypotheses in scientific models, and the existence of absolute space and time.

After several months in Bern, the job offer at the patent office was finalized. Einstein started out as a "technical expert third class." This was the lowest rank, but it paid reasonably well and he enjoyed the work, which challenged him to understand a large variety of technical inventions. The director nurtured Einstein's independent spirit; he encouraged everyone to always question everything, even what seemed obvious to everyone else. He also turned a blind eye when Einstein, after finishing up his required work, dedicated time to his own research.

About a year after moving to Bern, in January of 1903, Einstein married Mileva Marić. It was a civil wedding, with Habicht and Solovine acting as their only witnesses. Einstein's family especially had opposed the marriage. His mother had told him, "Like you, she's a book, but you ought to have a wife." Marić became his "wife," taking care of their household and giving him children. And although she lost the privileged position that she had once enjoyed as his main intellectual companion, she remained, for many years, his partner. In her quiet way, she supported him, and she appreciated his intellectual successes with a rare understanding of their importance.

Einstein enjoyed consistent fellowship with his reading group for a couple of years, but then Habicht graduated and moved away. Around the same time, a friend of Einstein's from the Polytechnic, Michele Besso, joined him at the patent office. Brilliant, spontaneous, disorganized, and with a good sense of humor, he was the perfect replacement for the Olympia Academy. Einstein and Besso walked to and from work daily in lively conversation about physics and philosophy. Einstein also read and reviewed new scientific books and articles to supplement his patent office income, thereby keeping up to date on the latest research. His life was filled with intellectual activity on all sides, and his independent research flourished.

One of the many problems Einstein was working on was attempting to resolve a contradiction that existed in electrodynamic theory. Specifically, the theory seemed to be at odds with the classical principle of relativity. According to this principle, there is no detectable difference between rest and motion at constant velocity. Galileo used it to argue against objections to a moving Earth that were based on the notion that if the Earth moved, we would feel it. To illustrate this idea, Galileo had considered the motion of objects inside the cabin of a boat — fish swimming in a bowl, water dropping, a person jumping or throwing something to a friend. In all of these examples, Galileo claimed, there is no difference between when the boat is at rest and when it is moving at any speed, as long as the motion is constant. Therefore, the Earth could be moving very quickly without its motion being felt by the people on it.

In this way, Galileo had established the relativity of motion — that there is no one reference frame "at rest" or "in motion" in any absolute sense. This was in contrast to Aristotle's theory in which all motion is in reference to the stationary center of the Earth, which is the center of the universe. Furthermore, Galileo established that constant motion is just as natural to an object as rest is, and an object that experiences no resistance will continue at the same velocity forever.

Newton named this tendency for an object to remain at rest or in motion at constant velocity the body's *inertia* and used it as the foundation for a new theory of motion. According to his law of inertia, an object at rest will remain at rest and an object in motion will continue in motion with a constant velocity unless a force acts to change its motion. This law applies in all inertial reference frames, reference frames that are moving at a constant velocity with respect to absolute space. Newton's other laws of motion and his law of gravity are valid in these same reference frames. Galileo's principle of relativity could now be phrased within the context of Newton's theory: the laws of motion and gravity are the same in all inertial reference frames.

However, this principle of relativity did not seem to apply to the new electrodynamic theory. As an example of this, Einstein considered the following thought experiment. If you are chasing a train with the same speed as the train, the train will appear to be at rest in your reference frame. Similarly, if you are chasing a light beam at the speed of light, the light should appear to be at rest. But light, according to electromagnetic theory, is an electromagnetic wave, and there is nothing in the theory that allows it to ever be at rest. The laws of electromagnetism seemed to fall apart in a reference frame moving at the speed of light.

Einstein wondered if perhaps one could just *assume* that the speed of light is the same in all inertial reference frames. This would save electromagnetic theory, in which the speed of light appears as a constant. But how could the speed of a train vary with reference frame but not the speed of light? If we run with a train, it appears to be at rest. But if we run with a light beam, it still appears to travel with the same speed, 300,000 km/s? How can this make sense?

Einstein considered modifying electromagnetic theory so that the speed of light changes depending on the reference frame, like the speed of a train. He didn't get anywhere with this idea, and besides, experiment seemed to contradict this. A couple of American physicists had repeatedly tried to measure the speed of light in different reference frames with an extraordinary amount of precision but could never measure any change.

One afternoon in May of 1905, after almost a year of fruitless thought, Einstein was walking home from the patent office with Besso and declared to him, "I give up." Then, suddenly, he had an idea. What if *time* was allowed to be relative? After all, absolute time was just a hypothesis with no experiential basis. If distance was also relative, the speed of light could be kept constant while the speed of a train varied according to reference frame. All the contradictions seemed to disappear. The next morning Einstein said to Besso, "Thank you. I've completely solved my problem."

Einstein then began writing. He established two postulates. First, the principle of relativity applies to the laws of electromagnetism as well as the laws of mechanics. Therefore, the laws of electromagnetism take the same form in all inertial reference frames. Second, the speed of light in a vacuum is the same in all inertial frames. With these two postulates, Einstein created a new theory of motion.

The mathematics, for the most part, had already been worked out. The Dutch physicist Hendrik Lorentz had developed a set of transformation equations that kept the speed of light constant, accounting for the experiments that were unable to measure any difference in its speed. Under these transformations, the laws of electromagnetism maintain the same form, independent of inertial reference frame. Einstein made small modifications to Newton's laws of motion, after which these too obeyed the principle of relativity. Einstein ended the paper by proposing an experiment to test the theory — a balance clock at the equator should run slightly slower than an identical clock at one of the poles. Einstein had completely solved the problem, and at the end of the paper he thanked Besso for his help.

Einstein worked furiously on the paper, writing it all up in less than a month. After finishing, he was so tired that he stayed in bed for a couple of weeks while Marić read the paper again and again, checking it over for errors. At the end of June, Einstein sent out the paper. And then he and Marić celebrated.

The paper gradually gained recognition, and it gradually gained a name. Einstein had called it the "Invariance Theory," referring to the invariance of the speed of light in different inertial frames. Max Planck, a physicist and the editor of *Annalen der Physik* who had reviewed Einstein's paper, called it the "Relative Theory." Eventually, Einstein began calling it the "Relativity Theory."

Within a couple years of its introduction, Einstein, Planck, and a few others had written articles extending Einstein's relativity theory. In September of 1907, Einstein was asked to write a review article on the subject. He responded that he would be glad to write the article but that he was not in a position to acquaint himself with all the recent literature since the library was closed during his free time. At this time, he was still working at the patent office eight hours a day, six days a week. They offered to send him relevant reprints, and Einstein began to write.

The article was for the *Yearbook of Radioactivity and Electronics*, due in December. As Einstein was writing, he became troubled by two inadequacies in the theory. It could not describe accelerated motion; the principle of relativity only applied to inertial reference frames. Furthermore, the theory did not include gravity. It was November, and he was sitting at his desk at the patent office thinking when suddenly it occurred to him that there might be a relationship between these two problems and a way to solve them both.

Einstein imagined a person falling freely, accelerating under the influence of gravity. To that person, it would seem as if gravity did not exist. If enclosed in a chamber, all loose objects would fall along with the person and appear to be floating. Alternatively, if a person were in an enclosed chamber in outer space, far away from any source of gravity, and the chamber were accelerated upward, it would feel as if the person were pushed downward. Any loose objects would fall, just as if gravity were pulling them down.

The experimental basis of these thought experiments was Galileo's law of free fall — that all bodies fall at the same rate in the absence of resistance. Galileo deduced this by observing falling bodies in different mediums, and Newton confirmed the result with pendulum experiments. More recently, the Hungarian physicist Loránd Eötvös had confirmed this result to such an extraordinary degree of precision that Einstein believed it was reasonable to assume that this independence of mass on gravitational acceleration is an exact law of nature.

If gravity accelerates all objects in the same way, Einstein reasoned, a reference frame at rest in a uniform gravitational field should be physically equivalent to a reference frame accelerating in the opposite direction without a gravitational field. Therefore, if we can understand accelerated motion within the context of relativity theory, we can begin to understand gravity.

This equivalence between gravity and acceleration was the starting point for a final section in Einstein's review article, "Principle of Relativity and Gravitation." He considered a body accelerating in the absence of gravity at a small rate such that most of the equations from his original relativity theory still applied. By analyzing this situation and assuming the equivalence between acceleration and gravity, he concluded that clocks should run slower in stronger gravitational fields.

Einstein then proposed an experiment to test this conclusion using radiation from the Sun. Because of the large gravitational field at the surface of the Sun, light waves emitted there should oscillate at a slower rate than if they were emitted on the surface of the Earth. This should cause the light to be shifted toward the red end of the color spectrum. It would be a small change, less than one part in a million, but theoretically testable.

In January of 1908, Einstein began looking for new employment. He wanted a job that would give him more time for independent research. He applied for a high school mathematics teaching position in Zürich, noting that he would "be ready to teach physics as well." He enclosed copies of his numerous publications along with the application. He did not even make the short list of applicants.

At the end of February, Einstein finally submitted his *habilitation* thesis, a paper showing independent scholarship that would allow him to teach at a university. The committee accepted the paper and after a successful sample lecture, he was given permission to teach. He taught his first course that summer, holding it at seven in the morning so that he could get to the patent office by eight. Besso and two other friends attended. In the winter term, he moved the class to the evening and a fourth student enrolled. The following summer, his three friends dropped out, and Einstein cancelled the class.

Eventually, his efforts toward gaining a full-time university position were rewarded. A non-tenured teaching position became available at the University of Zürich. The top candidate for the position was diagnosed with an incurable form of tuberculosis, and Einstein, next in line, got the position. In the middle of October in 1909, he ended his work as a patent clerk and began his university career.

Einstein grew as a teacher and developed his own style, disorganized but engaging. He did impromptu calculations on the board while explaining his thought process. He would stop in the middle of a lesson to ask if the students understood, encouraging them to interrupt him at any point if they were confused. He would often continue to talk with students during breaks and after class, answering their questions and sharing his current research with them. When the University of Prague offered Einstein a full professorship along with a significantly higher salary, his students in Zürich petitioned the administration to keep him, whatever the cost. The administration responded with an even higher salary, but in the end, they lost him to Prague.

Einstein's career continued to advance. In 1911, he was invited along with seventeen other prominent physicists to attend a meeting in Brussels organized by the Belgian chemist and industrialist, Ernest Solvey, to discuss the most important problems in physics. The meeting was chaired by Hendrik Lorentz; attendees included Max Planck, Marie Curie, Ernest Rutherford, and Henri Poincaré. The year before, Einstein had visited the aging, almost deaf Ernst Mach. He had now become acquainted with all the greatest physicists of his time.

Even before leaving for Prague, Einstein had talked with those in Zürich about returning as a professor to his old school, the Polytechnic, now a full university called the Swiss Federal Institute of Technology. Curie and Poincaré wrote recommendation letters, and Einstein was offered the job to begin in the summer of 1912. Not long after settling back in Zürich, he received an even more enticing offer from the University of Berlin, to join the university faculty and become part of the Prussian Academy of Sciences. He would be the director of a new physics institute with a large salary and no teaching responsibilities. He accepted and headed to Berlin in the spring of 1914, where he would remain until 1933 when he left Germany permanently.

Those who recruited Einstein to Berlin recognized the many contributions he had made to modern physics within the past decade. Here in Berlin more than anywhere, they appreciated the importance of his relativity theory, despite the lack of any experimental confirmation. He was not, however, recruited to work on relativity theory. In 1911, researchers in England had discovered that an atom is mostly space, with a dense, positively charged nucleus. Two years later, a Danish physicist created a mathematical model for the hydrogen atom. This was the future of science, an understanding of atoms, and the Germans wanted to play their part.

Einstein had already made great contributions to atomic theory, and everyone expected that he would make more, but by the time he arrived in Berlin he had turned his attention almost exclusively to gravity. He had realized that an additional consequence of the equivalence between acceleration and gravity is that a light beam should bend in the presence of a gravitational field, accelerating at the same rate as any massive object. He estimated how much a light beam would be deflected as it passed by the Sun, about 0.83 parseconds. This could be the first direct confirmation of relativity theory.

Einstein published a paper on the deflection of light in 1911 while in Prague, ending it with an invitation to astronomers to take up the question. Erwin Freundlich, a young astronomer from the University of Berlin, reached out to him, excited by the possibility. The observations would have to be made during a total solar eclipse in order for the stars in the vicinity of the Sun to be visible. The next suitable eclipse would be August 21, 1914.

The two worked together to raise money, and on July 19, 1914, Freundlich and his group left for Crimea in Russia. On August 1, Germany declared war on Russia. The astronomers, discovered with their photography equipment and telescopes, were captured as spies and their equipment confiscated. They eventually returned safely to Germany but with no data. The American and Argentinian astronomers who traveled to Russia to observe the eclipse did not fare much better, with clouds obscuring their view of the Sun during the eclipse. The expedition was, for all, a failure.

Meanwhile, Einstein continued to think about gravity. His starting point was the equivalence between acceleration and gravity, but he needed more. He began considering circular motion. If a person were enclosed in a chamber moving in uniform circular motion, that person would experience a constant inward acceleration, which would feel like an outward push. According to relativity theory, moving clocks run slow. Another consequence is that distances contract in the direction of motion. Soon, Einstein began thinking about space.

For an object moving in a circular path, its path length — the circumference of the circle — would be shortened due to the motion. However, the radius of the path, since it is perpendicular to the motion, would remain constant. The normal rules of calculating circumference and area in terms of radius would no longer be valid. With this thought, Einstein began to think of gravity in terms of geometry.

Geometry was not a particular strength of Einstein's, but it was a strength of his friend Marcel Grossmann. Grossmann was a friend from the Polytechnic; it was with his notes and help that Einstein was able to pass his mathematics exams. Grossmann excelled in geometry and after graduating, earned his doctoral degree and published several papers on non-Euclidean geometry, which could be used to describe the kind of curved space that Einstein was beginning to imagine. Grossmann had returned to their old school as a mathematics professor, and after Einstein arrived in Zurich in 1912 to join him as a colleague, he immediately reached out to his old friend and described his problem.

Grossmann was thrilled to help. He thought about it, consulted literature, and then told Einstein that he would need the non-Euclidean geometry developed by Bernhard Riemann. Riemann had developed a way to describe the geometry of a surface that contorted in any way — from flat to spherical or any other arbitrary curvature. Riemann also went beyond two-dimensional surfaces, developing a way to describe three- and even four-dimensional space. This fit well with the relativity theory as it had been reformulated by Hermann Minkowski in terms of four-dimensional space, with position taking three of the dimensions and time, the fourth.

In Riemann's geometry, distances are determined using something called a *metric tensor*, which describes the curvature of the space. In flat, two-dimensional space, the metric tensor can be used to derive the Pythagorean theorem and in three-dimensional space, the distance formula. In Minkowski's flat, four-dimensional space, it can be used to derive the law of inertia and the effects of motion on clocks and distances. In curved, four-dimensional space, if known, the metric tensor could be used to derive the path of a light beam or the trajectory of a planet.

With Riemannian geometry, the problem of gravity was reduced to finding an equation that could be used to determine the metric tensor for a particular physical system. The relativity principle could be applied naturally within this framework since a property of the metric tensor and other similar structures is that they preserve the same relationships among themselves under an arbitrary change in reference frame. Therefore, an equation written in terms of these tensors is covariant, meaning that it takes the same form in every reference frame, consistent with the relativity principle.

Einstein worked with Grossmann to learn the mathematics of Riemannian geometry, and then he struggled to find an appropriate equation to determine the metric tensor. The equation should be generally covariant, and it should also satisfy certain physical requirements. For example, it should reduce to Newton's law of gravity in the limit of a weak, static gravitational field, and it should satisfy classical conservation laws such as the conservation of energy and momentum. It should also be consistent with the equivalence between acceleration and gravity.

Einstein tried to determine the correct equation, alternating between mathematical and physical approaches. He started with a generally covariant equation and then checked to see if it was consistent with the physical considerations. Subsequently, he adopted a physical strategy, starting with the physical requirements and hoping that it would satisfy general covariance. He tried this and that, working forwards and backwards. He had all the pieces — he just had to get them to fit together. Finally, he found an equation that he thought could work. In 1913, he and Grossmann wrote up a paper with an "outline" of a general theory of relativity.

Soon after, Einstein began to have doubts. His equation, which he had arrived at using a physical approach, did not seem to be generally covariant. Furthermore, he and Besso used the theory to calculate the relativistic correction to the perihelion precession of Mercury, but the numbers didn't work out. According to observations, it should be 43 arcseconds per century. They arrived at half that value. Einstein went forward with the equation anyway. He concluded that general covariance was unattainable or at least physically uninteresting and ignored the discrepancy in the calculation of Mercury's precession. In 1914, Einstein wrote up another paper, not an "outline" but a formal foundation of *the* general theory of relativity. In the summer of 1915, he introduced this theory to the physicists and mathematicians in Göttingen.

Einstein lectured for a week at the University of Göttingen, about two hours each day. In the audience was the eminent mathematician, David Hilbert, who had once joked that physics was too difficult to leave to the physicists. Einstein was especially eager to convince Hilbert of his theory and spent much time explaining to him its details. He left Göttingen enchanted with Hilbert, a fellow pacifist, appreciating his passion and independence of thought. He was also delighted that he had been able to convince this great man of the correctness of his theory.

Hilbert had followed Einstein's explanation very well and had discovered an error. He let Einstein know that he was working on the theory now as well, and the two began a friendly exchange of letters. Meanwhile, Einstein paid a visit to Bern, where Marić and their children were now living. While there, he visited Besso to discuss the problem. Finally, Einstein realized his mistake. There was a clear contradiction in the theory. On November 4, at the weekly meeting of the Prussian Academy, he humbly announced that he had lost faith in the equations that he had previously derived.

For the next three weeks, Einstein worked frantically to try to figure out the correct equations. Finally, on November 17, Einstein reported his success in a letter to Besso: "My boldest dreams have now come true. *Generally covariant* gravitation equations. *Perihelion motions explained quantitatively.* The role of gravitation in the structure of matter." He had done it all. He had gotten all the pieces to fit together. With his new equations, he recalculated the precession of Mercury as 43 arcseconds, consistent with observations. On November 25, 1915, Einstein formally presented the final version of his theory to the Prussian Academy.

Hilbert had arrived at, essentially, the same equations as Einstein and wrote them up in a paper, "The Foundations of Physics," submitting his results to the Göttingen Society on November 20, five days before Einstein's presentation. Einstein feared that Hilbert would try to steal this victory from him, but he need not have worried. In his paper, Hilbert referred to the "magnificent theory of general relativity established by Einstein." Later he was heard to have said, "Every boy in the streets of Göttingen understands more about four-dimensional geometry than Einstein. Yet, in spite of that, Einstein did the work and not the mathematicians." This was Einstein's theory, his alone. And no one doubted it.

The war that had begun in Europe in 1914 had grown into a world war with battlegrounds extending throughout the land, air, and sea. Communication and travel were limited, but Einstein's theory still wanted confirmation. The deflection of light by the Sun recalculated using Einstein's new theory of gravity was 1.7 arcseconds, about twice as much as that predicted using Newton's theory of gravity. Another total solar eclipse would occur on May 29, 1919.

Willem de Sitter, a Dutch astronomer and mathematician, sent a copy of Einstein's paper on general relativity across the English Channel to Arthur Eddington, the director of the Cambridge Observatory. Einstein's theory was not well known in England, but Eddington studied the paper and quickly embraced it. England was at war with Germany, but Eddington was a pacifist. He did not want to fight, nor did he want to be jailed for refusing to fight. He came up with the bold idea, in the spirit of peace, that an English expedition should be sent to view the next solar eclipse to test the theory of a German. The authorities agreed and allowed Eddington, relieved of his duty to fight in the war, to lead the expedition.

By the time the expedition was preparing to leave, the war was over. Two separate groups were sent to view the eclipse; one traveled to the town of Sobral in the Amazon jungle of Brazil and the other, led by Eddington, traveled to Principe, a Portuguese colony off the western coast of Africa. It rained heavily in Principe on the day of the eclipse but cleared up sufficiently to get one usable photographic plate. The other expedition was also successful. They returned to England and spent the next several months developing the photographs, taking measurements, and performing error analysis.

In November, the group was finally ready to make an announcement. At a meeting of London's Royal Society, they reported that the data supported Einstein's theory. They called this "one of the greatest achievements of human thought." *The Times* of London trumpeted the news with a three-line heading: "REVOLUTION IN SCIENCE. New Theory of the Universe. NEWTONIAN IDEAS OVERTHROWN." *The New York Times* followed with a six-line headline that began: "Men of Science More or Less Agog Over Results of Eclipse Observations. EINSTEIN THEORY TRIUMPHS. Stars Not Where They Seemed or Were Calculated to be, but Nobody Need Worry." It ended with the line, "No More in All the World Could Comprehend It, Said Einstein When His Daring Publishers Accepted It."

On December 14, 1919, on the front page of *Berliner Illustrirte Zeitung*, was a large picture of Einstein. In the picture, his hair is dark and combed back. He has a thick mustache and is gazing downward with his chin on his hand. The caption of the picture reads, "A New Celebrity in World History."

# BOOK II

---

# The Story of Quantum

# 1

# The Skeptical Chemyst

And (concludes Carneades smiling) it were no great
disparagement  for a Sceptick to confesse to you, that as
unsatisfy'd as the past discourse may have made you
think me with the Doctrines of the Peripateticks,
and the Chymists, about the Elements and Principles,
I can yet so little discover what to acquiesce in,
that perchance the Enquiries of others have scarce been more
unsatisfactory to me, than my own have been to my self.

— Robert Boyle

On November 28, 1660, Christopher Wren gave one of his weekly astronomy lectures at Gresham College in London. Afterwards, the group in attendance moved to another room, as was the custom, for an informal discussion. On this day, they decided to create a new society. The society would be dedicated to the promotion of the new philosophy, to encourage collaboration in physical, mathematical, and experimental learning. It would accept members of different religions, countries, ranks, and professions, creating a diverse group joined together in the pursuit of scientific truths.

The English monarchy had just been restored and the new king, Charles II, approved the goals of the new society. Soon after his coronation, King Charles had been shown the moons of Saturn and Jupiter through a telescope. He, too, was interested in the new philosophy. He gave the society a royal charter and, in its final version, declared himself its founder and patron. He also gave it a coat of arms and a motto: *Nullius in Verba*, "Take nobody's word for it." The society became known as London's Royal Society. Central to this newly founded organization was Robert Boyle.

Born in Lismore Castle in January of 1627, Boyle was the fourteenth among the "hopeful children" of the Earl of Cork, an Englishman who had settled in Ireland. Robyn, as he was called, spent his first years in the home of an Irish family so that he could enjoy a healthier rustic lifestyle away from the pampering he would have received at the castle. His mother died while he was away, so he never knew her.

After returning to Lismore, the young Boyle began his education. He was tutored in Latin and French. At the age of nine, he was sent to England to attend Eton College. At twelve, he left for the continent with one of his older brothers, a French tutor, and servants. They traveled to Paris, Lyons, and Geneva and throughout Italy. They witnessed the conflict between Catholics and Protestants, which by then had escalated into war. They were in Florence near Galileo's home when he died in 1642. When they returned to England in 1644, they found a country torn apart by civil war.

Boyle was quiet, self-reliant, and studious. While still an adolescent, he chose the pseudonym *Philaretus*, "lover of truth." At the age of nineteen, he helped found the Invisible College in London, a society devoted to the development of scientific knowledge useful to society, especially knowledge related to agriculture and the extracttion of metals from ores. They sought knowledge through scientific experimentation and the experiences of artisans who practiced these trades.

Around this time, Boyle began his own research. His father had died, leaving him Stalbridge, an estate in the countryside of England. There, he set up a laboratory. He ordered an earthen furnace from a thousand miles away. He bought special glass containers, medicines, salts, metals, and an assortment of other equipment and materials. He mixed, boiled, and distilled. He felt a release of heat as he held mercury and gold together in his hand. He was interested in alchemy, especially as it applied to healing, and was intrigued by the idea that base metals such as lead could be converted into gold. However, he disliked the secrecy surrounding the subject. He wanted to bring the ancient subject of alchemy out into the open.

In 1654, Boyle moved to Oxford where several members of the Invisible College had migrated and begun their own branch of the society. They welcomed Boyle and later held meetings in his home. At Oxford, Boyle met Christopher Wren, who had studied there and then stayed on as a fellow. And he was introduced to Robert Hooke, who had arrived the year before at the age of eighteen.

Hooke was a gifted builder who made toy guns, toy boats, and toy clocks, which all worked as miniature versions of their larger counterparts. He was intrigued by the possibility of flight; he had worked out thirty ways of flying. He studied mathematics and astronomy, took organ lessons, and painted. He worked tirelessly, hardly eating or sleeping. When Boyle arrived at Oxford, Hooke was working as a laboratory assistant for one of the Invisibles, a doctor named Thomas Willis. Afterwards, upon Willis's recommendation, Hooke became Boyle's assistant.

Boyle set up his laboratory and began melting, burning, and distilling. His experiments then took a new direction after hearing of an air pump invented by the German scientist Otto von Guericke. Von Guericke's pump could remove the air from a closed container. With this pump, he demonstrated the enormous force of air. He pumped the air out of a copper sphere, which was actually two hemispheres that fit tightly together. He then had teams of eight horses pull on each hemisphere. The horses pulled and pulled but could not separate them. When air was allowed back in, the sphere easily came apart.

Boyle read about von Guericke's air pump and then set out

to make his own. He redesigned it with a glass globe so that he could perform experiments within the airless enclosure. He hired an instrument maker in London to build it for him, but the result was disappointing. He then turned to Hooke, who perfected the machine.

What an extraordinary thing, to be able to completely remove air from a space! A burning candle placed in the globe immediately went out after the air was removed, and the smoke from the candle *fell*. A mouse in the airless globe died almost instantly. A watch hung inside the globe ran normally — the second hand could be seen going around smoothly. However, as the air left the globe, the ticking of the watch became fainter and fainter until, finally, no sound was heard.

With his pump, Boyle demonstrated that air has weight, not "lightness" as Aristotle had claimed. He also demonstrated that air is compressible. He created a tall J-shaped glass tube where the upper end was open and the lower end closed. Mercury was poured into the upper end of the tube, trapping air at the closed end. As mercury was slowly added to increase the pressure on the air, the length of the air column decreased. Mathematical analysis indicated that the compression of air varies proportionally to the pressure applied to it.

In 1660, Boyle published a book describing his experiments regarding the nature of air. The book began as a letter to his twenty-year-old nephew, the Lord of Dungarvan, who wanted to reproduce some of the experiments. Afterwards, Boyle received so many requests for copies and much encouragement to share his work more broadly that he decided to publish the letter. He included additional experiments, forty-three in total. He added enough detail so that his nephew and anyone else who had the equipment could reproduce the experiments easily, and those who didn't could learn from them. The book was received eagerly by scholars throughout Europe and quickly brought Boyle fame.

It was around this time that Boyle, Wren, and others came together to form the Royal Society. Most of its earliest members were Boyle's friends, colleagues at Oxford, or former members of the Invisible College. Boyle could have been the president of the Society; he was asked to be on more than one occasion but refused. He never sought nor accepted a title of any kind, believing that titles would bring obligations that would impede his work. As a consequence, he remained throughout his life "Mr." Robert Boyle.

After the Royal Society was established, Hooke left for London to become its Curator of Experiments. Although Boyle remained at Oxford with his experiments, he attended occasional meetings and kept in close contact with the Society through its secretary, Henry Oldenburg. He also corresponded with Hooke, who sent him chatty letters about his own research. Boyle gave the Royal Society his air pump, which delighted the members and served as the centerpiece for their meetings.

By this time, Boyle had returned to his alchemical studies and written another book, *The Sceptical Chymist*. He wrote the book as a dialogue, with several characters representing different points of view. Themistius was an Aristotelian, who subscribed to the view that all matter on Earth is made up of four elements: earth, water, air, and fire. A second character, Philoponus, followed the teachings of the 16th century medical alchemist, Paracelsus, who held that all matter is made of three principles: salt, sulfur, and mercury. Although both Themistius and Philoponus were absent during much of the dialogue, their views were discussed. A third character, Carneades, hosted the debate in the garden of his home. He was the skeptic who challenged the ideas of Themistius and Philoponus as well as others. A fourth character, Eleutherius, acted as the judge.

The majority of the dialogue was centered on experiments. They discussed the decomposition of wood through burning. The resulting smoke, which can be analyzed by collecting the soot that sticks to the chimney, is made up of salt, oil, spirit, phlegm, and earth. The ashes can be separated into earth and salt. They also discussed the decomposition of blood into phlegm, spirit, oil, salt, and earth. With numerous examples, they demonstrated that matter is much more complicated than what was believed by the Aristotelians, Paracelsusians, and any other group who assumed a specific number of known elements.

Another experiment they discussed was one performed by the Dutch natural philosopher, Jan Baptist van Helmont. Van Helmont believed that there are only two elements, water and air, and that all of the other materials we observe, except for air, ultimately come from water. He subjected this idea to an experimental test according to the new philosophy. He planted a five-pound trunk of a willow tree in an earthen vessel with two hundred pounds of dry soil. He nourished the tree with only rainwater and distilled water, and he covered the soil to keep dust out. After five years, he again weighed the tree and found that it had gained one hundred sixty-four pounds, including its roots and fallen leaves. He dried out the soil and found that it had only lost three ounces. He concluded that the element of water had been transformed into roots, bark, leaves, and all of the other parts of the tree.

Boyle, as the narrator, challenged van Helmont's conclusion. Specifically, he questioned the notion that something considered an element can transform into another substance. Instead, he suggested that all matter is made up of tiny, indivisible particles. Such a theory had been proposed in ancient times by the Greek philosopher Democritus, who called these tiny particles *atomos* after the Greek word for "uncuttable." Aristotle had argued against this view, maintaining that smaller bodies are more easily destroyed than larger bodies, so it should be possible to divide matter indefinitely. But Aristotle was wrong about many things. Perhaps, thought Boyle, he was wrong about this too.

Boyle proposed that matter might be made of any number of different kinds of atoms, distinct with respect to their shape, size, and motion. Groups of these atoms could come together to form uniform, stable substances like water, metals, salts, and other materials. Furthermore, he suggested that fire and other processes might be able to break these particles apart and fuse them into other substances without distorting the atoms themselves. An observed change in a material may not actually be an alteration of the matter but rather a *rearrangement* of it. Regarding van Helmont's experiment, it might not be that water turned into wood, roots, bark, and leaves; rather, Boyle proposed, the water or something in the water broke apart and reformed to create these new substances.

Boyle also suggested that some kinds of arrangements of atoms are more stable than others. No one had ever succeeded in separating salt, sulphur, mercury, or gold. Gold, in particular, seemed to be extraordinarily stable. In one experiment, an ounce of pure gold and an ounce of pure silver were put together in an earthen vessel and then heated in a furnace for two months. Afterwards, the entire ounce of gold was retrieved and seemed to be unaltered.

After the publication of *The Sceptical Chymist*, Boyle continued to experiment and he continued to write. He wrote a philosophical book on the importance of science to humankind. He wrote a book detailing fifty experiments related to color, including experiments with hair dye made of a silver solution and tests using the purple juice of violets to determine whether a substance is acidic or alkaline. He wrote a treatise on hydrostatic paradoxes and the history of cold. He wrote about the cosmic qualities of things, considering both the subterranean region and the bottom of the sea.

Boyle had been at Oxford for well over a decade when his sister Katherine, Lady Ranelagh, invited him to live with her in London. She had taken him in years before when he had just returned from his trip abroad and introduced him to many intellectuals in London who would become the original Invisibles. Lady Ranelagh knew her brother well — he would not move without a laboratory. She offered him a back-house for this purpose, which Hooke helped convert into a suitable space, and ordered coal for the furnace. In 1669, Boyle left Oxford for good and settled into Lady Ranelagh's mansion at Pall Mall.

Since the age of twenty, Boyle had suffered from poor health of an unknown cause. He had colds, fevers, and chills. He also had problems with his vision; he could see, but it was as if he were looking through a thin layer of mist. In the summer of 1670, his health problems reached a climax with a paralyzing stroke. Initially, he could not even raise his hand to feed himself. He was only forty-three years old.

Boyle was well cared for at Pall Mall. He was taken outside every day for fresh air. He had his legs and feet massaged and his legs and arms exercised. He tried various remedies; the most effective seemed to be the dried flesh of vipers. Within a year, he was well enough to give a demonstration at a Royal Society meeting for a couple of visiting Italian nobles.

Boyle returned to his research, hiring assistants to help conduct the experiments and record his observations on account of his poor health and eyesight. He designed a new air pump and continued to explore the properties of air. He distilled seawater, separating it into crystals of salt and a liquid that tasted like pure, saltless water. He wrote books about these experiments and others about gems, glass, and celestial magnets.

In addition to published books, Boyle wrote letters about his work in progress to Henry Oldenburg. Oldenburg was a friend, and he and his wife were frequent guests at Pall Mall. It was not for Oldenburg's sake that he wrote, but rather so that Oldenburg could share the letters with the members of the Royal Society in his *Philosophical Transactions*. Among his many readers was Isaac Newton.

In 1676, Newton read with concern Boyle's latest letter, in which he spoke of an "uncommon experiment about the incalescence of gold and mercury." Newton recognized in this an attempt to create the philosopher's stone, a mythical elixir of life with healing properties and the ability to turn base metals such as mercury into gold. Newton had also been tempted to seek such an object, but he worried about the impact it would have on the world. He urged Boyle through Oldenburg to keep silent about this work until he better understood its consequences should he succeed. Boyle could see the wisdom of this and, with some reluctance, followed Newton's advice.

The next year a German merchant paid Boyle a visit. He brought a substance that had been obtained by an alchemist experimenting with urine in his attempt to make gold. In a darkened room, he put the substance into a glass globe. It glowed. He scattered bits of it about the room and they twinkled. The substance was called *phosphorus*, meaning "bringing light."

After this visit, Boyle and his assistant worked to produce their own supply of phosphorus. Boyle then observed how it reacted with various liquids including water, turpentine, and clove oil. He used a magnifying glass to concentrate a beam of light on it — it burst into flames. Boyle performed many varied experiments on the substance, which he summarized in two books.

The years passed. Boyle continued to experiment and he continued to write. He wrote books on the history of human blood and the experimental history of mineral waters. His bed-chamber was filled with glassware, pots, chemical instruments, mathematical tools, books, and papers. His health was failing, but he still had so much to do. He needed more time, but large portions of most days were consumed entertaining visitors. Finally, in 1689, he posted a note on the door of Pall Mall with an announcement: Mr. Boyle, upon recommendation of his "skillful and friendly" physician and seconded by his best friends, will reserve Tuesday and Friday mornings as well as Wednesday and Saturday afternoons for himself, to regain his spirits and put his papers in order.

In July of 1691, after decades of poor health, Boyle finally wrote his last will and testament. In addition to bequeathing his property and possessions, he requested that certain papers be burned. He died five months later, on the last day of the year. With his death, as one friend lamented, "England has lost her wisest man, wisdom her wisest son, and all Europe the man whose writings they most desired."

# 2

# The New French Chemistry

The word ought to bring about the birth of the idea;
the idea should depict the fact;
they are three imprints from the same stamp.
And since it is words that preserve and transmit ideas,
the result is that it would be impossible to improve science
without improving its language.

— Antoine Lavoisier

Antoine-Laurent Lavoisier's great-great-great-great-grandfather was a postal rider for the king. From these humble origins, his ancestors rose through the economic and social hierarchy. His great-great-grandfather worked in the local court system. His great-grandfather became a successful merchant. His grandfather and father both became lawyers, and it was expected that he, too, would become a lawyer. In 1761, at the age of eighteen, Lavoisier was sent to Paris Law School. Three years later, he earned his law degree and was accepted into the *Parlement de Paris*, but he never practiced law.

By this time, Lavoisier had set his sights on another goal — to be admitted into France's *Académie des Sciences*. He had begun working toward this goal while still in law school. He attended public lectures on chemistry held in the *Jardin du Roi*. He apprenticed with a mineralogist and a botanist, both Academy members. He did all this and more while also keeping up with his legal studies. He was serious and solitary, so much so that he received the following advice from a friend, delivered with a bowl of gruel: "regulate your studies, and believe that one more year on earth is worth more than a hundred in the memory of men."

After finishing his legal studies, Lavoisier devoted himself more completely to science. He worked on a mineralogical atlas of France and wrote four scientific memoirs. He entered a competition sponsored by the Academy on street lighting, shutting himself in a blacked-out room for six weeks while doing the research. He didn't win the competition; the prize was shared by three men who had more practical solutions. The king, however, impressed by Lavoisier's theoretical analysis, presented him with a gold medal, an unprecedented honor.

The Academy of Sciences was founded in 1666 by Louis XIV's chief minister to provide a formal community for French scientists. It was royally funded and only allowed a fixed number of members in each scientific category. When a spot for a chemist opened up in 1768, Lavoisier was nominated and won the majority vote by a small margin. The appointment was then arbitrated by the king, who decided in favor of another more senior scientist, consistent with the wishes of the committee. Lavoisier was given immediate membership as a "supernumerary adjunct" with the promise that he would fill the next vacancy.

Membership in the Academy brought Lavoisier a small income but not enough to support him in any reasonable fashion in Paris. However, only weeks before he was admitted into the Academy, he had secured for himself another source of income. With money he had inherited from his mother, he bought shares in the General Farm. This investment, combined with the salary he received from the Academy, brought him enough money that he was free to pursue science without having to fall back on his law degree.

As a member of the General Farm, Lavoisier took on the role of inspector for the Tobacco Commission. In this capacity, he developed a chemical test to detect the corruption of tobacco and devised a precise system of weighing. The position also gave him the opportunity, through Jacques Paulze, the director of the Commission, to mingle with many of France's intellectuals who strove to reform or abolish archaic social institutions. And it gave him the opportunity to meet Paulze's daughter, Marie-Anne.

Antoine Lavoisier and Marie-Anne Paulze were married in December of 1771. Marie-Anne was young, barely fourteen at the time, but she was intelligent and hardworking. Madame Lavoisier took her first chemistry lessons from her husband. As the years passed, she became a passionate and knowledgeable supporter of his work. She joined him in the laboratory and took notes during his experiments. She translated articles and sometimes entire books for him. She also helped draft his texts and drew illustrations of his instruments and laboratory.

Furthermore, with Madame Lavoisier's enthusiasm for science and social graces, the Lavoisier home became one of the most important gathering places for scientists in Paris, welcoming such prominent international scholars as Joseph Priestley, Joseph Black, James Watt, and Benjamin Franklin. This is where they ended up, but Marie-Anne was also there at the beginning, when Lavoisier began a program of research that would result in a complete overhaul of the subject of chemistry.

Chemistry at this time was disorganized. Experimentally it had flourished, but theoretically it was a hodgepodge of Aristotelian-influenced ideas without the coherence of Aristotle's original theory. Furthermore, there was a large disconnect between theory and experiment. Lavoisier saw this especially in the chemistry course he took in the *Jardin du Roi*. The lecturer was accompanied by a demonstrator who would, often dramatically, illustrate the ideas being discussed with an experiment. As often as not, the experiment would contradict the theory being presented.

Nonetheless, there was one theory that experiment seemed to prove and that almost all scientists held as true: the phlogiston theory. The phlogiston theory was introduced in 1667 by the German physician Johann Belcher. At this time, Robert Boyle was experimenting with his air pump and found that neither a flame nor a mouse can survive in an airless space. He reasoned that there must be something in the air that allows a flame to burn and a mouse to live. Meanwhile, Robert Hooke discovered that certain combustible substances that contain saltpeter burn in a vacuum. Therefore, whatever it is in the air that allows a flame to burn must also be a part of saltpeter.

Belcher had another idea. He suggested that earth can be subdivided into three parts, and one of these parts contains the principle of combustion. He called the combustible part *terra pinguis* or "fatty earth." This combustible substance became known as phlogiston, from a Greek word meaning "to burn." Georg Stahl, a student of Belcher, formalized and popularized the theory. Others extended it, applying it successfully to a large and diverse range of physical phenomena. According to the theory, materials such as wood, charcoal, and sulfur burn easily because they are rich in phlogiston. As they burn, phlogiston is released into the air. The burning stops once all the phlogiston has left the burning object. The phlogiston in the air is then reabsorbed by plants and trees.

The phlogiston theory could also explain calcination, a process in which a metal's metallic properties are removed with the application of heat — the resulting material is a powdery substance called the "calx" of the metal. This process can be reversed by heating the calx with charcoal. It was common practice for miners to heat ores with charcoal to eliminate any calcined portion of the metal, thereby refining it. The phlogiston theory could explain both of these processes. During calcination, phlogiston escapes into the air. It is then returned to the metal when heated with phlogiston-rich charcoal. The phlogiston theory accounted for the corrosion of metals in the same way; a metal rusts because the phlogiston is slowly leaving the metal. The theory also explained some reactions involving acidity in terms of a transfer of phlogiston.

In 1766, Lavoisier purchased a Latin translation of a treatise on sulfur by Stahl, which included his presentation of the phlogiston theory. Lavoisier was impressed with this treatment: "For the first time in the history of chemistry," he wrote, "a theory was embodied in the facts it aimed to explain." He filled the pages of the book with annotations, especially surrounding combustion and calcination.

By 1772, however, Lavoisier had begun to doubt Stahl's theory. In April of that year, he gave a presentation to the Academy titled, "Memoir on Elementary Fire." He began with a summary of existing theories and experiments relating to fire and its interaction with other matter. His conclusion was simply that the nature of phlogiston remains unknown. He then went on to propose a series of experiments that could be done with a Tschirnhausen lens, a huge lens owned by the Academy that had been used to publicly demonstrate the burning of diamonds.

The burning of diamonds was a practice revived from a hundred years earlier in the court of Cosimo III, the Grand Duke of Tuscany. In the Tuscan experiment, six thousand florins worth of diamonds and rubies were subjected to intense heat for twenty-four hours. The rubies suffered no change while the diamonds disappeared without a trace. Now, this demonstration was repeated in the *Jardin de l'Infante*, with two great lenses focusing the midday rays of the Sun on a crucible filled with jewels. To the amazement — and sometimes horror — of the audience, the diamonds again vanished without a trace.

Lavoisier had other ideas for the use of these lenses. With a Tschirnhausen lens, he proposed, it would be possible to subject an object to intense heat in an enclosed container. Specifically, it would be possible to perform calcination and combustion reactions in an enclosed space in order to understand the role of air in such processes. He then listed a series of experiments to be performed on metals, stones, crystals, and fluids. He didn't offer any guesses as to what might be discovered. He merely suggested that a great deal of work remained to be done.

In October, Lavoisier was finally permitted to use the Academy's lens and began experimenting. First, he burned phosphorus under a bell jar. He collected the condensed liquid, an acid, and found that it weighed more than the original phosphorus. He did the same with sulfur and obtained a similar result; the sulfur had gained weight. Finally, he took a quantity of minium, a calx of lead, and heated it with a small amount of charcoal under the bell jar. He observed that a significant amount of some kind of gas was released in this process.

Lavoisier scribbled down a note describing these experiments. In the note, he proposed that the observed weight gain in the case of phosphorus and sulphur was the result of a great quantity of air fixing itself to these substances during combustion. Furthermore, he suggested that the weight gain of metals during the process of calcination that had been observed by Boyle was due to the same cause, air attaching itself to the metal during the heating process. Lavoisier's experiment with minium confirmed this hypothesis; great amounts of gas were released as the metal calx was returned to its metallic state.

In short, Lavoisier was suggesting that the phlogiston theory had gotten it backwards. Rather than phlogiston *leaving* a substance as it burns, is heated, or rusts, air is *entering* the substance. When phlogiston was thought to enter a substance, for example when a metal calx is heated with charcoal, air is actually leaving. On November 1, 1772, Lavoisier deposited his note with the secretary of the Academy of Sciences, sealed in an envelope, securing his claim to this discovery.

A few months later, Lavoisier began a new laboratory notebook. In it, he sketched out a plan. He wrote that everything in the past was only indications; history had presented separate sections of a great chain and only joined a few links. He proposed that every experiment now be repeated with greater precautions, paying particular attention to air that fixes itself to or is released from other matter. In his notebook, he anticipated a "revolution in physics and chemistry."

Lavoisier would have an opportunity to make a presentation at an upcoming Academy meeting in April, but his ideas wanted experimental support. He tried to capture the vapors emitted from diamonds burning under a glass vessel, but the vessel shattered. He heated lead in a furnace holding a retort, but the retort cracked. He then attempted a calcination experiment using glass jars, a washing basin, and a crystal pedestal designed for holding fruit. He was puzzled when he observed that the calcination stopped on the surface. The date for the April meeting came; what he had done so far would have to be enough.

At the meeting, Lavoisier presented a guess. He believed that the increase in weight that occurs during calcination should equal the weight of the air that is absorbed and somehow fixed to the metal. The reverse should be true as well: when a metal calx is heated with charcoal, the decrease in weight of the metal calx as it returns to its metallic state should equal the increase in weight of the surrounding air. It is air, he claimed, that is entering and leaving the metal, not phlogiston leaving and entering. On May 5, Lavoisier opened the sealed letter in front of his Academy colleagues. Lavoisier had staked his claim, but no one was rushing to claim this discovery for themselves.

That summer, Lavoisier continued his experimental work, increasing the precision in his measurements and taking greater precautions. He also began to write a book, compiling the history of air that is released from or fixed to matter. Lavoisier called this air "elastic vapor." Paracelsus and other alchemists had called these vapors *spiritus silvestris* meaning "forest spirit." Van Helmont named them *gas* from the Dutch word meaning "ghost." Boyle referred to these same substances as "artificial air," and others simply called them "airs." The confusion surrounding their name reflected the confusion about their nature, a confusion that Lavoisier aimed to correct.

Lavoisier began his historical survey of elastic vapors with van Helmont, who found that vapors are released from matter during many processes — the fermentation of wine, the rising of bread, the ignition of gunpowder, the burning of charcoal, and the effervescence of metals when they are combined with acids. He developed a theory of digestion to account for the release of gas out of the body. He also recognized that the gas released in these situations is different from the air that we breathe and can be fatal, thereby accounting for deaths in mines, cellars where spirituous liquors are fermented, and air saturated with the vapors from burning charcoal.

Boyle had repeated many of van Helmont's experiments in the airless globe of his air pump and also in condensed air, confirming many of van Helmont's conclusions. However, while van Helmont only acknowledged the release of vapors, Boyle demonstrated that there are also some processes, such as the burning of sulfur and amber, that consume the air in which they burn.

The next significant advance in the study of elastic vapors was made by the English clergyman, Stephen Hales. Through his experiments, Hales determined that air is an essential component not only in the decay and fermentation of plant matter, but also in its growth. He found that plants process a large amount of air, which then becomes fixed to the plant matter and loses its elasticity. He suggested that the "fixed air" can be released to regain its original elastic, vaporous state. He also attributed the weight gain in calcination to air attaching itself to the metal. Furthermore, Hales invented a device to capture and measure the bulk of airs that are released or absorbed in various situations. With this device, he studied reactions involving such substances as blood, oyster shells, coal, vinegar, wax, gold, and bones. He also studied various combustion reactions and the respiration of small animals.

Lavoisier praised Hales's numerous and carefully performed experiments along with his accompanying reflections. He urged the reader to consult Hales's writing in the original; in it, the reader will find "a most inexhaustible fund of meditation." Lavoisier also pointed out that the German Stahl, who extended and popularized the phlogiston theory, did not appear to be at all acquainted with Hales's work although they were contemporaries. Stahl maintained that air takes no part in chemical reactions, despite Hales's numerous experiments to the contrary. Moreover, Lavoisier suggested that Hales's work did not produce the reformation in chemistry that it should have, overshadowed as it was by Stahl's phlogiston theory.

Nevertheless, the study of chemistry continued to advance. In the 1750s, the Scottish chemist Joseph Black, a professor who taught chemistry to medical students, began to experiment in an attempt to find something powerful enough to dissolve stones that form in the urinary tract. As he worked, Black consistently kept track of the various substances using a balance scale. When he heated quicklime, he found that in forming a calx it lost weight. Through a series of combination reactions, first with an acid and then with an alkali, he was able to restore it to its original form, in which it had precisely the same weight as it had originally.

Black concluded that the loss in weight in forming the calx was due to elastic vapors leaving the substance, and the gain in weight, which was accompanied by a restoration of the original properties, was due to the elastic vapors being reabsorbed. He named these elastic vapors "fixed air," borrowing the term from Hales. However, while Hales used this term more generally for any kind of air that can be fixed to matter, Black used it specifically in reference to the air involved in his experiments with quicklime and other similar substances. He then began to study the properties of this fixed air. He found that it does not support combustion. It is released in the burning of charcoal, fermentation of beer, and respiration. He also determined that fixed air does not attach equally to all bodies; it has a greater affinity for some substances than others. Finally, it is only a small part of atmospheric air.

Following the publication of Black's work, there arose a great interest in this "fixed air" he had identified. The Irish physician David MacBride discovered that it is released during the decay of animal matter. Moreover, when he applied fixed air collected from fermentation to partially decayed flesh, the firmness of the flesh was restored! Henry Cavendish, an English natural philosopher, demonstrated that cold water can absorb great quantities of fixed air, creating bubbly, acidic-tasting water. He also discovered that water in this form can dissolve almost any metal. Meanwhile, the English minister Joseph Priestley discovered that although fixed air cannot support combustion or respiration, these properties can be restored to the air through exposure to a growing plant.

Although many were inspired by Black's experiments and copied his experimental techniques, practically no one copied his use of a balance scale to track the movement of substances during chemical processes. As a result, arguments regarding whether the observed physical and chemical changes are caused by the movement of air or the movement of phlogiston remained unresolved. Lavoisier believed that careful measurements of weight were key to resolving this question. At the end of the book, he shared his own experiments in which he made consistent use of a balance scale, noted temperatures, and repeated experiments that he had performed in air now in a vacuum, trying in every way possible to remove doubt from his conclusions.

With that, Lavoisier finished his book *Opuscules Physiques et Chimiques*. He sent copies to the Royal Society of London, where he knew it would reach Priestley and Cavendish, and the Royal Society of Edinburgh with an extra copy especially for Black. Lavoisier wanted to create a revolution in physics and chemistry. He wanted to create something as magnificent as Newton's theory — a complete theory of matter. There was still much work to be done, many missing links to be found and connected. Lavoisier was not yet ready to announce any great discovery, but at least now he had organized the pieces.

When the Academy had reviewed *Opuscules*, they commended Lavoisier for his restraint in not drawing hasty conclusions. In the book, Lavoisier had even conceded that the phlogiston theory might be compatible with his experiments; while air entered a substance causing it to gain weight, phlogiston might simultaneously be leaving. Nonetheless, Lavoisier was confident in his observations and his interpretation of them. An increase in the weight of a substance when burned or heated is always accompanied by a loss of air. In this case, air is fixing itself to the substance. Similarly, a decrease in weight is caused by air being released.

However, Lavoisier didn't have any explanations. He couldn't explain why some objects burn and others don't. He couldn't explain why some substances absorb elastic vapors when heated while others release them. He had no explanations, only observations, and these were not enough to form a theory.

Soon, new clues emerged. In 1774, the French military pharmacist Pierre Bayen, who created mercury calces for medical purposes, discovered that the calx of mercury could be reduced to pure mercury by heating, *without* charcoal as well as with it. When it was heated with charcoal, it gained weight, as was typical for a metal calx. When it was heated without charcoal, it lost weight and released a gas. Bayen interpreted this in terms of Lavoisier's theory, that the gain or loss of weight is due to air fixing itself to or being released from the mercury.

Meanwhile, in England, Priestley continued to experiment with various methods of producing different kinds of airs and made this same observation, that it is possible for mercury calx to be reduced without charcoal. He then began to study the air that was released. A candle lit in this air burned even more brightly than in ordinary air. A mouse enclosed in a container filled with this air lived more than twice as long as it would have in common air. He inhaled the air himself; his chest felt light and his breathing especially easy for some moments afterwards.

According to the phlogiston theory, a fire cannot persist and an animal cannot survive for long in a completely enclosed space because the phlogiston released saturates the air. Priestley reasoned that since this air is better for both combustion and respiration, it must be pure air without any phlogiston. He therefore called it "dephlogisticated air." Carl Scheele, a Swedish chemist who had discovered this same air and, like Priestley, interpreted it in terms of the phlogiston theory, called it "fire air." Lavoisier initially referred to it as "eminently breathable air."

Lavoisier spent the next several years experimenting with this newly discovered air. In addition to confirming the properties of the air discovered by Priestley, he performed his own experiments. He reduced mercury calx without charcoal to obtain eminently breathable air and then performed the reverse reaction, combining this air with mercury to form a calx. He discovered that after the mercury had absorbed all of the respirable air in forming the calx, there was another elastic vapor remaining: a non-combustible, non-breathable air different from fixed air, which he called "mofette." When he combined the mofette with the respirable part of the air, he obtained ordinary atmospheric air.

Lavoisier also performed a reduction of mercury calx with charcoal. When the reduction was complete, the only remaining substances were pure mercury and fixed air. Therefore, the fixed air must be charcoal combined with the respirable part of air that had been part of the calx. The pieces were beginning to fit together.

Even more pieces came together when considering respiration. The breathable part of the air, Lavoisier determined, combines with the blood, giving it its red color. It is then exhaled as a part of fixed air while the non-respirable part, the mofette, is exhaled unchanged. Furthermore, since combustion also consumes the respirable part of air while producing fixed air, he hypothesized that respiration is a kind of combustion that produces heat, which would explain how it is possible for a body to maintain a constant temperature even while heat flows from the body into the cooler environment.

Lavoisier performed more experiments, trying to understand the relationship between the eminently breathable air and acids. He found that it combines with phosphorus during combustion to produce the acid of phosphorus. It also combines with sulfur to produce vitriolic acid. In both cases, more than half of the weight of the acid comes from air. He found that eminently breathable air is also a component of nitrous acid, carbonic acid, oxalic acid, and others. From these results, he came to the conclusion that it must be a component of all acids. With this, he gave the air a name. "I shall henceforward designate dephlogisticated air or eminently respirable air," he announced at an Academy meeting in November of 1779, "by the name of acidifying principle, or if one likes better the same meaning in a Greek word, by that of *le principe oxygine*."

In 1775, Lavoisier assumed a post in the Gunpowder Commission, an organization created to replace a private company that had been unable to satisfactorily supply the country with gunpowder. Under Lavoisier's direction, both the quality and quantity of gunpowder rose to such an extent that, rather than having to import gunpowder from other countries, it became a profitable export. In this position, Lavoisier was given both housing and a laboratory.

Hence, in April of 1775, Antoine and Marie-Anne Lavoisier moved into a private apartment in the Royal Arsenal, where they would live and work for almost two decades. Lavoisier used his income from the General Farm to furnish the laboratory with the most advanced equipment available. Eventually his laboratory included a scale accurate to one part in four hundred thousand. With this scale he verified with an enormous degree of precision, in experiment after experiment, what he had been assuming for years, that the total weight before and after every combustion, calcination, reduction, and any other type of reaction was always the same.

In the late 1770s, with more and more pieces coming together, Lavoisier turned his laboratory into a school. It had its own teachers, French chemists who had been converted to what they called the "New French Chemistry," in which chemical changes are accounted for by the movement of different kinds of elastic vapors in and out of substances. The teachers instructed in laboratory techniques as well as theory. Among its first students was Madame Lavoisier.

During these years, Lavoisier woke up early in the morning, often spending three hours in private research before beginning his workday, which would be filled with business related to the General Farm, Gunpowder Administration, and the Academy of Sciences. He then returned to his private studies for several hours in the evening. On Saturdays, he and Marie-Anne would happily spend the entire day in the laboratory, even taking their lunch there, with science friends, assistants, and students. Madame Lavoisier ran a social salon at the Arsenal to complement the school and laboratory, welcoming French intellectual and political leaders as well as numerous international scholars visiting Paris.

Meanwhile, the English chemists continued to innovate, adding fresh experiments to the growing body of chemical knowledge. Years before, Cavendish had isolated an inflammable elastic vapor that could burn even without any air or fuel. In 1781, Priestley began to experiment with this same air. When he ignited a mixture of inflammable air and dephlogisticated air, a violent explosion resulted that left mist on the walls. He told Cavendish about his experiments, who then began his own experiments. Cavendish discovered that the mist was plain water. He also found that when the airs were combined in a ratio of two volumes of inflammable air to one volume of dephlogisticated air, all of the air was converted into water.

Priestley and Cavendish then both attempted to apply the phlogiston theory to these observations. Priestley suggested that water must be pure atmospheric air that has been saturated with phlogiston. The inflammable air, therefore, is simply phlogiston. Cavendish disagreed, claiming inflammable air must equal water plus phlogiston and dephlogisticated air equals water minus phlogiston. The "plus phlogiston" and "minus phlogiston" cancel out when combined, forming pure water.

While Priestley and Cavendish were creating water out of air, Lavoisier was experimenting with heat. When an object is heated, something flows into the body. He considered this an element, something like the matter of fire, but he needed a new word for the new chemistry. He decided to call it "caloric." The caloric has no weight; Lavoisier had demonstrated repeatedly that the weight of an enclosed system is the same before and after a reaction even when great amounts of heat are applied. Looking for a way to quantify heat, he recalled Black's work demonstrating that the temperature of ice remains constant while it melts. Therefore, since all of the heat, or caloric, goes into melting the ice, the amount of ice melted can act as a way to measure the amount of heat released during a reaction.

Lavoisier then worked with Pierre-Simon Laplace, a French scholar and fellow member of the Academy, to invent what they called an "ice calorimeter," a device that uses ice to measure the amount of heat released. With this instrument, they found support for Lavoisier's hypothesis that respiration is a form of combustion by comparing the heat released during the burning of coal to that released during the respiration of a guinea pig.

Upon hearing about Priestley and Cavendish's work with inflammable gas, Lavoisier and Laplace turned their attention toward this new elastic vapor and began to do their own experiments. They soon verified the English scientists' results, that the inflammable air combines with oxygen to produce pure water. Then by shooting streams of the two vapors together in different proportions and observing the brightness of the explosion, they determined that the ideal ratio is a little less than two parts of the inflammable gas to one part oxygen, similar to Cavendish's results. Later, Lavoisier and

Claude-Louis Berthollet, an Academy colleague who had worked with Lavoisier manufacturing gunpowder, engineered a way to separate water by dripping it into a red-hot iron gun barrel. They then collected the inflammable air, much lighter than ordinary air, for use in balloons.

Lavoisier published memoir after memoir, which, although not attacking the phlogiston theory directly, put its supporters on the defensive. For every result he published that seemed to contradict the theory, its defenders made up new explanations to fit the new observations. Finally, Lavoisier launched a public attack on the phlogiston theory in his paper, "Reflections on Phlogiston," which he presented before the Academy in June of 1785.

Lavoisier suggested that phlogiston is imaginary, that its existence in matter is an unjustified assumption. All the known facts of combustion and calcination can be explained better and more simply without phlogiston than with it. Furthermore, the phlogiston theory is vague. It has no laws and makes no predictions. Phlogiston is sometimes free fire and other times, fire with earth. It could have weight, no weight, or negative weight. In some situations it can move through the pores of vessels, and other times it cannot. Its existence in matter was used to explain acidity, opacity, and color as well as their opposites. Phlogiston was a "Proteus," the Greek prophetic man of the sea who constantly changed shape to avoid answering questions.

Lavoisier concluded, "I see with much satisfaction that young men, who are beginning to study the science without prejudice, and geometers and physicists, who bring fresh minds to bear on chemical facts, no longer believe in phlogiston in the sense that Stahl gave it." From the vantage point of the Arsenal, with chemistry courses and laboratory work all confirming and building on Lavoisier's ideas, this statement must have seemed true. In reality, outside of this small arena, there were few scientists embracing the new chemistry. Even among those teaching chemistry who were sympathetic to these ideas, very few, if any, were willing to revise their entire courses to reflect this new school of thought. Therefore, the young chemists, with very few exceptions, were still being instructed in the old phlogiston theory.

Opposition to Lavoisier's attack arose abroad. Lavoisier was accused of merely reciting Stahl's theory backwards and substituting phlogiston for oxygen. In England, Priestley was still sore over the credit given to Lavoisier for the discovery of oxygen. He had not only isolated it before Lavoisier but had even traveled to Paris to demonstrate it for him. And then Lavoisier had renamed Priestley's dephlogisticated air "oxygen" and taken the credit!

The Irish chemist Richard Kirwan rallied the opposition in the book *Essay on Phlogiston and the Constitution of Acids*. Loaded with sarcasm, Kirwan discredited Lavoisier's theory by calling it arbitrary and cited experiments that "proved" the existence of phlogiston. Furthermore, he identified inflammable air as phlogiston and attempted to establish its existence in all combustible substances. Kirwan's counterattack was successful, strengthening opposition to the antiphlogistic theory, especially in England, Germany, and Sweden.

In the meantime, Lavoisier had begun to assemble his own army. Already on his side were Madame Lavoisier and Antoine Fourcroy, who taught at the Arsenal. Berthollet, who had worked with Lavoisier producing gunpowder and inflammable air, declared his support publically at an Academy meeting, and his influence helped bring others to Lavoisier's side, including the mathematician Gaspard Monge. Lavoisier, Fourcroy, Berthollet, and Monge then traveled to Dijon to visit Guyton de Morveau, a prominent chemist who was working on a dictionary of chemistry, to try to win him over. Despite Kirwan's efforts to keep him loyal to the phlogiston theory, de Morveau eventually joined Lavoisier's side.

Lavoisier had gathered a small but very capable group by the time Kirwan's book was published. Madame Lavoisier led the response by translating Kirwan's book into French. Lavoisier, Fourcroy, Berthollet, Monge, and de Morveau added essays refuting Kirwan's theory, each presenting arguments in their own area of expertise. Kirwan responded with another edition that included the French essays along with rebuttals to their arguments, but his efforts were obviously weakening.

By the time Kirwan's second edition was published, the French chemists had launched another offensive. De Morveau had already been working on a new chemical dictionary. Now Lavoisier, Fourcroy, and Berthollet joined him with the goal of creating new chemical nomenclature to reflect the new chemistry. In April of 1787, Lavoisier introduced the project to the Academy and shared with its members the principles that motivated their decisions.

Chemical terminology had developed haphazardly, claimed Lavoisier. Some substances were named based on their appearance or properties. Others were named after the discoverer or place of discovery. Still others were given names based on their alchemical association with a planet. Names given in this way made memorizing difficult. Furthermore, the names were often misleading; chemical nomenclature was filled with oils, butters, and flowers that had nothing to do with the biological substances whose names they bore. That names were given in this way was excusable when so little was known. Now that chemistry has advanced, its terminology should also advance.

The name of a substance, Lavoisier proposed, should indicate its chemical composition. But here too there was confusion. There was no consensus regarding the elements, the simple substances that make up all matter. Lavoisier and his colleagues made the decision to use the definition Boyle had proposed in the *Sceptical Chymist* but never applied: "We shall content ourselves here with regarding as simple all the substances that we cannot decompose, all that we obtain in the last resort by chemical analysis." With this, they offered a starting point for a new theory of elements.

However, as Lavoisier emphasized, any list of elements must be tentative. What is now regarded as simple may later, with more powerful means of chemical analysis, be found to be composed of even simpler elements. Therefore, any attempt at a system of naming should be flexible enough to adapt to future discoveries. To this end, he and his colleagues proposed a system that was not meant to be set in stone, but rather one that was a living entity that could grow and evolve with the growth of chemical knowledge.

De Morveau and Fourcroy followed up Lavoisier's introduction in later meetings, applying the system and demonstrating it with tables that connected the old terminology with the new. They offered a long list of elements including oxygen, sulfur, phosphorus, iron, gold, silver, lead, and mercury. Inflammable air was renamed "hydrogen." Charcoal was renamed "carbon." Lavoisier's mofette became azote meaning "no life." Later it was renamed "nitrogen" after it was discovered to be a component of saltpeter, also called niter. Elastic vapors were called "gases."

For compound substances, they devised a classification system with different naming conventions depending on the type of substance. Suffixes were used in naming acids to indicate the amount of oxygen. For example, acids formed with sulfur were called "sulfuric" or "sulfurous acid," the former having more oxygen than the latter.

Similarly, phosphoric acid has more oxygen than phosphorous acid. Fixed air became "carbonic acid." The calces of metals were renamed "oxides," so that the calx of lead became lead oxide. Names of salts included the metal and the acid from which they were formed, so that a salt made from copper and sulfuric acid was called "sulphate of copper."

The initial reactions to the new terminology were mostly negative, but Lavoisier and his colleagues persevered. Two other Academy members, Jean Hassenfratz and Pierre-Auguste Adet, introduced chemical symbols to go with the new terminology. Adet also founded a journal for the new chemistry, *Annales de Chimie*, which published its first edition in April 1789. In that same year, Lavoisier published a treatise on chemistry, *Traité Élémentaire de Chimie*.

When Lavoisier began writing, his intention was only to explain and extend his earlier work on the reformation of chemical nomenclature. However, it grew and evolved into a complete chemistry textbook. It was an introductory textbook, but at this point practically everyone was a beginner. He wrote about the discovery of oxygen and presented his theories of acids, combustion, and the formation of metal salts. He included his law of mass conservation: the total mass of all the substances involved in a chemical reaction remains constant. The book contained his theory of heat, including the formation of gases due to the absorption of the caloric.

Most of this was old, a summary of memoirs already delivered to the Academy. However, Lavoisier also presented a new theory of fermentation. He explained that the fermentation of fruit juices and other sweet substances was accompanied by the production of carbonic acid and the spirit of wine, which he restored to the old Arabic name "alcohol." He also included his recent discovery that sugar is made up of hydrogen, oxygen, and carbon in specific proportions.

Moreover, Lavoisier included detailed descriptions of experiments so that others could replicate and confirm his results, complete with warnings on how to avoid explosions and shattered glass. And he included descriptions of his instruments accompanied by drawings done by Madame Lavoisier. The book contained everything a student would need to know, and even everything a teacher would need to know, for a complete course in chemistry.

Lavoisier's book was so well received that it was hastily translated into English the following year — in one month! — so that it would be ready for the new school year. Soon after, it was translated into German, Italian, Spanish, and Dutch. The book had been another offensive, another strategy to promote the new chemistry. And it was successful; practically every opponent laid down their arms to receive Lavoisier's book, a treasury of chemical knowledge. The only notable holdout was Priestley, who remained loyal to the phlogiston theory for the rest of his life.

While Lavoisier was winning the chemical revolution, another revolution was gaining momentum. Tensions between the aristocracy and the commoners reached a breaking point, and King Louis XVI was compelled to call upon the States-General, a governing body in France somewhat like a popular parliament, which had not convened since 1614. Furthermore, the Third Estate, who spoke on behalf of the commoners, were to have voting power equal to or greater than the other two estates, who represented the nobility and clergy.

Lavoisier welcomed the summoning of the States-General. He had been a strong advocate for progressive reforms throughout his career. As a member of the Academy, he wrote up long, detailed proposals for the improvement of sanitary conditions in hospitals and prisons. He helped reduce some of the unfair tax collection practices of the General Farm, in part by abolishing special taxes placed on Jewish persons and businesses. He gave greater rights to the villagers whose excrement-laden soil provided the saltpeter for gunpowder. Lavoisier looked forward to a better France, a constitutional monarchy that would be even more receptive to progressive change.

The meeting of the States-General began on May 5, 1789 and lasted until the middle of June, at which point the Third Estate declared itself the National Assembly of France. The king closed the meeting hall; the Third Estate then reconvened in an indoor tennis court, where they began to draft a constitution for France. By the end of June, King Louis had surrendered and ordered the nobility and clergy to join the National Assembly.

Soon after, riots broke out. Demonstrators stole thousands of guns from *Hôtel de Invalides* and *Hôtel de Bretonvilliers*, the headquarters of the General Farm. The commander of the Royal troops then ordered that the gunpowder held at the Arsenal be transferred to the fortress at the Bastille for safekeeping. Protesters began tearing down the customs wall built around Paris by the General Farm. Lavoisier, who had been appointed to keep order, spent many hours in conversation with the National Assembly, trying to justify the move of gunpowder to the Bastille. He assured them that it would be available for popular interests should that become necessary. This became irrelevant when the Royal troops defending the Bastille surrendered to the thousands of protesters. Three days later, on July 17, Louis XVI arrived in Paris and pinned the new revolutionary cockade to his hat in front of the National Assembly.

Lavoisier was now placed in charge of the destruction of the Bastille. He found a contractor and made arrangements for payment out of his own income. They debated various techniques to ensure that the destruction would be done orderly and safely. All of this became irrelevant as workers finished the job with pickaxes, tearing down the Bastille stone by stone.

Things eventually settled down, and the National Assembly began to build a new government, a constitutional monarchy. In these efforts, Lavoisier's expertise was called upon again and again. He was elected to be a deputy to the Paris Assembly. He was asked for assistance regarding the technicalities of issuing paper money as well as the detection of counterfeit. He was consulted by committees that dealt with issues of health and taxes. He served on the committee of weights and measures, acting as its treasurer. He was also chosen to be one of six board members for the newly organized Royal Treasury, which upon Lavoisier's suggestion became called the National Treasury. In this role, he developed a careful system of accounting so that one could know the condition of the treasury daily.

In June of 1791, the royal family attempted to escape France. They were caught and dragged back to Paris. After this, Louis XVI attempted to maintain his power, but the radicals became stronger and fiercer. Eventually, they took over the government and imprisoned the royal family. They established the Revolutionary Tribunal for the judgment of anyone deemed an enemy of the Republic. They set up a guillotine, a newly invented device intended for humane execution, in a public square that had recently been renamed *Place de la Révolution*. There, on January 21, 1793, they beheaded the king.

Foreign nations had tried to save King Louis, and after his execution, other nations stepped forward. They were met by armed revolutionaries who defiantly stood up against the other monarchies. Celebrating the king's execution, the radical journalist Jean-Paul Marat shouted, "Allied kings threaten us, and we hurl at their feet as a gage of the battle the head of a king. We must establish the despotism of liberty to crush the despotism of kings." This is what they did, and in the name of "Liberty, Equality, and Fraternity," a multitude of supposed enemies of the Republic shared the fate of the king.

By this time, Lavoisier had distanced himself from politics as much as possible. He had left his position in the Gunpowder Commission, and he and Marie-Anne moved out of their apartment at the Arsenal, leaving behind their precious laboratory. Lavoisier was no longer working for the General Farm, which had by then been abolished, accused of mismanagement; its assets were in the process of being liquidated. He had resigned from his position in the National Treasury but was still on an advisory board of the Bureau of Arts and Crafts. In this capacity, he proposed a free, universal educational system open to both boys and girls regardless of their means or social class. He also outlined curricula for primary and secondary schools and proposed the creation of trade schools.

On September 10, 1793, officials visited Lavoisier's home on behalf of the Committee of Public Safety. They searched and confiscated many of his papers, especially foreign correspondence that included letters from Benjamin Franklin and Joseph Black. A couple of weeks later, the papers were returned, and Lavoisier was given a certificate of civil virtue, with a note asserting that everything in the documents "does honor to your civic spirit and is susceptible to dissipate any sort of suspicion."

On November 24, an order went out for the arrest of all former members of the General Farm. Lavoisier was out when the officials arrived; his father-in-law, Jacques Paulze, was also away when they came to his home. A few days later they went together to turn themselves in. The specific accusation being made was that the General Farm had defrauded the nation of 400 million livres. Although the committee assigned to review the charges lowered the amount to 130 million livres, they proposed that the accused be tried as enemies of the state before the Revolutionary Tribunal, which practically guaranteed their execution. Eventually, a more careful audit showed that the state actually owed 8 million livres to the General Farm. In light of this new information, all of the accused were exonerated, but by then it was too late.

Based on all the surviving documents, it appears that Lavoisier faced his final days with calm. He wrote to Marie-Anne from prison, trying to console her. He assured her that he had enjoyed a happy life since he had known her. When she had visited the day before, he noticed that she looked sad. "Why should you be," he wrote, "since I am resigned to everything and since I regard everything I will not lose as won?" Lavoisier was content knowing that he had left his mark, that he would be remembered.

# 3

# On the Nature of Light

I had always felt so unable to understand what light is,
that I would have gladly spent all my life in jail,
fed with bread and water, if only I was assured
that I would eventually attain that
longed-for understanding.

— Galileo Galilei

Christiaan Huygens was one of the lucky ones who began his scientific career during the glorious period after Galileo had effectively broken Aristotelian physics but before Newton established his laws. So many problems left unsolved! Every scientist had the right to dream of making the next discovery. They cooperated and competed, all wondering what would come out of Galileo's "New Science."

Huygens was Dutch by birth, born in 1629 into a wealthy and well-connected family. His father was a classical scholar and diplomat, friends with such prominent thinkers as Rene Descartes and John Donne. Christiaan's early schooling was done at home through a private tutor. He learned to sing, play the lute, and compose Latin verses. He also enjoyed drawing and making mechanical models. At sixteen, he entered the University of Leyden, where he studied law and mathematics, acquiring great skill especially in the latter.

Huygens published his first work at the age of twenty-two, a treatise about "squaring the circle," the ancient Greek problem of constructing a square whose area is equal to that of a given circle using only a straightedge and compass. Gregory of Saint-Vincent, a Flemish mathematician, had attempted a proof of its impossibility. Huygens demonstrated that Gregory's proof was incorrect. He followed up with another longer work on the same problem. With these two papers he gained widespread recognition as a mathematician of the highest caliber; he was compared to Apollonius and Pappus, two of the greatest geometers in the ancient world.

Around the same time, with the encouragement of his father and the assistance of his elder brother, Huygens began to build telescopes. He ground his lenses by hand and experimented with different shapes. His first telescope was twelve feet long. With this, he discovered a moon around Saturn. He also looked for moons around Mars and Venus but could not find any. He then built an even larger telescope with twice the magnification and used it to more closely study Saturn's unusual appearance. Galileo thought that it might be three spheres joined together. However, with his improved telescope, Huygens could see a band extending across and beyond the planet. He interpreted the band as a ring around Saturn and created a mathematical model that could predict its future appearance. With these discoveries, he established himself as an astronomer of the highest rank.

Huygens also began tinkering around with different designs for a pendulum clock. Galileo and others had had ideas about making such a clock, but no one had actually carried them through to completion. Huygens cleared up all practical challenges. His clockmaker reproduced and sold clocks built on his design, and soon pendulum clocks appeared throughout the world. Huygens followed up his construction with a theoretical work on the subject, the first thorough mathematical treatment of any dynamic system. He included details on the construction and calibration of the clock along with his theoretical analysis in the book *Horologium Oscillatorium*. This work assured him a place among the greatest physicists of his age.

Huygens first went to Paris in 1655 at the age of twenty-six. He enjoyed the music, drama, and company of the intellectual and artistic people in the city. He talked with astronomers about Saturn and was pleased to find that his telescope was as good as any that could be found in Paris. He talked with mathematicians about mathematical problems. He talked with clockmakers and lensmakers about their crafts. He attended informal meetings dedicated to the new philosophy that had sprung up around Paris. He enjoyed these too but felt that they spent too much time on philosophical arguments.

In the early 1660s, Huygens's father and younger brother traveled to Paris on a diplomatic mission, bringing with them one of Christiaan's pendulum clocks, which they presented to King Louis XIV. At this time, the king's chief adviser was conspiring to turn Paris into the cultural center of the world. A scientific academy similar to the Royal Society in London, which was just getting started, would fit nicely with this plan. Toward this end, the king invited Huygens to Paris to assist in creating such an institution. Although he was at home in Holland for most of the intervening years leaving the details of the organization to others, Huygens became the nominal head of Paris's new *Académie Royale des Sciences*, which gathered officially for the first time in December of 1666.

By this time, Huygens had moved into an apartment in the *Bibliothèque du Roi*, where the headquarters of the new Academy had been established. He lived and worked there for the next fifteen years, even remaining there after Louis XIV waged war on Holland. Throughout this time, Huygens was a central member of the Paris Academy. Even so, when he became severely ill and feared for his life, he entrusted his unfinished work not to his colleagues in Paris, but to London's Royal Society.

Huygens visited London for the first time in 1661, just before the coronation of Charles II. He conversed with many who would become the original members of the Royal Society, which was in its final stages of organization. Afterwards, he kept in touch with the group primarily through its secretary Henry Oldenburg. When Huygens became sick in 1670, he sent his papers to Oldenburg with an explanation. He felt that the seat of science was in England. Furthermore, its members promoted the new philosophy not out of vanity, ambition, or self-interest, but out of "natural principles of generosity, inclination to Learning & a sincere Respect and love for the truth." He felt that the Paris Academy had been infected with envy because its existence was based on the whim of a king and his minister.

Not everyone shared Huygens's admiration for the Royal Society. Apparently Charles II laughed at the group for wasting its time with the weighing of air. The Paris group, rather, occupied itself mainly with astronomy. They had Huygens with his enormous, powerful telescope and his discoveries of Saturn's moon and ring. In 1669, Italian astronomer Giovanni Cassini joined the Academy and, with a grant from Louis XIV, set up an observatory in Paris. The French Academy was distinguishing itself as the leader in astronomy. It must have been with great interest that they heard about a new kind of telescope that had been built in England.

Huygens received the news from Oldenburg in January of 1672. A professor of mathematics from Cambridge, Isaac Newton, had built a telescope using mirrors rather than lenses. It was a stubbly little telescope — only half a foot long — and yet it magnified forty times, as much as the best telescopes in London and Italy. Huygens called the design "beautiful and ingenious" and set to work constructing one for himself. However, he soon gave up, unable to produce a mirror with sufficient polish.

In March, Oldenburg sent Huygens a copy of the most recent *Philosophical Transactions*, which included a letter written by Newton proposing a new theory of light and color. This was Newton's first scientific paper, but already Oldenburg knew him well. He cautioned Huygens that Newton never put forth any idea lightly and asked for his opinion of the theory.

Newton's letter began by describing experiments that he had performed with a prism. He had darkened a room, allowing a beam of sunlight to pass through a hole in the window frame. He then placed the prism in the path of the light, and a rainbow was projected on the wall behind the prism. This observation was not surprising. It had been known since ancient times that light shining through pieces of broken glass produces various colors. Glass prisms were designed to display the colors most clearly, creating a rainbow. At that time, it was assumed that the colors came *from* the glass.

It was also known that light bends when it moves from one medium into another, such as when light passes from air to glass or water. This bending, called "refraction," was used to explain the movement of rays through lenses, important in the theoretical understanding of telescopes, microscopes, and eyeglasses. Through his experiments, Newton now attempted to understand refraction as it pertained to colors.

He used a board with a small hole in it to isolate the blue, and then the red, part of the beam. With careful measurements, he found that the blue beam bent more than the red beam. He then sent these beams separately through a second prism. The blue again bent more than the red. Furthermore, the blue beam remained blue as it passed through the second prism, and everything it shined on appeared blue. The same was true for red. Once a color had been separated from the others, the prism could not alter it or create new colors.

A prism doesn't *create* colors, concluded Newton, it *separates* them. Color must be an inherent property of a ray of light that is accompanied by a greater or lesser tendency to refract, which Newton referred to as its "refrangibility." And sunlight is made of rays of varying refrangibility that separate when passed through an angled or curved transparent object like a prism or lens.

Just as rays with different refrangibility can be separated when passed through a prism to create a rainbow, these same rays after being separated can be recombined to create new colors. The most surprising and wonderful of these combinations, wrote Newton, was the composition of whiteness. There is no simple ray that can create white. Rather, all of the primary colors mixed in the proper proportions creates light that is perfectly white.

All of this was the background to understanding Newton's telescope that magnified with mirrors rather than lenses. He created it to solve problems that were known to lensmakers but not understood. First, there was the problem that spherical lenses blurred their images — the larger the lens, the greater the blurring. In addition, the larger the lens, the more rings of color were seen through it. Lensmakers tried different techniques and lens shapes but couldn't find a way to eliminate the blurring and the unwanted rings of color. Now, Newton could explain this. The problem was not a matter of construction but rather an inherent problem due to the composite nature of white light. Rays of light with varying amounts of refrangibility bend by different amounts — this causes the observed blurring and extraneous rings. The only way to get around this problem was to avoid using lenses altogether.

Huyens's immediate reply was that the proposed theory was "very ingenious," but he needed to see if it was compatible with all the experiments. Three months later, he composed a more complete reply. He agreed that Newton's experiments proved the compound nature of white light. However, he wondered if Newton might be content with only two colors, yellow and blue, with the other colors regarded as these two with varying depth. With only two colors, he suggested, it would be much easier to find a hypothesis to explain color. Until Newton could explain the difference between these two colors, Huygens wrote, he would not have taught us the nature of colors but only the accident of their different refrangibilities.

Newton replied to Huygens denying that all colors could be created from only yellow and blue. He also denied that it would be easier to form a hypothesis with only two colors rather than an infinite variety, unless it would also be easier to assume only two shapes, sizes, and amounts of speeds or forces for the particles or pulses of light rather than an infinite variety, which he considered to be a "harsh supposition." Besides, it was not his purpose to create a hypothesis explaining the nature of light, but only to share his experiments. Furthermore, Newton declared, if Huygens thought that he could paint the world with only blue and yellow, he should become an artist.


Huygens and Newton had both become interested in light at around the same time in the mid-1660s after reading Robert Hooke's *Micrographia*. While Newton wiggled a needle around in his eye socket and shined light through prisms, Huygens adopted a more mathematical approach in his attempt to understand the nature of light.

Huygens enjoyed Hooke's descriptive and experimental approach to light and also Boyle's experiments with colors. However, he believed that the key to understanding the nature of light and color was in the study of refraction. He thought that neither Hooke nor Boyle had paid sufficient attention to refraction, one of the few optical phenomena that could be described mathematically.

According to the law of refraction, every transparent medium can be associated with a quantity called an index of refraction. With the index of refraction of the medium it was possible, using the law of refraction, to determine how a ray of light coming from the air at any angle will bend when entering the medium. This was extremely important in the optical analysis of lenses.

Pierre de Fermat, a French mathematician whom Huygens had met during his early visits to Paris, derived the law of refraction using a "least time principle." According to this principle, a ray of light takes the path between any two points that minimizes its time of transmission. Huygens felt that this principle had an Aristotelian feel to it; it seemed to imply that light has both knowledge and intent. However, as he repeated Fermat's calculations, he became more and more convinced that Fermat's conclusion at least was correct: the index of refraction of a medium is equal to the ratio of the speed of light in air to that in the medium. Because of this, Huygens agreed with Fermat that the speed of light is finite, contrary to the commonly held belief that light travels instantaneously.

Huygens was in Holland in November of 1676 when Ole Rømer, a Danish astronomer who had been working as Cassini's assistant in the new Paris observatory, made the announcement that the speed of light, although large, is finite. He had noticed that the time between the eclipses of one of Jupiter's moons was shorter as Earth approached Jupiter than when it receded and used this information to estimate the speed of light, about 210 million meters per second.

The news of Rømer's discovery rekindled Huygens's interest in light. He was sick at the time but began to work even while recovering. Shortly after his return to Paris in the middle of 1678, he presented to the Academy a treatise on light. He held off on publishing it, though, wanting to first translate it into Latin so that it could be more widely read.

Huygens and Newton met for the first time in 1689, shortly after Newton published *Principia*. By this time, Huygens had moved back to Holland permanently on account of his poor health. He planned a trip to England "only to make the acquaintance of Mr. Newton," as he wrote to his brother, "whom I exceedingly admire for the beautiful inventions that I found in the work that he sent me." In June, Huygens arrived in London, where he met Newton at a Royal Society meeting. Later, the two traveled together by stagecoach from Cambridge to London. Huygens also met with Boyle on several occasions during this visit and had the opportunity to witness some of his chemical experiments. He returned to Holland with many regrets at his own isolation.

Huygens recorded nothing of interest about the time he spent with Newton. There is some evidence that they spoke of gravity. Before his trip to England, Huygens still clung to Cartesian vortices as an explanation for gravity. Afterwards, he began to distance himself from this view while Newton reconsidered it. It is unlikely that they argued out their views on light in any meaningful way.

After returning to Holland, Huygens finally published his treatise on light. He still hadn't translated it into Latin, and so it remained in the original French, *Traité de la Lumière*. In this work, Huygens proposed that light is a kind of pulse that results from the motion of the particles in a burning object. The pulses spread out in all directions as spherical surfaces like sound spreads through the air. He called these pulses "waves" because of their resemblance to the circular ripples that form, for example, when a stone is thrown into the water. He then questioned whether the spreading of light takes time, as does that of sound and water waves. Appealing to Rømer's discovery, he answered that it does.

With this model, Huygens could explain the observation that light rays can pass through each other, as when people look into each other's eyes. The light rays reflected off the two sets of eyes pass through each other without distortion. The pulses will briefly interact as they pass through each other, but afterwards they regain their original shapes. Similarly, rays of light coming from many diverse places and then intersecting will not hinder each other.

Huygens also demonstrated mathematically that light entering a medium in which its speed changes will bend, consistent with the law of refraction. Furthermore, he reproduced the result that Fermat had obtained using the least time principle, that the index of refraction of a medium equals the ratio of the speed of light in air to the speed of light in the medium. Using known data on refraction indexes, he determined the ratio of the speeds in different mediums. For example, he predicted that light moves three times slower in water than in air. Using Rømer's value for the speed of light, Huygens then estimated the speed of light in water to be about 70 million meters per second.

In his treatise, Huygens proposed that light is the motion of some kind of matter. This is evident, he claimed, in the generation of light by fire, whose rapidly moving particles melt even the most solid of substances. However, light does not move through the air. Boyle demonstrated this with his air pump; when air is removed from a glass globe, a sounding body inside it can still be seen even though it cannot be heard. Huygens reproduced Boyle's experiment, verifying this result. He concluded that light must be the motion of whatever is left in the globe after the air is emptied out. He called this substance "ethereal matter" or just "ether." Ether, he reasoned, fills the universe, occupying the spaces between particles of ordinary matter.

Huygens speculated on the nature of ether, most importantly asserting that it must have "spring" or elasticity. Air is compressible, and the amount by which it is compressed is proportional to the force it must exert to regain its natural volume, just as with a spring exerting a force to return to its equilibrium size. As a consequence of this springiness, a large impulse oscillates with the same frequency as a small impulse, similar to the periodic motion of a watch or clock. This same property applied to sound implies that a loud sound travels at the same speed as a soft sound. In the case of light, a bright light travels at the same speed as a dim light.

Therefore, the speed of light should be constant through the ether, independent of the strength of the impulse that produces the light. Huygens suggested that light slows down in substances like glass and water due to collisions with ordinary matter. With this, he was able to explain an optical illusion that had been observed but never understood: the Sun, Moon, and stars are visible *before* they have actually risen above the horizon. Since air has weight, its density should not be constant, but rather greatest closest to the Earth. As a ray of light enters the atmosphere, it gradually bends as the light collides with more and more ordinary matter. The bending causes the light to take a curved path, allowing a person to see beyond the horizon.

Finally, Huygens attempted to explain a strange kind of transparent substance from Iceland through which objects appear double. Although it was referred to as a crystal, Huygens identified it as a kind of talc due to its hardness and ease in being broken apart. He experimented extensively on the substance and found that a ray entering it splits into two rays. One of the rays follows the usual laws of refraction with a constant index of refraction consistent with other similar materials. The other ray acts as if its index of refraction varies with incident angle. Huygens suggested that two waves form in the substance, one that spreads out as spherical surfaces and the other, representing the unusual refraction, spreads out in the form of ellipsoidal surfaces. He offered rigorous mathematics to support this suggestion. However, he could not explain why, if another crystal is held behind the first, the rays that split upon entering the first crystal do not split again when they encounter the second.

Throughout his treatise, Huygens ignored color. He had come to believe that color could not be explained mathematically. He also held firmly onto what he called the "true Philosophy," in which the causes of all physical phenomena are conceived in terms of mechanical motions — light, gravity, and all other forces must result from collisions between particles, during which impulse can be transferred between objects but not created nor destroyed. "This, in my opinion, we must necessarily do," he wrote, "or else renounce all hopes of ever comprehending anything in Physics."

Newton had no philosophy within which to understand light. He had no mathematics to describe light. And he had no hypothesis for the nature of light. He only had his experiments. After both Huygens and, more persistently, Hooke, criticized him for not being able to offer a compelling hypothesis concerning the nature of light, Newton retreated. It was not until 1704, after both men had died, that he finally compiled his treatise on light, which he published in English as *Opticks*.

It was a comprehensive treatise on optics, which included all the usual topics associated with the subject — the laws of reflection and refraction, the anatomy of sight, and a discussion of telescopes. But more importantly, it was a treatise on color. Practically every proposition, application, and even every question he posed and discussed had at its heart an attempt to understand color.

Sunlight is made of rays with different refrangibility; each ray *separately* obeys the law of refraction with its own index of refraction for a given medium. And the refrangibility of a ray is associated with its color. Newton listed the colors in order of increasing refrangibility: red, orange, yellow, green, blue, indigo, and deep violet. When sunlight shines through a prism, the different rays bend according to their various refrangibilities to form a rainbow.

Newton made it clear that when he spoke of a colored ray, he was speaking carelessly. Rays are not themselves colored but rather create the perception of color when they interact with the nerves in the eye. Furthermore, two rays that appear the same when projected on a screen may not actually be the same kind of light. For example, although the color orange can be created by mixing red and yellow light, when this light is passed through a prism, it splits into its components. However, when orange light that has been isolated by a prism is passed through a second prism, it remains orange. Moreover, mixing rays from opposite ends of the rainbow — red and deep violet — creates a purple color unlike any kind of pure light.

Newton mixed light, and he mixed paints. He discovered that although white light can be created through an appropriate mixing of colored light, white paint cannot be. The reason is that all colored powders suppress some of the rays with which they are illuminated. They appear colored because they reflect some rays more than others. Red pigment, for example, appears red because it primarily reflects red light. Black pigment reflects almost no light. Since all colored powders suppress at least some of the rays, certain combinations of paints can produce a dusky white, but not a full, bright white.

Newton also experimented with thin transparent bodies, such as panes of glass or thin layers of water. He collected extensive data, demonstrating the regular cycling of colors as the thickness of the transparent body is increased. He layered transparent surfaces on top of each other with a small air gap, producing rings of color. From all of these studies, he came to the conclusion that the permanent color of objects is determined in some way by the size of the particles comprising it. He then drew connections between his data and the colors of various substances.

Newton's stated purpose in his treatise was not to explain the properties of light by hypothesis, but rather to propose and prove them "by Reason and Experiments." However, toward the end of the work, he added some conjectures by way of "Queries." His optical studies had been interrupted, and he did not think he would ever return to them. He included a list of questions for others who wished to build on the work that he had started.

Throughout the first twenty-four Queries, Newton considered light as a wave traveling through the ether: Do not rays excite sensations of color, similar to the way vibrations in air cause different musical notes? And so wouldn't deep violet, which has the most refrangible rays, be perceived when the most rapid vibrations travel through the optical nerve to the brain? Doesn't light propagate outward like the ripples excited by a stone thrown into stagnant water? Isn't this ethereal matter more rare and elastic than air, pervading all bodies and filling the heavens?

In the twenty-fifth Query, Newton asked, "Are there not other original Properties of the Rays of Light, besides those already described?" He then began to discuss the strange substance from Iceland, referred to as a crystal, that Huygens had studied and written about in his *Traité de la Lumière*. With several chemical and physical tests, Newton demonstrated that the substance is some kind of talc. He observed what Huygens had observed, that a ray entering the crystal splits into two rays, one of which follows the usual laws of refraction while the other exhibits an unusual refraction, which he described in detail. He also observed that if another crystal is placed behind the first, the light that had divided when entering the first crystal does not divide again.

Newton noticed something else as well. If the second crystal is placed parallel to the first, the angles for the second refraction are identical to those of the first. However, if the second crystal is placed *perpendicular* to the first, the angles for the second refraction are reversed. Newton concluded that every ray of light must have two opposite sides. The orientation of the sides of the light with respect to the orientation of some kind of internal structure in the Iceland crystal must cause a usual refraction in one case and an unusual refraction in the other.

Newton continued with his Queries, challenging the view that light is a motion through the ether: "Are not all Hypotheses erroneous, in which Light is supposed to consist in Pression or Motion, propagated through a fluid Medium?" He then challenged the notion of the ether entirely. First, such a medium that is supposed to fill the heavens would hinder the regular and lasting motion of the planets and moons. Second, there is no evidence for such a substance. In conclusion, we should reject the existence of the ether along with the hypothesis that light is a motion through that medium. For support, he appealed to the oldest and most celebrated philosophers of Greece and Phoenicia who believed in the existence of atoms in a vacuum.

Newton believed, based on his experiments with light, that matter is made up of atoms — hard particles, perhaps indestructible, with great spaces in between them that sometimes hinder the passage of light and other times allow light to pass through. Now he wondered whether light, too, could be made up of particles: "Are not the Rays of Light very small Bodies emitted from shining Substances?" In this way, rays of light would have several different properties and conserve these properties as they pass through different mediums, consistent with observations that light keeps its refrangibility and tendency to bend this way or that when passing through the strange Iceland crystal.

Newton went further by proposing that light might be composed of particles of different sizes. Rays of light entering glass bend toward the glass as if the rays were attracted to it. If the same force acts on all the particles, those with the smallest mass would bend the most. Therefore, the particles of violet should be the smallest since they are the most refrangible. In this way, the different colors with their varying refrangibilities could be explained simply by attributing different masses to them.

Finally, Newton addressed the question of the origin of light. If light is not a motion in the ether caused by the vibration of matter, where does it come from? Light is produced from burning matter, and light can heat up matter, causing it to burn. In his thirtieth Query, he proposed that light and matter might be able to turn into each other: "why may not Nature change Bodies into Light, and Light into Bodies?" He then questioned whether the particles of matter and light might interact, not through the medium of the ether but rather have certain forces by which they can act directly on each other at a distance. With these last questions, Newton ended his treatise.

Newton died in 1727 at the age of eighty-four. He was given a full state funeral, the first commoner to receive such an honor. Voltaire, a French writer who had been exiled to England, wrote of the event: "I have seen a professor of mathematics, simply because he was great in his vocation, buried like a king who had been good to his subjects." Voltaire became captivated by Newton, who had dissected light "with more dexterity than the ablest artist dissects a human body." After returning to Paris, he became captivated by the elegant Émilie du Châtelet, one of the few mathematicians who could comprehend Newton.

Voltaire and du Châtelet moved in together, collected books, set up a laboratory, and began writing. Voltaire wrote a book in a popular style expounding Newton's theories of light and gravity. He wrote poetically and included relevant history, all the way back to the ancient Greeks. He presented Newton's theory of light, explaining that light is made of particles with different weights — particles of violet light have the smallest weight and therefore bend the most when encountering another medium, and red the greatest, hence bending the least. When white light is viewed through a prism, the different colors bend different amounts, creating a rainbow! He encouraged his readers to test this for themselves, creating their own rainbows with prisms. He shared the story of Newton's discovery of gravity, his realization that the force that pulls an apple toward the Earth is the same as that which causes the Moon to fall around the Earth and the planets to fall around the Sun.

Meanwhile, du Châtelet published a series of works focusing on some of the mathematical and philosophical aspects of Newton's theory of gravity. She published a paper in the *Journal des Sçavans* arguing against Descartes' vortex explanation for the motion of falling bodies and celestial orbits. She wrote a textbook of physics in which she proposed a metaphysical foundation for Newton's theory of gravity. Finally, she embarked on the daunting project of translating Newton's *Principia* into French, adding original commentary that offered different ways to understand Newton's celestial mechanics.

Paris, which had been the center of Cartesian thought, was soon converted to Newton's theories. As one observer commented, "All Paris resounds with Newton, all Paris stammers Newton, all Paris studies and learns Newton." From Paris, Newton's influence spread outward. Francesco Algarotti, an Italian friend of Voltaire, wrote *Il Newtonianismo per le dame*, a popular science book on Newton's theories written for women. Poets celebrated Newton in their poems. Artists paid tribute to him with color wheels and rainbows radiating from prisms. Newton's theories were taught in universities throughout Europe, his theory of gravity hand in hand with his particle theory of light.

In 1807, the Royal Society published a two-volume set of lectures on natural philosophy written by the English physician Thomas Young. Young was a lecturer at the newly founded Royal Institution, an establishment that promoted science education and research especially for practical applications. After two years, his position was terminated with two years' advance pay. It was during this time that he began to compile and extend his lectures into a collection, *A Course of Lectures on Natural Philosophy and the Mechanical Arts.*

The lectures were organized as a student might work through them, beginning with important theoretical background information in an area and then moving to applications. For example, after a general treatment of motion and forces, he considered applied topics such as architecture, carpentry, machinery, printmaking, and timekeeping. He included Newton's theory of gravity followed by lectures on practical astronomy, geography, and meteorology. He also included several histories, including the history of optics and the history of terrestrial physics.

Along with his lectures, Young included tables summarizing all known physical constants and conversions between units of all types, ancient and modern, found throughout the world. He included a detailed index so that one could easily search for small pieces of information without reading the entire text. The most impressive resource that he included, however, was a catalog of references. Here, he listed the best sources of information on the various branches of mathematics followed by references to accompany each lecture. The depth and breadth of his research is reflected in the number of resources he listed — about twenty-thousand in all.

Young's lectures were geared toward practical applications, consistent with the Royal Institution's mission to allow craftsmen and artisans to benefit from scientific knowledge. Nonetheless, he also included a lecture on the nature of light, although he admitted that such an inquiry served no practical purpose. He justified this inclusion on the grounds that it is "extremely interesting." Moreover, understanding the nature of light may help us understand the nature of sensations and the composition of the universe.

Young's interest in the nature of light grew from his medical research on the physiology of sight and sound. In 1800, he had published a letter in the Royal Society's *Philosophical Transactions* summarizing these inquiries. He didn't claim to be finished with his research, but it had gotten so large and had begun to go in so many directions that he thought he should share it in case, for any reason, he could not continue with his work. Toward the end of the letter he included the section, "Of the Analogy between Light and Sound."

In this part, Young maintained that although Newton's particle theory of light was universally accepted in England, there had been some who disagreed. Notably, the Dutch astronomer and physicist Christiaan Huygens had suggested that light is a wave that propagates through the ether. At this point, Young did not take any definitive stance on the matter, though he did offer some objections to Newton's particle theory of light. He also offered additional support for a wave theory of light that stemmed from an analogy between light and sound.

In 1801, Young was chosen for the second time as the recipient of the annual Bakerian medal, an award given by the Royal Society to recognize outstanding achievement in physical science. In addition to a medal, the recipient was awarded a sum of money and required to give a talk on any topic in the field. Young chose as his topic the nature of light and colors.

Although little is to be gained, he began, toward the advancement of science by offering a hypothesis without experimental support, there is much to be gained from a discovery that unifies a diverse number of observations. In this regard, Young did not have anything completely new to propose. Rather, his purpose was to review what others had proposed, giving credit to those who originated the ideas, and then to apply them to a great number of diverse phenomena. It was not even necessary to produce original experiments since much work had already been done in this area, notably by Newton whose optical experiments remained unrivaled.

The bulk of Young's lecture referred to Newton, quoting many long passages from his optical treatise. True, Newton had advanced a theory in which light is made of particles that are emitted from a luminous body. But he had also suggested the alternative, that light might be excitations in the ether from the vibrations of a burning object. Furthermore, he had offered an explanation for color in the context of a wave theory of light.

Newton suggested that the sensation of color might depend on the frequency of the vibrations excited by light on the retina. The most rapid vibrations, which have the shortest distance between one vibration and the next, excite the color violet, while the least rapid vibrations excite the color red. Using this, he explained the different refrangibilities of the colors of light. Since red light has the longest distance between vibrations, it bends the least; since violet light has the shortest distance between vibrations, it bends the most.

Newton had used this idea — that the color of light depends on frequency — to explain the colors produced by thin plates. He also suggested an analogy between color and sound: "May not the harmony and discord of colors arise from the proportions of the vibrations propagated through the fibres of the optic nerve into the brain, as the harmony and discord of sounds arise from the proportions of vibrations in the air?"

Young picked up where Newton had left off. If light is a wave, then when light rays combine, they can add or cancel depending on whether similar or dissimilar parts of the waves combine. This effect had been demonstrated already with the phenomenon of "beats" in music. When two musical notes with slightly different frequencies are played together, their sound gradually increases and decreases in loudness as the two notes alternately add and cancel. The same idea should apply to light waves; when they overlap, they add or cancel depending on whether their motions coincide or oppose each other.

Young then explained Newton's experiments with thin plates. He assumed that rays combine by adding or cancelling depending on whether the motions of their vibrations are together or opposed. He also assumed, in some cases, a reversal of the direction of the wave's vibration at the surface of reflection. In this way, Young could conceptually explain the patterns of colors resulting from Newton's numerous and meticulously recorded observations of the interaction of light with thin plates.

With a quantitative analysis of Newton's observations, Young was able to obtain consistent values for the distance between vibrations in a wave for each color, called its wavelength. He determined that the wavelength of red light is about 680 billionths of a meter. For violet, it is about 440 billionths of a meter. With an estimate for the value of the speed of light, he then calculated the frequencies of the various colors. Violet, for example, vibrates about 707 trillion times a second.

Following its publication, an anonymous writer from the influential *Edinburgh Review* wrote a devastating review of Young's lecture. He criticized it as being "destitute of every species of merit." He admonished the Royal Society for giving its authority to "dangerous relaxations in the principles of physical logic." He ridiculed Young's hypothesis as "the unmanly and unfruitful pleasure of a boyish and prurient imagination, or the gratification of a corrupted and depraved appetite."

The Royal Society apparently ignored this attack on their own body, for they awarded yet another Bakerian medal to Young, who again devoted his commemorative lecture to the wave nature of light. He built upon his previous lecture by analyzing his own original experiments as well as additional experiments found in Newton's *Opticks*. He also added clarity to his conclusions: homogeneous light is made up of opposite qualities that repeat at regular intervals in the direction of motion. These opposite qualities are capable of neutralizing each other when united, extinguishing the light. He called this the general law of the interference of light.

The author from the *Edinburgh Review* returned to attack. After endeavoring to explain away the most decisive of Young's experiments, he concluded that the law of interference was absurd and illogical. He entreated that his readers dismiss this work on the grounds that it lacks all traces of merit. While again reproaching the Royal Society for their continued support of Young, he showed contempt for the Royal Institution, which had employed him: "Has the Royal Society degraded its publications into bulletins of fashionable theories for the ladies of the Royal Institution? *Proh pudor!* Let the Professor continue to amuse his audience with an endless variety of such harmless trifles, but in the name of science, let them not find admittance into the venerable repository which contains the names of Newton, and Boyle, and Cavendish..."

This attack was effective. Most of those who learned of scientific ideas from the more accessible *Review* condemned Young as a scientist and felt that his optical studies were a deplorable assault on the venerable Newton. Finally, Young responded in the pamphlet *A Reply to the Animadversions of the Edinburgh Reviewers*. Unable to find a publisher, he published it himself. He sold only one copy.

In his response, Young objected to the author's claim that he was the only champion of the wave theory. He also objected to the accusation that he had not read the "plainer" parts of Newton's work and had altered his opinions. In addition, he objected to the accusation that he considered Newton a "sorry philosopher." Young had read the entirety of both his *Opticks* and *Principia* with reverence, attention, and delight. However, he wrote, "much as I venerate Newton, I am not therefore obliged to believe that he was infallible." He urged those who doubted his experiments to try them out for themselves.

While responding to numerous arguments raised by the author from the *Review*, Young gave some context for his own work, explaining that he had become interested in the nature of light by way of his medical research. He took the job at the Royal Institution hoping that he could be of some use in teaching natural philosophy, but after two years it became clear that the duties of such a job were not compatible with his profession as a physician, so he resigned. However, wanting to create something permanent out of his work, he began to compile his lectures with a mass of reference works so that every student and researcher could know what had been done and what was left to do. Afterwards, he planned to end his pursuit of general science.

Young published his *Lectures* in 1807, which included one lecture devoted to the nature of light. Afterwards, he abandoned this line of inquiry, turning his attention back toward medicine and the study of languages; he eventually played a leading role in deciphering the enigmatic Rosetta Stone. The general population remained hostile toward Young as a scientist and toward the wave theory he promoted. However, word of his work traveled across the English Channel, and there, some became interested.

In December of 1807, the Paris Academy announced their latest competition. The winner would be whoever could explain the strange refraction of the Iceland crystal using Huygen's wave theory of light. At the time of the announcement, Etienne-Louis Malus was in Egypt. He had been pulled out of *École Polytechnique* to fight with Napoleon. There, he began to study light in order to escape the misery of his life, which included dysentery and the Bubonic Plague.

When he returned to Paris in 1808, Malus heard of the competition and began to experiment with the Iceland crystal and calculate. After a year, he submitted his entry and won. Afterwards, he went back to his studies. He looked at sunlight reflecting off a window through the crystal. He looked at the flame of a candle reflecting off of water through the crystal. He expected to see a double refraction but only saw a single refraction. He concluded that light must have "sides," that a ray of light must be asymmetrical. Only the rays oriented a certain way pass through glass or water. The others are reflected back, and these rays refract when passing through the crystal only one way, thereby eliminating the double refraction. Malus announced his findings to the Academy in 1811, calling this new property of light "polarization."

In March of 1817, the Academy announced another challenge: to determine how light travels around an obstruction. They only received two submissions, one anonymous and another from a civil engineer, Augustin-Jean Fresnel. Fresnel had become interested in light while building roads and bridges in a remote part of France. Studying optics from textbooks and performing his own experiments, he had become convinced of the wave nature of light. In his contest entry, he applied Huygens's wave concept, updating it using the more modern method of calculus.

One of the judges for the competition was Simeon Poisson, a distinguished mathematician and a staunch advocate of Newton's particle theory of light. Having reviewed Fresnel's entry, Poisson did his own calculations. He concluded that if Fresnel were correct, there should be a bright spot right in the center of the shadow produced by the obstruction. The judges performed the experiment and, just as Poisson had predicted, there was a pinpoint of light right at the center of the shadow! Even then, Poisson would not waver, stubbornly clinging to the Newtonian view. The other judges, however, were convinced and awarded the prize to Fresnel.

More and more scientists were won over to the wave theory of light as additional experimental and theoretical confirmation appeared. Soon practically the entire scientific community subscribed to the view that light is a wave, and Newton's particle theory of light faded into the background of history.

# 4

# Prelude to Electromagnetism

To you alone, true philosophers, ingenuous minds, who not only
in books but in things themselves look for knowledge,
have I dedicated these foundations of magnetic science
— a new style of philosophizing.

— William Gilbert

The year after the publication of Copernicus's *De Revolutionibus*, William Gilbert was born. At the age of fourteen, Gilbert left his birthplace in Colchester, England to attend St. John's College in Cambridge. He earned a bachelor's and master's degree and then went on to study medicine, obtaining his medical degree in 1569 at the age of twenty-five. He then spent three years in Italy, where the best medical schools of his day were located. When he returned to London, he began practicing as a physician and rose to become the personal physician of Queen Elizabeth. He was accepted as a member of the Royal Society of Physicians, eventually becoming its president.

Though Gilbert was a physician by trade, his passion was science. He hosted monthly scientific gatherings at his home in London. He was one of the earliest Copernicans, at least with regard to the rotation of the Earth. Even before the invention of the telescope, he proposed that there may be many fixed stars beyond the limits of our vision. His earliest scientific investigations were alchemical, but he quickly turned his attention almost exclusively to the study of magnets and a new type of substance he called "electrics," bodies that attract small bits of matter in the same way as amber.

Gilbert researched magnets and electrics for almost two decades and then summarized his work in the book *De Magnete*, which he published in 1600. He didn't quote the ancients for support, for they had little to say about the subject. They knew that magnets, also called lodestones, attract iron. They also knew that amber attracts straw and chaff. These same observations existed in ancient literature from all over the world accompanied by various conjectures and lore.

More recently another property had been discovered: a freely moving magnet aligns itself north or south. With this knowledge came the invention of the mariner's compass. Gilbert talked with sailors to learn what they knew about the compass. He talked with miners to learn what they knew about magnets and iron. But most of all, he experimented.

Gilbert learned that every magnet has two poles, like the Earth, one pointing north and the other south. He devised a way to determine which pole is which for any magnet: put the magnetic stone in a wooden vessel in water — the stone will orient itself such that the north end of the magnet points north and the south end points south. He then found that the north end of one magnet attracts the south end of another. If the like ends are put together, they repel. Furthermore, a cut lodestone results in two magnets, each with a north and south pole oriented the same way as the original stone.

Gilbert also determined that although the attraction or repulsion of a magnet is strongest at its poles, it is not confined to them. Rather, the magnetic force surrounds the lodestone; if a compass is slowly brought around from one pole to the other, the force seems to follow a continuous line connecting the two poles. He also discovered that all forms of iron, whether iron ore or smelted iron, share these magnetic properties, if only weakly. By comparing various properties of lodestone with those of iron, Gilbert came to the conclusion that lodestone and iron are the same thing. Furthermore, since an iron compass needle aligns itself with the Earth's poles, the terrestrial globe itself must be a magnet! Gilbert devoted a large portion of the book applying this model — the Earth as a large magnet — making sense of reports made by sailors as they used compasses throughout the world.

Gilbert also studied amber, which was known to attract straw and chaff. He discovered that many substances share this property, such as diamonds, gems, rock crystals, glass, sulfur, and sealing wax. Furthermore, he found that these materials not only attract straw and chaff, but also metals, wood, leaves, and stones. They even attract water and oil. He called the attracting bodies "electrics" and the attractive force "electrical." He invented a simple device to detect the electrical force between two objects — a long needle balanced on a sharp point, which would move when attracted to a nearby object. He called this device an "electroscope."

Magnets and amber were often discussed together in writings, both ancient and recent, linked by their ability to communicate an invisible force. Gilbert, however, drew a sharp distinction between them. The magnetic force is strong and enduring. It cannot be quenched by water or flame, and its attraction is strong enough to influence objects even through thick slabs of wood, marble, and stone. The electric force, on the other hand, is weak and seems to be trapped inside matter. It can be awakened by friction but is just as easily hidden again, even by a breath of moist air. The electric force can only attract small pieces of matter and is easily shielded by a single sheet of paper or piece of cloth.

Gilbert proposed that the magnetic and electric forces differ not only in strength but also in what and how they influence. The magnetic force only attracts other magnetic objects and does so in a structured way. When bits of iron are scattered around a magnet, they align themselves in a pattern as if tracing the path of the force from one pole to the other. The electric force attracts everything. Moreover, when objects are scattered around an electrified object, they orient themselves randomly.

Although most of Gilbert's book dealt exclusively with his experiments and the connection between these experiments and others' observations, he also included some philosophical speculations, specifically in regard to the nature of the electric and magnetic forces. He prefaced this discussion with the regret that the word "attraction" had crept into magnetic philosophy. Although he, too, used this word, he emphasized that he does not mean attraction in the sense spoken by Orpheus in hymns, that iron is drawn to lodestone as a bride embraces her spouse. Similarly, he found fault with the word "force," which calls to mind tyrannical violence. He nonetheless continued to use these words but emphasized that when he spoke of the force of attraction, he simply meant the tendency for bodies to unite.

Gilbert attributed the electric force of attraction to a flow of matter out of the electric body. This would explain why even a piece of paper can disrupt the force between two objects. He questioned whether the magnetic force could be mediated the same way. Evidence seemed to contradict this, since the magnetic force can even pass through thick pieces of stone. Based on his research, the only way to take away the magnetic powers of a lodestone is with prolonged exposure to intense heat. Even then, the stone's magnetic power is not completely removed but only confused. Another magnet placed nearby can restore the lodestone's power, though the poles will not be oriented as before. Gilbert concluded, therefore, that magnets must have some kind of innate strength.

Furthermore, Gilbert proposed that the force exerted by magnets to attract other magnetic objects is similar to whatever holds the Moon, Sun, and stars together. The invisible influence of a magnet is like the influence of the Moon on the waters of the Earth, creating tides of equal strength whether the Moon is visible above the horizon or hidden on the other side of the Earth. With these thoughts, he proposed that the whole universe is animate, endowed with a soul — all the globes, stars, and the glorious Earth! He condemned Aristotle's universe, calling it a "monstrous creation, in which all things are perfect, vigorous and animate, while the earth alone, luckless small fraction, is imperfect, dead, inanimate and subject to decay." He ended the book with a defense of the Copernican explanation for days and nights, insisting that they are due to the rotation of the Earth about its axis.

Gilbert did not expect his conclusions to be widely accepted. In the preface to his book, he wrote that he was aware of how difficult it was to initiate any change and especially for new ideas that challenge established doctrines to win approval. He told his readers that he had made every attempt to write clearly rather than fill the book with rhetoric; although he sometimes used new words, he did not do so as alchemists did to obscure the subject, but rather to give names to the nameless in order to make the subject more plain so that others could continue to discourse on it. Finally, he urged his readers, even if they didn't agree with his opinions, to take note of his experiments, which had been performed with much care at great personal expense and many sleepless nights.


As a dissenter opposing the Church of England, Joseph Priestley was barred from their educational system, including the great institutions of Oxford and Cambridge. He acquired a good education nonetheless through dissenting schools and self-study, with the generous support of his aunt who raised him from the age of nine. He studied to become a minister at the Daventry Academy, a dissenting college where students and teachers alike were nearly equally divided on every important question. He then took a position as a minister at a small church in Needham.

In Needham, Priestley gave lectures on the theory of religion. In his next position at a church in Nantwich, he established a small school for boys and girls. With the income from the school, he was able to purchase some books and instruments, including a small air pump and electrical machine. He entrusted this equipment to the students in the highest class, encouraging them to do experiments to entertain their parents and friends. Next, Priestley took a position as a language tutor at the academy in Warrington.

Priestley was gifted at languages. He studied Latin, Greek, and Hebrew as part of his formal education. He studied French, Italian, and High Dutch on his own, acting as a translator in both French and Dutch. He also learned Chaldee, Syriac, and some elementary Arabic. He was well-qualified to be a language instructor, but he would have preferred to teach mathematics or natural philosophy.

After settling into the position at Warrington, Priestley had the opportunity to travel to London, where he met Benjamin Franklin. Priestley was a good writer; he had already written books on theology, education, and political history. Priestley now proposed that he write a book on the history of electricity. Franklin approved the project, and Priestley went forward, keeping in constant communication with Franklin and other philosophical friends from London. Before a year had passed, he sent a completed draft to Franklin.

In the preface to the book, Priestley shared his delight with the subject of natural philosophy and more specifically the subject of electricity. He also celebrated the writing of histories that trace the development of ideas. When we see continual progress and improvement we can't help but imagine that it will continue indefinitely into the future, a prospect he considered "boundless and sublime." More practically, collecting all of the widely dispersed studies can provide a better foundation for further inquiries into electricity as well as the other branches of natural philosophy. In addition, Priestley suggested that electricity, light, and chemistry offer an inlet into the internal structure of matter, on which sensible properties depend. By following this new path, even the glory of Newton might be eclipsed by the new worlds that are opened up.

Priestley began the history of electricity with the writings of the ancients and then quickly moved on to William Gilbert. He called Gilbert the father of modern electricity but claimed that he had left his child in its infancy. Robert Boyle built on the work of Gilbert, discovering new electrical substances and new ways of bringing out electrical properties. He also showed that an electrified body both moves and is *moved by* other bodies. Otto von Guericke, a German contemporary of Boyle who invented the air pump, discovered that after a body approaches and touches an electrified object, it is then repelled by it. Furthermore, bodies in an electrified atmosphere acquire a charge opposite to that of the atmosphere. Unfortunately, von Guericke's work was overlooked, and his discoveries had to be rediscovered.

Priestley next considered the achievements of Francis Hawksbee, an instrument maker who began giving weekly demonstrations at Royal Society meetings in 1703 after Newton assumed its presidency. Hawksbee conducted a large variety of experiments, most of which he performed with a self-designed air pump. The tube attached to the pump could rotate so that it could easily be electrified with friction. Hawksbee did experiments with various substances inside the globe. In doing so, he found several ways to produce light both inside and outside the globe, light that was bright enough to be used for reading. He discovered that when the light was on the outside of the globe, it was often accompanied by a "snapping" sound. He also found that all electrical phenomena were intensified by warmth and weakened by moisture.

Priestley then reviewed the accomplishments of several other English scientists, noting that for a while only the English were interested in the study of electricity. The first foreigner to take a serious interest was Charles du Fay, the superintendent of the French king's garden and a member of the Paris Academy of Sciences. Du Fay performed many electrical experiments that he presented in six long memoirs to the Academy. A summary of these experiments was also included in the Royal Society's *Philosophical Transactions* in December of 1733.

Du Fay discovered that all solid bodies, with the exception of metals, could be made electric by some combination of heat and friction. Suspended by silk cords, he electrified himself and observed that if another person or a piece of metal approached within an inch of him, a kind of pricking would shoot out of his body accompanied by a crackling noise. In the dark, he could also perceive a small light. Furthermore, du Fay rediscovered what von Guericke had observed, that after a body approaches an electrified object, it is repelled by it. Through further experimentation, du Fay identified two distinct types of electricity. The first is acquired by substances such as glass, rock, crystal, precious stones, and wool. He called this type of electricity "vitreous." The second type is acquired by amber, silk, and paper, which he called "resinous." Each type of electricity repels itself and attracts the other.

In the 1740s, the Germans began experimenting with electricity, making great advances in the extraction of electrical power. Their new machines were so powerful that a shock could send a jolt through a person's entire body and kill a small bird. They could set fire to various spirits and even ignite hot oil or sealing wax. With the invention of the Leyden phial, large amounts of electricity could be stored.

After finishing up the history of electricity in Europe through about 1750, Priestley set his sights across the Atlantic on the discoveries of Benjamin Franklin. Franklin had been studying electricity in isolation and communicating his work to one of the fellows of the Royal Society in a series of letters dated between 1747 and 1754. These were compiled in the book *New Experiments and Observations on Electricity, made at Philadelphia in America.* The letters were read and admired across Europe, translated into practically every vernacular European language as well as Latin. Franklin's principles of electricity became accepted as the true principles, just as Newton's principles were considered the true principles of nature in general. Foreign publications on electricity were filled with references to "Franklinism," "Franklinist," and the "Franklinian system."

Through his experiments, Franklin discovered that friction does not *create* electrical matter, but rather it moves electrical matter from one object to another. In charging a Leyden phial, for example, electrical matter from one side is passed to the other side. It was therefore impossible, as had been observed, to be electrified by touching a Leyden phial if standing on wax or glass, since it could not give more electricity than it received. If two people stood on wax, each touching opposite sides of the phial, both would be electrified. Then, if the two of them touched each other, they would both receive a shock much greater than if they touched some other person.

In order to communicate these ideas more clearly, Franklin introduced new terminology. He called one side of the phial "positive," assuming that it had gained electrical matter relative to its natural state, and he called the other side "negative," assuming that it had lost electrical matter. He confirmed that the inside and outside of the phial always have opposite charges using a cork ball suspended by silk. If the cork was attracted to the outside, it was repelled by the inside, and if it was attracted to the inside, it was repelled by the outside.

Franklin also observed many similarities between electricity and lightning. Lightning appears crooked and waving in the air, as do sparks. Lightning strikes the highest and most pointed object in an area, such as tall trees, towers, and masts of ships. The same had been observed with electricity, that sparks are especially drawn to pointed objects. Moreover, lightning and electricity can burn, melt, and split apart objects. They can even blind and kill animals. Lastly, magnets had been observed to lose their power after being struck by both lightning and electricity.

Franklin was not alone in observing these similarities. The

French clergyman Abbé Nollet went so far as to explain the mechanism by which lightning strikes: the clouds become electrified with a charge opposite to that of some terrestrial object through heat and friction; as the clouds discharge, they spark and crackle as lightning and thunder. The connection between lightning and electricity was obvious enough that several people independently noted their similarity. But it was Franklin alone who devised a way to test this idea.

In doing so, Franklin made use of another discovery that he had made regarding electricity, that it takes the path of least resistance. And so, for example, although a dry rat can be killed by electricity, a wet rat cannot. The electricity flows through the water on the outside of the rat, leaving its body unharmed. Franklin also knew that different materials conduct electricity with varying degrees of ease. Hempen string, for example, is a very good conductor of electricity while silk string is not. Metal is an excellent conductor that draws in electricity, especially when it is pointed.

With these principles in mind, Franklin designed an electrical kite. He attached a wire to a kite, tied a hempen string to the wire, and then, on the other end, hung a key. He also tied the hempen string to a piece of silk, which he held in one of his hands. As a thunder storm approached, Franklin, soaking wet, raised the kite into the air. Electricity streamed out of his finger as it approached the key! He then collected the electricity in phials and used it to kindle spirits and carry out all the other experiments typically done with electrical machines.

Franklin also devised an inexpensive way to protect buildings from lightning by simply placing a metal rod higher than any other part of the building. A wire attached to the rod could safely conduct any electricity down to the ground or a water source, where it would disperse. In this way, Franklin was able to bring electricity into his house through an insulated rod. In order to avoid missing out on an opportunity, he hung two bells that would ring any time the rod was electrified. He then collected the electricity in jars, which he could use for his electrical experiments.

Priestley's history of electricity went many hundreds of pages beyond his account of Franklin's discoveries. He wrote about improvements in electrical apparatus, the conducting power of various substances, and medical applications of electricity. He included descriptions of the most entertaining electrical experiments. He gave a brief summary of all the important conclusions regarding electricity, including clarification of some of the terminology. For example, "electrics" like glass or wax, which can hold a charge, are poor conductors of electricity, while "non-electrics" like metals and fluids are good conductors. He also included "Queries and Hints calculated to promote further discoveries in electricity," which included sections on the electric fluid, electrics and conductors, and the electricity of the atmosphere.

In the last chapter of the book, Priestley wrote of an experiment that he had performed upon Franklin's suggestion. Franklin had observed that cork balls suspended entirely inside an electrified metal cup seem not to be affected by the electricity. He asked Priestley to verify this observation and make it public in his book. Priestley electrified a tin quart vessel that stood on top of baked wood. He attached a pair of pith balls to a glass rod and hung them so that they and the strings that connected them to the glass rod were entirely within the cup. If at any point the strings came above the mouth of the cup, the balls would separate, but as long as the balls and strings stayed within the cup and did not touch the sides, they appeared not to be affected at all by the electricity of the cup. Priestley then posed the question: May we not conclude that electricity follows the same inverse-square law as gravity, since, as can easily be demonstrated mathematically, a body inside a shell of mass would feel no force?

With that, Priestley ended his book. It was published in 1767 as *The History and Present State of Electricity, With Original Experiments*. That same year, Priestley left his teaching position in Warrington, having obtained a position as a minister of the Mill Hill chapel in Leeds. During his first year in Leeds, he lived next to a brewery and became interested in the fixed air produced in the process of fermentation. He later came up with a method to produce his own supply of fixed air and a means of impregnating water with it. After this, his scientific inquiries became increasingly centered on the subject of chemistry.

In March of 1800, the Italian scientist Alessandro Volta wrote a letter to the president of London's Royal Society to share a new apparatus that he had designed. It was a combination of good conductors in a specific arrangement. Any two metals could be used, though copper and zinc work especially well together. Also needed were layers of a conducting liquid, such as water or brine, or pieces of pasteboard soaked in such a liquid. The three conductors should be arranged in alternating layers always in the same order.

In his letter, Volta described the physiological effects that this instrument has on various parts of the body. It creates a "very disagreeable quivering and pricking" when touched by fingers and a convulsion and painful prick when applied to the tongue. When he applied it to his brain, the result was incredibly unpleasant; fearing that it might also be dangerous, he would not try it again. The instrument functions like a Leyden phial, generating a spark and crackle when the outer layers are touched at the same time. However, it does not need to be charged or recharged. It seems to provide a continuous and unlimited supply of electricity.

Volta's invention was inspired by research done by a fellow countryman, Luigi Galvani, who had observed a violent convulsion in the leg of a dead frog after being touched with a scalpel. After much experimentation, Galvani determined that the limbs convulsed any time a metal arc, generally made of two different metals, served as a connector between the nerves and muscles. He believed that electricity came from the nerves and that the metal arc was merely serving as a conductor. Volta disagreed, suggesting instead that the electricity came from the metals and that the organs only acted passively. The proof was in his invention, which created electricity without any animal matter. He referred to his apparatus as an "artificial electric organ." It soon became known as a battery pile, Voltaic pile, or simply a battery.

Hans Christian Ørsted was a scientist and philosopher. He earned a degree in pharmacology at the University of Copenhagen and then continued his studies in philosophy, earning a doctorate for his research on Kant's philosophy of nature. After learning about Volta's invention of the battery pile, Ørsted began experimenting with alternate designs. In 1801, he published a paper describing a new kind of battery along with a method to determine the electrical current it produces by measuring the gas released when it breaks water into hydrogen and oxygen.

Subsequently, funded by the Danish government, Ørsted went abroad to study in France and Germany. In Germany, he was introduced to the philosophy of nature advocated by Friedrich Schelling. Schelling subscribed to Kant's philosophy of nature that all matter interacts through fundamental forces of attraction and repulsion. Taking this a step further, Schelling claimed that the fundamental forces could take different forms in different circumstances. Therefore, heat, light, electricity, magnetism, and gravity can turn into each other. Ørsted agreed with much of Schelling's philosophy, especially his idea regarding the unity of all the forces. However, he did not share Schelling's disdain for experimental work. He returned to Denmark and in 1806 obtained a position as a professor of physics at the University of Copenhagen.

In 1807, Ørsted announced his intention to examine the effect of electricity on a magnetic needle, although it was more than a decade before he actually performed such an experiment. He had initially entertained the idea of using the electricity from a thunderstorm. He later considered that he might observe an effect using a battery connected to a thin wire; perhaps some kind of magnetism would radiate from the wire along with heat and light. He could then observe any magnetic effects on a compass. He tried this for the first time during a lecture, but he only observed a small twitch of the needle.

Ørsted repeated the demonstration three months later with a thick wire replacing the thin wire to increase the current running through the wire. The wire did not noticeably heat up, nor did it glow. Regardless, the deflection of the compass needle was clearly visible. Over the next several months, he confirmed these results with a more powerful setup. In July of 1820, Ørsted publicly announced his result: he had created magnetism with electricity, sending a shockwave through the scientific community.

# 5

# Faraday's Lines of Force

The man who is certain he is right is almost sure to be wrong; and he has the additional misfortune of inevitably remaining so. All our theories are fixed upon uncertain data, and all of them want alteration and support. Ever since the world began opinion has changed with the progress of things, and it is something more than absurd to suppose that we have a certain claim to perfection; or that we are in possession of the acme of intellectuality which has or can result from human thought.

— Michael Faraday

Michael Faraday's father was a blacksmith who moved his family to London, hoping to give them a better life. Michael was born there, and when he was five, the family moved into a small apartment above a stable in central London. He went to school only for a short time. His mother pulled him out after the schoolmistress tried to correct his speech impediment with a cane. Young Faraday had trouble with his R's; he would, for example, pronounce his brother Robert's name as "Wobert."

At the age of thirteen, Faraday took a job as an errand runner for a book and newspaper shop run by a progressive émigré from France, George Riebau. Riebau quickly recognized his worth and within a year offered him a seven-year apprenticeship as a bookbinder. Here, Faraday learned the skills of the trade while enjoying the fellowship of Riebau and the other apprentices. He also enjoyed the eclectic collection of books that went in and out of the shop, which he could read during his off-hours: picture books, adventure stories, philosophical works, nature books, and many others.

One of the books that had a great impact on Faraday was *Improvement of the Mind* by Reverend Isaac Watts. "Even the lower orders of men have particular callings in life," wrote Watts, "wherein they ought to acquire a just degree of skill." Inspired by this book, Faraday embarked on a program of self-improvement. He began elocution lessons two hours a week and invited friends to correct his grammatical errors. He took drawing lessons from a French artist who was living in a room above Riebau's shop. He started a commonplace book titled "The Philosophical Miscellany," filling the pages with facts about light and color, electric fish, meteorites, lightning, oxygen, water spouts, and the formation of snow.

Watts also recommended that one should "be not too hasty to erect general theories from a few particular observations, appearances or experiments." Faraday took this to heart as he began to read about electricity from an *Encyclopedia Britannica* article that he was binding. The author had confidently put forth Benjamin Franklin's one-fluid theory of electricity, in which positive charge is an excess of some electrical fluid and negative charge is due to the lack of the same fluid. But Faraday wasn't so sure; many French scientists preferred a two-fluid model in which one fluid gives rise to positive charge and another, negative charge. He began to question and experiment, charging up a homemade Leyden phial and using it to shock himself and others.

In February of 1810, Faraday attended his first scientific meeting, a lecture on electricity at the City Philosophical Society. While the Royal Society's meetings were exclusive and the Royal Institution's meetings were expensive, the City Philosophical Society welcomed everyone for the small price of a shilling. It had been founded a couple years earlier by John Tatum, a silversmith who gave weekly lectures and demonstrations on a variety of scientific topics. Faraday began attending these meetings regularly, taking notes and drawing sketches of the equipment. He then went home and rewrote his notes more carefully. He also repeated many of the demonstrations.

The same year Faraday began attending meetings at the City Philosophical Society, he read the book *Conversations on Chemistry* by Jane Marcet. It was written in the form of a dialogue between Mrs. B and her two students, diligent Emily and flighty Caroline. They conversed about heat, light, metals, acids, and elements. They performed experiments with items that could be found around the house. Faraday cross-examined Marcet's assertions with the experiments he could perform himself and in every case found them true to the facts as he understood them. He felt, therefore, that he had found "an anchor of chemical knowledge."

Through Marcet's book, Faraday was introduced to the Cornish chemist Humphry Davy. In his teens, Davy had abandoned a medical apprenticeship with his godfather to work in the Pneumatic Institution in Bristol where he became internationally famous for his experiments, mostly on himself, with laughing gas. In 1801, at the age of twenty-two, he was hired to run the chemical laboratory at the newly established Royal Institution and act as an assistant lecturer.

Davy was charismatic, knowledgeable, and passionate about science. In addition, he was a good showman with dramatic, often explosive, demonstrations. The fashionable elite of London, both men and women, flocked to his lectures. He became so popular that Albemarle Street had to be made one-way to accommodate the carriages that lined up to hear his talks. During an illness, so many people were inquiring about him that the Royal Institution posted hourly updates outside their headquarters.

Faraday and his friends at the City Philosophical Society could never hope to afford tickets to Davy's lectures, but they read and heard about his achievements. And then one day, Faraday was offered a ticket to hear Davy speak. The ticket was given to him by a customer, Mr. Dance, a member of the Royal Institution whom

Riebau had shown Faraday's notes from the City Philosophical Society meetings, beautifully written and illustrated. The ticket was for Davy's farewell lectures, his final lectures at the Royal Institution. Afterwards, he planned to devote himself more fully to research.

Davy was a showman but also a free-thinking scientist. One of his earliest papers was an attack on Lavoisier's theory of heat. Lavoisier had suggested that heat is a material fluid called "caloric." Davy disagreed, proposing instead that heat is a form of energy. His paper was filled with speculations on matter and light, and his work was strongly criticized. Subsequently, Davy continued to follow his instincts in challenging accepted doctrine, but he began backing up his conclusions with carefully planned and executed experiments. In particular, Davy put forth the idea that all chemical forces are electrical, that it is electricity that binds elements together to form compounds. He used his public lectures to advance this idea. Marcet, who frequently attended his lectures, popularized this view through her book.

In his final lecture series given from February to April of 1812, Davy launched an attack on Lavoisier's theory of acids. He spoke from memory, animated by the joy of discovery, backing every theoretical assertion with an experiment that he performed on the spot. He applied a current to muriatic acid, transforming it into a pungent, greenish gas. Others who had observed this gas had assumed that it was a compound that included oxygen. But Davy, through careful experimentation, eliminating every conceivable source of error, demonstrated that the gas is composed of hydrogen and an element that he named "chlorine." Lavoisier was wrong; it is not the presence of oxygen in a substance that gives it acidic properties, for this pungent, greenish gas exhibits fully the properties of an acid without containing oxygen. Instead, Davy proposed, the acidity of a substance is a result of its electrical properties.

Davy's attack on Lavoisier's acid theory was criticized, especially by chemists on the Continent, but Faraday became a believer. He rewrote the notes from Davy's lectures carefully at home and saved up money to buy some chemical supplies. With seven copper half-pennies, seven discs of zinc, and paper soaked in saltwater, he created his own battery. He placed the ends of the battery into a solution of Epsom salt and observed a stream of tiny bubbles coming from the negative wire. After a couple hours, he observed magnesia suspended in the solution. He made a bigger battery and attempted to decompose water. When he took the layers of the battery apart, he noticed that the zinc discs were coated with copper and the copper was coated with zinc oxide. The metals must have passed by each other in the saltwater.


Faraday's apprenticeship was over. He knew that he needed employment, but he did not want to spend the rest of his life binding other people's books. He wanted to learn science, and he did not want to just read about it. As he wrote to a friend, "how terrified I should be to set about learning science from books only." He wrote a letter to Joseph Banks, president of the Royal Society, asking for any scientific position no matter how menial, but he received no response. He took a bookbinding position with another French émigré. Soon after, another opportunity came his way.

After an explosion in his laboratory, Davy was temporarily blinded. He needed an assistant so that he could continue with his work, and Mr. Dance recommended Faraday. Faraday took a few days off from his job to work as Davy's assistant. After returning to work, he carefully bound the notes from Davy's lectures and sent them to him, asking whether there was any chance of a permanent position at the Royal Institution. Davy wrote back almost immediately. Although the Royal Institution had no open positions, he would happily give Faraday the next one available. Shortly after, he fired his bottle washer for fighting and gave the job to Faraday.

In March of 1813, Faraday began working at the Royal Institution. He was paid twenty-five shillings a week, given lodging in the attic, and had access to the laboratory. He washed bottles, swept floors, and cleaned fireplaces. Soon he was extracting sugar from beetroot and working alongside Davy with explosive compounds. When Davy planned a tour to visit scientists and cultural centers around the Continent, he brought Faraday along as a chemical assistant.

By this time, Davy was the most important scientist in all of Europe. In 1805, the Royal Society had honored him with the Copley medal, their most prestigious award. Two years later, he received the Volta Prize from the *Institut de France*, even while France and Britain were fighting a bitter war. In 1812, he married a rich, young widow and retired as a lecturer from the Royal Institution. He now had the means and time to travel; he wanted to go to Paris to collect his prize. Others thought him a traitor, but Davy insisted it was the countries, not the scientists, who were at war. In a show of broad-mindedness,

Napoleon arranged special passports for Davy, his wife, a valet, a maid, and Faraday to travel anywhere within his empire.

They set out in October of 1813, Faraday doubling as a chemical assistant and valet; Davy's valet backed out at the last minute, fearing for his life as Britain and its allies prepared to invade France. After returning eighteen months later, Faraday received a modest pay increase and a new title, "Assistant and Superintendent of the Apparatus and Mineralogical Collection." He continued as Davy's assistant, keeping his experiments going and neatly writing up his disorganized research notes. Davy gradually gave Faraday more freedom, allowing him to publish small research projects under his own name. Faraday also continued with his plan of self-improvement — reading, taking evening classes in elocation and oration, and going to City Philosophical Society meetings, sometimes giving lectures himself.

In October of 1820, Davy announced to the Royal Institution the discovery made by Danish scientist Hans Christian Ørsted that a compass needle placed near a current-carrying wire is deflected at right angles to the current. Davy and Faraday repeated the experiment, verifying Ørsted's results. When they oriented the current vertically, they could see the compass needle held horizontally follow a circular path as it was moved around the wire.

The news of this discovery had reached Paris as well. Inspired by Newton's law of gravity, André-Marie Ampère tested whether a force exists between parallel wires. He found that such a force does exist; wires that carry current in the same direction attract and wires that carry current in opposite directions repel. Ampère then determined a law relating the strength of the force to the current and separation between the wires. This law could be used to determine the force between wires of any shape and orientation by adding up the forces between pairs of current elements.

Another French scientist, Augustin-Jean Fresnel, suggested that magnetism in iron might come from tiny currents circulating around each particle of matter, which, when aligned, create a permanent magnet. Ampère tested this idea by winding a wire in the form of a helix and then passing a current through it — he assumed that this arrangement would be equivalent to many separate currents all circulating in the same direction. In fact, the helix of current acted just like an ordinary magnet with a north pole on one end and a south pole on the other. With this and his force law for currents, Ampère had a full theory of electromagnetism that quickly gained almost universal acceptance within the scientific community.

Faraday initially took little interest in this problem; he was busy courting the sister of one of his friends at the City Philosophical Society, Sarah Barnard. In June of 1821, he and Sarah married and moved into an apartment at the Royal Institution. Around the same time, another friend from the City Philosophical Society, the editor of *Annals of Philosophy*, asked Faraday to write a historical account of electromagnetism. As a consequence, Faraday began to read all that he could on the subject.

Faraday read, repeated experiments, and tried to follow the reasoning of Ørsted and Ampère as they theoretically interpreted their observations. He couldn't understand Ørsted's vague theory of conflicts, in which the electric current was a wave of chemical disruption and constitution. Ampère's work seemed more clear, but Faraday, who had never been schooled in mathematics, could not understand his equations. And so he began to develop his own theory.

Ampère had focused on the force between two currents, ignoring whatever it was that caused a compass needle to follow a circular path around a wire. But Faraday thought that this circular force was important. Davy had been working with another scientist, William Wollaston, who predicted that a magnet should cause a current-carrying wire to spin. Faraday believed, rather, that a current-carrying wire should revolve around the pole of a magnet.

In order to test this idea, Faraday stuck a magnet to the bottom of a basin using wax, aligning its poles vertically. He filled the basin with mercury so that only the top of the magnet was exposed. He then suspended a wire above the magnet, placing one end of the wire freely in the mercury. He connected the other end of the wire to a battery. When a wire was connected to the other end of the battery, completing the circuit, the wire moved in rapid circles around the magnet. He then reversed the arrangement, and this time the magnet revolved around the wire. "There they go! There they go! We have succeeded at last!" he shouted to his fourteen-year-old brother-in-law, who had accompanied him to the laboratory; the two danced around the table and then went off to the circus to celebrate. Faraday commemorated the event in his journal: "Very satisfactory, but make a more sensible apparatus."

Within a few months, the Royal Institution set up a large rotator in the lecture hall and sent smaller versions to scientists throughout

Europe. Through his apparatus and the series of review articles he wrote for *Annals of Philosophy*, Faraday became famous for his work on electromagnetism. However, in the decade following his discovery, he had little time to extend this research. He was chosen as the Royal Institution's director with the immediate task of rescuing it financially. To this end, he began a Friday evening lecture series in the laboratory. These lectures became so popular that they were moved to the upstairs theater where Davy had given his lectures years before. He also introduced a special Christmas lecture series for children. The membership and income of the Institution swelled.

In addition to his responsibilities as director of the Royal Institute, Faraday was called upon to help with several commercial projects, the most important of which was to improve the manufacturing of optical glass. Shortly after being elected to the Royal Society, he had been asked to lead this project and felt obliged to accept. After three years of effort, his only tangible result was weakness and a nervous headache. He set aside the work for a couple months, rambling around the countryside with Sarah until his symptoms went away. After two more years, he finally produced a piece of glass of acceptable quality. The committee wanted more of it so that they could produce bigger lenses, but Faraday felt that he had done enough. In July of 1831, he handed six volumes of experimental notes on the manufacturing of optical glass to the Royal Society and resigned from the committee.

Faraday accepted a part-time teaching position at the Royal Military Academy in Woolwich. Together with his salary from the Royal Institution, this new job allowed him to live comfortably while still have time to do independent research. He continued to think about electromagnetism. If electricity can create magnetism, can magnetism create electricity? Many had tried, but no one had yet succeeded. In August, Faraday began a new section in his laboratory journal, "Expts. On the production of Electricity from Magnetism, etc. etc."

Through a friend, Faraday had heard about the work of the German scientist and musician Ernst Chladni. Chladni had filled a plate with a thin layer of sand. By stroking the rim of the plate with a violin bow, beautiful patterns of standing waves formed in the sand. These same patterns could be created in a sand-filled plate by stroking the edge of another nearby plate — sound waves carry the vibrations from one plate to the other. Faraday wondered, could electricity and magnetism be transmitted in a similar way?

Inspired by Chladni's research, Faraday sent current through one circuit speculating that it might induce current in another nearby circuit, as vibrations in one plate induce vibrations in another plate. He wrapped one side of an iron ring with a wire formed in the shape of a helix, connecting the ends of the wire to a battery. He did the same to the other side of the iron ring except instead of connecting the wires to a battery, he placed their ends over a delicately balanced horizontal magnetic needle. This acted as a homemade galvanometer to detect current; whenever current flowed through the wire, the magnetic needle would be deflected.

Faraday found that the needle was deflected for a moment right when the battery was connected to the primary circuit, indicating a current in the secondary wire. It was deflected the other way as the battery was disconnected. Any time the current was stable in the first circuit, no current was induced in the secondary circuit.

This was something. Faraday had created electricity, which creates magnetism, from electricity. But could he do the same thing without the primary circuit, using only a magnet? He attempted this by placing an upright iron cylinder with a wire wrapped around it inside a V-shaped "jaw" formed from two magnets; he connected the ends of the wire to a galvanometer to detect current. When the jaws opened and closed, the galvanometer needle moved, first one way and then the other way. When the jaws were stationary, the galvanometer indicated no current. He observed a similar result when he moved a magnet quickly toward and then away from a helix of wire. A current was induced in one direction when he brought the magnet toward the helix and then in the other direction when he moved it away. When the magnet was stationary, no current was induced in the helical wire.

Now Faraday understood why others had not been able to create electricity from magnetism. Electricity is induced only when some kind of *change* occurs in the magnetism — when there's relative motion between the magnet and the wire or, as in his first experiment with the electromagnet, the strength of the magnetism changes. Now he wondered, was it possible to create a *continuous* current? He then spun a copper disc in the presence of a magnet. As expected, a continuous current was produced in the disc.

Ten years earlier, Faraday had surprised the scientific community by creating motion with electricity and magnetism. Now he surprised them again, creating electricity with motion and magnetism. They applauded him, this son of a blacksmith with almost no formal education who had risen to become one of the most important experimentalists in all of Europe. And they saw nothing in his experiments that caused them to question their own theories.

The scientific community was content with their principle of "action at a distance," which they now applied to electricity and magnetism as it had been applied to gravity. They were content with their equations, which described electric and magnetic attraction and repulsion along straight lines connecting the charges or currents, similar to Newton's law of gravity that describes the straight line attraction between masses. They had no hypothesis to explain *how* these forces are mediated, but they had confidence in the equations and their connection to experiment.

Faraday didn't understand their equations, and he found nothing in their theories that could explain why the induced current disappears once the magnet stops. He had been led to this discovery through an analogy with sound. Might not electromagnetic forces be conveyed progressively through space, like sound travels through the air or waves travel through the water? And what about the force that seems to trace a circular path around a current-carrying wire? Or the force that seems to come out of one pole of a magnet and then curve around and go into the other? Ampère had ignored these curved forces, focusing only on the resulting attraction and repulsion between the wires. But Faraday thought that, somehow, these curved forces were important.

Faraday shared his experiments through the Royal Society's *Philosophical Transactions* and the Royal Institution's Friday Night Discourses. He induced electricity with the Earth's magnetism by quickly flipping a large wire. He showed that this induced current is the same as regular current by using it to make a frog leg twitch. He gave a name to this new phenomenon, the induction of a current due to a change in magnetism, calling it "magneto-electric induction." He also offered a theoretical explanation.

In his explanation, Faraday introduced the concept of "lines of magnetic force," curved lines as indicated by a compass needle or iron filings sprinkled around a magnetized object. As an example, the lines of magnetic force around a magnet are concentrated at the poles of the magnet and then curve outward, drawing continuous lines from one pole to the other. He also demonstrated that the lines go *through* the magnet, forming continuous loops that pass through the poles. If the magnetic force changes, as when current is applied to an electromagnet, the lines of force also change and travel outward through space. With this idea, Faraday explained magneto-electric induction: a current is induced whenever a wire "cuts through" magnetic lines of force.

Scientists throughout Europe excitedly worked to confirm and extend Faraday's experiments. Meanwhile, Faraday turned his attention to another area of electrical research: electro-chemical decomposition. He had made the observation that while liquid water conducts electricity, ice does not. He then found many other substances — various oxides, chlorides, iodides, and salts — that share this property with water, acting as conductors when liquid but non-conductors when solid. Furthermore, these same substances can be decomposed by electricity when in their liquid state. Many other substances he tested such as sugar, caffeine, and resin do *not* acquire conducting powers when they become liquids. These same substances also could not be decomposed by electricity.

Faraday surmised from these observations that there must be an intimate relationship between electricity and chemical decomposition. He conducted a series of experiments to explore this relationship, decomposing various liquid substances using a battery. When wires connected to the ends of the battery were submerged in the liquid, oxygen and acids would collect near the positive wire, while hydrogen, combustibles, metals, and bases would collect near the negative wire.

Decomposition experiments like these were commonly interpreted by assuming that the poles of the battery are centers of electrical power, acting at a distance on the particles in the fluid to break them apart. Therefore, the strongest force should be near the poles and the weakest within the decomposing substance. Faraday tested this view and found it incorrect. He varied the separation between the battery wires and found no difference; the amount of decomposition was constant given a fixed amount of electricity. He also used a galvanometer to determine the current in the liquid at different locations. He found that the current was the same whether he measured it in the center between the poles or near one pole or the other. He concluded, therefore, that the poles are *not* the center of the force; rather, the force is more or less constant throughout the fluid within which decomposition occurs.

With this observation and his earlier one that a substance's ability to be decomposed by electricity is correlated with its electrical conductivity, Faraday developed a new theory of electro-chemical decomposition. He suggested that the force of decomposition is not external, emanating from the poles, but *internal*, within the decomposing substance. To explain these internal forces, he introduced the idea of "lines of action" that follow the electric current in the decomposing fluid from one pole to the other. Similar to the lines of magnetic force, these lines of action are not assumed to be straight; even in the simplest case of two wires immersed in a drop of fluid, the lines curve significantly at the poles.

In Faraday's theory, the poles are not the source of the force but rather points at which the current enters or leaves the fluid. The lines of action weaken or neutralize chemical affinity in one direction and strengthen it in the other in such a way that fragments such as oxygen and acids move toward the positive pole and others such as hydrogen, bases, and combustibles move toward the negative pole. This process occurs through a series of successive decompositions and recombinations between complementary fragments. When the fragments eventually reach the poles, finding no partners with which to recombine, they are expelled. In this way, the poles are also the endpoints of the decomposition process.

With the help of the Cambridge historian William Whewell, Faraday created new terminology to accompany his new theory, introducing such words as "anode," "cathode," "electrolyte," and "ion" to describe the process of electro-chemical decomposition, now renamed "electrolysis." He invented a way to measure Volta-electricity using an instrument he named a "Volta-electrometer." After extensive research with numerous substances in various conditions, he formulated two quantitative laws of electrolysis.

Faraday continued his experimental studies in the messy area of chemistry. All the while, he kept his sights on the big questions. He looked for connections, believing that electricity, magnetism, chemical decomposition, voltaic excitement, and "many other things more or less incomprehensible at present" would eventually come together, perhaps under one general law. He maintained his skepticism toward the prevailing view that all forces act in straight lines instantaneously at a distance, continually looking for new ways to disprove one aspect or another of this long-held belief.

To this end, Faraday turned his attention to the subject of electrical induction. He began his inquiries in this area with a thought experiment regarding electrolysis. He considered the moment right when the battery is connected *before* the decomposing particles begin to break up and recombine. At this moment, he reasoned, all the particles should be polarized, their positive side toward the negative electrode and their negative side toward the positive electrode. Opposite sides of adjacent particles should face each other so that the particles line up to form chains from one electrode to the other.

This led Faraday to speculate that all electrical induction might be caused by action of contiguous particles. If this were true, the inductive effect between a charged body and an insulator should depend on the insulator's ability to form chains of polarized particles, which should be different for different substances. Faraday then constructed a device that allowed him to determine an inductive constant for various insulators and found, as he had expected, that this constant is different for different substances.

In Faraday's view, these experiments seemed incompatible with the principle of action at a distance since the induction occurs through connecting particles. Moreover, the paths of the polarized particles are curved, contrary to the prevailing assumption that forces are transmitted along straight paths. Faraday called these curved paths "lines of electric force" in analogy with the lines of magnetic force and lines of action. Furthermore, with his studies of induction in insulators, he was able to demonstrate that induction does not occur instantaneously, as was assumed, but rather takes time.

Faraday continued to consider the induction of metals. He applied charge to various metals and found that in every case the charge settles on the outside. He concluded that electricity cannot penetrate a closed metal container. He then demonstrated this with a twelve-foot-side cube covered in tin foil and insulated from the ground. Faraday stepped inside the metal cage, which was then charged with a generator. Sparks shot out from the corners of the cage while he stood calmly inside, completely unaffected by the electricity.

Then using a metal sphere, Faraday again challenged the assumption that the electrostatic force always travels in straight lines. He charged a rod and placed it near a brass sphere. Beyond and in line with these, he placed various objects such that the metal sphere intercepted any straight-line paths between the charged rod and the objects. If the electrical force only travels in straight lines, the metal sphere should effectively create a shadow beyond it, shielding objects from the effect of the rod. However, the rod was able to induce a charge on the objects, demonstrating that the electric force must be able to travel in curved lines.

Between 1831 and 1838, Faraday published fourteen series of experiments that he compiled in the book *Experimental Researches in Electricity*. He accomplished this even while teaching, organizing the Friday Evening Discourses, and taking occasional analytical or consulting jobs. He had received an honorary degree from Oxford and had become Fullerian Professor of Chemistry at the Royal Institution. The Royal Society had given him their highest award, the Copley medal. In addition to membership in England's scientific societies, he had been honored with membership in the Royal and Imperial Academies of Science in Paris, Florence, Berlin, Göttingen, Copenhagen, Stockholm, Petersburg, and many others.

Meanwhile, Faraday's health was deteriorating. Increasingly, he suffered from nervous headaches and had trouble maintaining concentration. His memory began to fail, and he had difficulty writing and even speaking; by the time he got to the middle of a sentence, he couldn't remember how it had begun. The doctor ordered him to take a month off and then another and then another. He enjoyed the company of his family and friends, spending time by the seaside and hiking in the mountains. He found comfort in his wife, Sarah, who was like "a pillow to his mind." He loved running out in a thunderstorm and could walk thirty miles in a day. But without work, Faraday's life was incomplete. In his journal, he wrote, "I would gladly give half my strength for as much memory." He then quickly reminded himself to be grateful for what he had.

Faraday did no work in his laboratory for two years. When he returned, he did a little research and gave occasional lectures. He gave advice on ventilation in lighthouses, led government inquiries into explosions at a gunpowder factory, and took other assignments that he accepted out of social responsibility. He performed small studies on the electrification of steam and the liquefaction of gases. It seemed that his days of making great discoveries were over.

In June of 1845, Faraday went to the annual meeting of the British Association for the Advancement of Science, held at Cambridge that year. Between sessions, the university student William Thomson introduced himself. Four years earlier, in his first year at Cambridge, Thomson had read Faraday's *Experimental Researches*. At first, he was appalled that in all of the dense prose, Faraday had not included a single equation. However, after continued efforts to understand his meaning, he became in his own words, "inoculated with Faraday fire." It was in this spirit that he approached Faraday.

Soon after meeting Faraday, Thomson attempted a mathematization of Faraday's ideas, demonstrating that his electric lines of force around a charged object could be represented by the same set of equations that had been used to describe the flow of heat in a metal bar. In August, he wrote Faraday a letter. He wondered, could the electric lines of force be detected with light? Specifically, would it be possible to detect a change in the polarization of light shined on a transparent substance when it is subjected to an electric force? Perhaps Faraday had already tried this...

Faraday *had* tried this and obtained a negative result. However, Thomson's letter motivated him to try again, casting aside all other work. He used a variety of sources of electricity, transparent substances, and various arrangements without success. He decided to try applying a magnetic force, rather than electric, which could be made much stronger. He had powerful electromagnets in the laboratory at the Royal Institution. He also had the piece of optical glass that he had manufactured years earlier, made of highly refractive lead borate. With these, he finally found an arrangement that indicated a small change in polarization through the flickering of the light of a candle.

Faraday obtained an even stronger electromagnet from the Military Academy in Woolwich. He observed the same effect, except this time he could verify something that he had long suspected, that it takes time for an electromagnet to reach full intensity. Also, by varying the number of loops of wire in the electromagnet, he demonstrated that the rotation of the angle of polarization is proportional to the strength of the magnet. Furthermore, he discovered that magnetism does not act directly on light but requires matter as a mediator. Faraday performed test after test, recording in his notebook "very fine effect" followed by "effect was best yet" and so on.

He performed experiments with a large assortment of transparent substances and found them all, to some degree, to be affected by magnetism. Faraday conjectured that every substance was in some way magnetic. To verify this, he needed an even stronger magnet. He found half of a huge iron link from the anchor of a ship and had it made into a gigantic electromagnet, wound with 552 feet of wire and weighing 238 pounds in all.

Faraday suspended a glass rod between the magnet's north and south poles. The rod swung, aligning itself at a *right angle* to the lines of magnetic force. He tried many other substances: crystals, powders, various liquids, wood, beef, apple, and bread. All of these, like the glass, aligned themselves at right angles relative to the lines of magnetic force. Most metals also aligned themselves in this same way. Only iron, cobalt, and nickel aligned themselves with the magnetic force lines.

Faraday had discovered a new magnetic property of matter. His lines of force allowed him to offer a theoretical explanation for what he had observed. With the help of Whewell, he created a name to describe this new property. Substances that align themselves at right angles to the lines of magnetic force were named "diamagnetic." Those that align themselves parallel to the lines, materials that had previously been called magnetic, he renamed "paramagnetic."


On April 3, 1846, the inventor Charles Wheatstone was scheduled to give the Friday Evening Discourse at the Royal Institution. Though an extremely able scientist, he was scared of public speaking and just as the lecture was about to begin, he panicked and quickly departed. Not wanting to disappoint the audience, Faraday improvised a lecture based on Wheatstone's latest invention: the electromagnetic stopwatch. However, he quickly ran out of things to say before the hour was up. In the remaining time, he shared some of his private thoughts on electromagnetism, matter, and light.

Faraday described a universe crisscrossed by lines of force — electric, magnetic, and probably others. When disturbed, these lines vibrate like waves along a rope, sending ripples with a rapid, but finite, speed. The universe, aside from these lines, is empty. The points at which the lines of force meet are where we *perceive* matter to exist. Faraday expressed doubt in the existence of the luminous ether that supposedly pervades the universe and acts as a medium for light. Instead, he suggested that light is probably one manifestation of the vibrating lines of force. Gravity, too, might be conveyed in the same way, by gravitational lines of force. If this were true, the force of gravity should take time to communicate its influence rather than act instantaneously at a distance.

Faraday shared these thoughts with his audience and then

wrote them up in a letter published in the Royal Society's *Philosophical Transactions* titled "Thoughts on Ray-Vibrations." In closing, he wrote, "I think it is likely that I have made many mistakes in the preceding pages, for even to myself, my ideas on this point appear only as the shadow of a speculation."

The response to Faraday's paper was almost unanimously negative. Faraday received many letters asking him to clarify, exactly *what* is a line of force? Many publications followed with rebuttals and criticisms. One attack was particularly blunt: Faraday should go and learn some mathematics and leave theoretical physics to those who were properly trained.

But Faraday had stuck his neck out and would not pull it back in. He pushed for the reality of his lines of force, which he alternately referred to as fields. He argued against the luminous ether as a medium for light and against the principle of action at a distance applied to any force. He appealed to Newton for support, quoting a letter in which he had referred to the idea of action at a distance as "an absurdity."

Faraday also continued his experimental research. He returned to his work on magnetic induction, establishing a quantitative law relating the concentration of magnetic lines of force intercepted and the quantity of electricity induced. He dropped blocks made of various substances through a 350-foot-long vertical helix — from the ceiling of the Royal Institution's lecture hall onto a cushion on the floor — to test whether a change in gravitational field might induce a current. The experiment was a failure; despite all of his efforts, he never detected any current. But this did not shake his belief that gravity, along with electricity, magnetism, and light, were all somehow unified, embodied in lines of force crisscrossing the universe.

# 6

# Maxwell's Equation's

Now, my great plan, which was conceived of old, and quickens and kicks periodically, and is continually making itself more obtrusive, is a plan of *Search and Recovery*, or Revision and Correction, or Inquisition and Execution, etc. The Rule of the Plan is to let nothing be wilfully left unexamined. Nothing is to be *holy ground* consecrated to Stationary Faith, whether positive or negative.

— James Clerk Maxwell

James Clerk Maxwell spent his early years in southwestern Scotland among the gently rolling hills of Galloway in Glenlair, an old family estate that his father had inherited and his parents, together, had lovingly restored. We can glimpse into his childhood through a letter written when he was almost three years old. His mother begins the letter, addressed to her sister, and then turns the pen over to the "abler hands" of her husband: "He is a very happy man... he has great work with doors, locks, and keys, etc., and 'Show me how it doos' is never out of his mouth. He also investigates the hidden course of streams and bell-wires, the way the water gets from the pond through the wall and a pend or small bridge and down a drain into the Water Orr, then past the smiddy and down to the sea, where Maggy's ship sails."

Maxwell spent his childhood days following his father about the farm, knitting with his mother, playing with frogs, catching sunshine in a tin plate, reading, drawing, and running around with the estate children. His mother was his teacher, in charge of his education until he was eight. Around this time, she became sick and died. His father then employed a highly recommended sixteen-year-old to live with them as a tutor, but young Maxwell rebelled against the tutor's rigid teaching style. The interaction between pupil and teacher is nicely captured in a sketch drawn by his cousin: a boy sits in a washtub in the middle of a duck pond with his tutor on the side trying to reel him in with a rake.

Reluctantly, the elder Maxwell sent his son away to school — to the academy in Edinburgh, where he lived with his father's sister. Maxwell was ten when he began, entering in the second month of the second year in an odd-looking tunic and shoes with square toes, which inevitably provoked a group of older boys. He responded to his tormentors with ironic verse. He returned home with his tunic in rags and his frill torn, but his spirit was intact. In the end, his courage, good nature, and agile strength won the respect of his peers, and his quirky sense of humor and rustic, eccentric ways were accepted.

Maxwell got on well enough academically for the first few years. Then, stimulated by a course in geometry, he began to excel. He rose to the top of his class, mastering his coursework while maintaining the playful spirit for learning and discovery that he had as a child. He became intrigued with geometric solids, as he wrote to his father: "I have made a tetrahedron, a dodecahedron, and two other hedrons whose names I don't know." A year later, he included detailed instructions on how to draw an octahedron in a letter to his aunt.

Maxwell continued to innovate, and without even realizing it he rediscovered and extended work done by Descartes on shapes called "Cartesian Ovals." He also devised simple ways to construct the curves with the help of a pencil, string, and pins. His proud father brought this work to the attention of his friend, a professor at Edinburgh University, who recognized its scholarly worth. And so, only fourteen years old, Maxwell published his first academic paper, which was read for him to Edinburgh's Royal Society since he was considered too young to read it himself.

At sixteen, Maxwell enrolled at Edinburgh University to complete his general studies. His father's plan was that he become a lawyer, as he himself had studied to be but never practiced. Although Maxwell had already become interested in science by this time, he was also open to the law profession. He wanted to know everything about everything — studying law might be interesting too. He enrolled in philosophy and logic courses as well as science, leaving himself open to either option.

The coursework at the University of Edinburgh was sufficiently light that Maxwell had abundant time for independent study. During this time, he spent about half of his time in Edinburgh and the other half in free exploration at his home in Glenlair. He set up an improvised laboratory in a room over the washhouse and equipped it with such items as salt, sulphuric acid, broken glass, various metals, sealing wax, clay, beetles, and spiders. He performed optical experiments, trying out different ways to polarize light. He created batteries, explored the relationship between chemical action and electricity, and attempted to make an electric motor. He also created a centrifugal pump and conducted various aquatic experiments, making use of one of the ponds on the estate.

And Maxwell read. He read classics in Greek and Latin, Young's *Lectures*, and Newton's *Opticks*. He read mathematical and engineering books by Fourier, Monge, and Cauchy. He read philosophical works by Kant and Hobbes. His self-imposed workload was so great that he got in the habit of waking up early, around 5:45 am, bringing up the water barrel and doing whatever other strenuous work needed to be done around the estate so that he would be "saddened down," able to sit and concentrate for long periods of time afterwards. He read, absorbed, and innovated. Soon he had two more theoretical papers accepted by Edinburgh's Royal Society, both read for him as he was still deemed too young to read them himself.

During this time, Maxwell wrote long letters to his best friend, who had left the University of Edinburgh after a year to study at Oxford. Most of his other friends had also left to continue their education at one of the great universities in England. Maxwell had begun his third year at Edinburgh, and his father still held out hope that he would stay in Scotland and study Scottish law. He knew about his son's growing passion for science and had even encouraged it; the two had been going to scientific meetings together since the younger Maxwell was twelve. Nonetheless, a scientific career wasn't *practical*.

Finally, as Maxwell cheerfully reported to his friend, "the Cambridge scheme has been howked up from its repose in the regions of abortions, and is as far forward as an inspection of the Cambridge *Calendar*." His father relented, enrolling him at Cambridge to study natural philosophy. And so, in the autumn of 1850, with vast amounts of knowledge, physical experience, and an independent spirit nurtured by philosophy, Maxwell packed his bags and left for England.


Maxwell thrived in the vibrant intellectual atmosphere at Cambridge, reuniting with old friends and making new ones. After spending his first term in the elite Peterhouse, he transferred to the larger and more social Trinity College, where Newton had studied two hundred years earlier. He took his school work seriously, taking advantage of private tutors to meet the intense demands of his classes. He enjoyed outdoor activities, especially walking, swimming, and rowing. He wrote poems, some serious and others to amuse his friends. He was invited to join an intellectual twelve-member discussion group called "The Apostles."

Even while participating in Cambridge life to the fullest, Maxwell continued his independent investigations, publishing two more scientific papers. He also experimented with his daily schedule, trying in various ways to make the best use of his waking hours. In one such experiment, he got in his exercise in the middle of the night, from 2-2:30 am, by running through the corridors and up and down the stairs. He continued this until the residents along his course began taking shots at him with boots and hairbrushes as he passed.

By the time Maxwell attended Cambridge, the university had

developed a highly competitive exam culture where the top students, coached by private tutors, battled for prizes in standardized examinations as if they were sporting events. These exams were scattered about the undergraduate curriculum, called "tripos" presumably after the stools students would sit on to be grilled orally in the days before the written exams. The most intense of these was the Mathematical Tripos, which everyone in the university had to pass to earn any kind of degree. It was taken in a student's fourth year in chilly January in the Senate House, a building without either a fireplace or a stove.

Students sat for these mathematical exams five and a half hours a day for three days. Those attempting to graduate with honors would take exams for an additional four days. The questions were contrived puzzles that had little to do with physical reality. This was not Maxwell's particular strength. Although he had almost perfect physical intuition, he typically made careless analytical errors. However, he worked hard with one of the best tutors, learning systematic analysis, tricks, and shortcuts to improve his speed and accuracy in solving the types of problems on the exams. In the end, he placed second in the Mathematical Tripos, earning the prestigious title of "Second Wrangler." He then went on to compete for the Smith Prize and tied for first.

Maxwell's years as an undergraduate at Trinity were happy ones. Afterwards, he stayed on at Cambridge for a couple years, first as a bachelor scholar and then as a fellow. He had some responsibilities that went along with these positions — supervising examinations and teaching classes at the university. He also taught at a working men's college one evening a week. Even with these occupations, he was left with large amounts of time to spend as he wished. He remained an active member of the Apostles and was elected to the Ray Club, a group devoted to the discussion and promotion of natural philosophy. Additionally, he continued his general education, reading works by Chaucer, Bacon, Pope, Berkeley, and many others. All the while, he kept in good physical shape, working out at the new gymnasium, swimming, and skating.

During this time, Maxwell's research interests turned toward optics. He discovered that by combining red, green, and blue light, it was possible to make any color, even white. He first demonstrated this with a "Color Top," a top covered with strips of red, green, and blue colored paper; when it was spun, the colors blurred together and appeared as one. By adding a black strip, he could adjust the color's brightness. He also mixed colors using light from the Sun with a "Color Box," which he created using a prism to separate the differently colored rays. With these devices, he tested the color perception of various subjects and found that most people perceive color the same way. He tested people with color blindness and discovered that most of them lacked a red color receptor. He also invented an instrument to look into the eye through the pupil.

Next, Maxwell began to study electricity and magnetism. He had toyed around with the subject before, performing various experiments in his laboratory above the washhouse in Glenlair. Now he tackled the subject in earnest, wanting to learn everything that was already known. He especially wanted to read books that dealt with observed phenomena, those that avoided "old traditions about forces acting at a distance." He had written to Professor William Thomson, a good friend of his cousin's husband, asking for a reading list. Naturally, Thomson directed him toward Faraday's *Experimental Researches in Electricity*; it was here that Maxwell began his studies.

In Faraday, Maxwell found a kindred spirit, one whose thinking matched his own. He admired the openness and integrity of Faraday's work, that he included his failed attempts as well as his successful ones, sharing not only the results but also the process of discovery. He witnessed the concept of lines of force grow over decades of experimentation and thought.

Maxwell came to believe in these lines of force, but he also believed the results of other scientists who formulated the laws of electricity and magnetism mathematically in a way that could be tested quantitatively. Familiar with both conceptions, he set out to unify the approaches. He had been known to take a board full of equations used to solve a physical problem and reduce them to a simple picture. Now, he took Faraday's pictures and began to translate them into equations.

Thomson had shown the equivalence between Faraday's electrical lines of force and the equations of static electricity using the analogy of heat flow. Maxwell picked up where Thomson left off, making use of an analogy between the electric lines of force and the flow of a weightless, frictionless, incompressible fluid. In his model, positive charges were the source of the fluid and negative charges, the sinks. He conceived of lines of force as flexible tubes that transfer the fluid from the sources to the sinks throughout space.

The tubes fill all of space and act like real walls, consistent with Faraday's lines of force that cannot cross. The flow of the fluid is caused by pressure differences, just like in an actual fluid. The strength of the electric force is proportional to the speed of the fluid at any point in space. The tubes coming from a source expand as they get farther from it; as the area of a tube expands, the speed of the fluid decreases. A more specific analysis showed that the speed decreases as the square of the distance from the source increases, consistent with the inverse-square law of electricity.

Making use of this analogy, Maxwell was able to model all the static laws of electricity and magnetism, including the effects of materials with various electric and magnetic properties. Metal surfaces were the sources and sinks of the fluid. Insulators were given various resistances to fluid flow to represent their inductive capacities. Magnetism was modeled as having the fluid flow in continuous loops. In this way, Maxwell showed the equivalence between Faraday's methods and the mathematical treatments of electricity and magnetism by Ampère, Gauss, and others.

It was a strange model, with the strength of the force proportional to the speed and fluids flowing in continuous loops without either a source or a sink. As Maxwell emphasized, however, it was an *analogy*, not meant to represent physical reality. Rather, its use was to provide a means for thinking about a problem physically without committing to any specific physical picture. He wrote up this model in the paper "On Faraday's Lines of Force," published in the *Transactions of the Cambridge Philosophical Society* in December of 1855.

In January, Maxwell learned of a professorship that was opening up in Scotland: the Chair of Natural Philosophy at Marischal College in Aberdeen. Although he was happy at Cambridge, he would need to look for a new position in the next few years anyway. And, as he wrote to a friend, it would be good to be out, to feel "the rubs of the world." He discussed the opportunity with his father, who was thrilled at the idea of his son being closer. Also, the vacation schedule would allow him to spend the whole summer in Glenlair. His father, who had been in poor health for the past year, cheered at this prospect and even appeared to get better. However, while Maxwell was with him during the Easter holidays, his father took a turn for the worse, and the morning after a troubled night spent with his son by his side, Mr. John Clerk Maxwell passed away.

Maxwell wrote to friends and relatives, organized his father's funeral, and assumed his role as the laird of Glenlair. After returning to Cambridge, he received the news that his application to Marischal College had been successful. He finished up his work in Cambridge and then returned to Glenlair for the summer, where he took care of estate business, wrote more letters, and continued his scientific work. And he mourned his father's death. When apart, Maxwell and his father had exchanged letters almost daily. He now took his pen, as he often did in times of joy and sadness, and wrote poetry:

> *Oh! Those signs of human weakness, left behind for ever now,*
> *Dearer far to me than glories round a fancied seraph's brow*
> *Oh! the old familiar voices! Oh! the patient waiting eyes!*
> *Let me live with them in dreamland, while the world in slumber lies!*
>
> *Let me dream my dream till morning; let my mind run slow and clear,*
> *Free from all the world's distraction, feeling that the Dead are near,*
> *Let me wake, and see my duty lie before me straight and plain.*
> *Let me rise refreshed, and ready to begin my work again.*

In November of 1856, Maxwell packed up his things, which included a new color box, built sturdy enough to withstand the journey, and traveled northward to assume his position at Marischal College. He was not yet twenty-six and the average age of his colleagues was fifty-five. They kindly welcomed him; he was invited out so often, he rarely ate in his own lodgings. He settled into academic life in Aberdeen as well, teaching a full load at Marischal and evening classes at the local mechanics school.

That winter, Maxwell sent Faraday a copy of his paper, "On Faraday's Lines of Force." He must have been at least a little nervous about how this great man, who had been breaking new ground in electromagnetic theory for decades, would receive his work. Maxwell had boldly entered the field, as he admitted at the beginning of the paper, having hardly done a single electrical experiment. But Faraday was only grateful. He wrote back, sharing his experience while reading it: "I was at first almost frightened, when I saw such mathematical force made to bear upon the subject, and then wondered to see that the subject stood it so well." He thanked Maxwell, writing that his words gave him much encouragement.

Soon after, perhaps emboldened by Maxwell's support, Faraday published a paper formally proposing gravitational lines of force, extending an idea that he had put forth earlier for which he had so far received only criticism and doubt. He sent Maxwell a copy of the paper, asking for his opinion. He followed up almost immediately with another letter in which he apologized for asking his opinion, perhaps before he was ready to give it. Faraday needn't have worried. Maxwell took his time answering, but when he did, his response was long, thoughtful, and supportive.

Maxwell wrote to Faraday that the idea was sound, that it could work if the lines of force were not attractive but repulsive. The lines would emanate from all matter in the universe; two nearby objects would be in each other's shadow and therefore be *pushed* together. The resulting force of attraction would obey an inverse-square law, indistinguishable from Newton's law of gravity. These gravitational lines of force could "weave a web across the sky" and "guide the stars in their courses." Faraday, again, was grateful. He wrote back, "Your letter is the first intercommunication on the subject with one of your mode and habit of thinking. It will do me much good, and I shall read and meditate on it again and again."

In this same letter, Faraday included a plea. He asked whether it were possible, when mathematicians working on physical problems have finished their analysis, if their conclusions could be expressed in common language as "fully, clearly, and definitely" as with mathematical formulas. Might not the mathematical physicists translate their results out of their "hieroglyphics" so that others might also work on the problem by experiment?

Maxwell did not get a chance to continue his research in electricity and magnetism for many years. He was busy with his classes and optical studies, and not far into his first term at Marischal, he became engaged in another problem: Saturn's ring. It was the subject of the biennial competition for the Adams Prize, given by St. John's College in honor of John Couch Adams's discovery of Neptune. Was the ring solid, fluid, or was it composed of many separately moving pieces of matter? This was a fiercely difficult problem that had been attempted by some of the best minds in the world, including the great French mathematician Simon Pierre Laplace, but no one had yet been able to solve it.

Maxwell devoted his best energy to the problem for over a year. He performed simple experiments with corks and rings. He built a more elaborate device to model the motion of a ring of satellites with two large wheels turning on parallel parts of a cranked axis with thirty-six little cranks, each carrying a tiny ivory satellite. He applied mechanical principles and rigorous mathematics. In the end, in an essay weighing twelve ounces, Maxwell demonstrated clearly that the ring of Saturn must be composed of many separately orbiting bodies. His entry, which was the only entry, won the prize.

Even with his research, teaching responsibilities, and numerous correspondences, both personal and scientific, Maxwell still found time for general reading — Buckle's *History of Civilisation*, astronomical essays by Herschel, Elizabeth Browning's *Aurora Leigh*, and many others. Work was good and reading was good, as he wrote to a friend, but friends are better. Joy began to radiate throughout his correspondences. He shared with another friend what he believed to be his greatest achievement by way of discovery: "the method of converting friendship and esteem into something far better."

Among the many colleagues who generously welcomed Maxwell into their homes was the president of Marischal College, Reverend Daniel Dewar. He and his family, which included his thirty-three-year-old daughter Katherine, became very fond of Maxwell. They were impressed by his great knowledge and admired his excellence of heart. One visit led to another, and soon they began to treat him as one of the family, even inviting him on a vacation. There, he proposed marriage to Katherine, and she accepted.

After keeping it a secret for several months, they announced their engagement to friends and family. Happiness spilled out of Maxwell's letters as he shared the news. He assured his aunt that they needed each other and that they understood each other better than most couples he had seen. She is not mathematical, he continued, but she "certainly won't stop the mathematics." He boasted to a friend that their eyes were both so good that they could see the spot on the Sun without a telescope and that they planned to do optical experiments together over the summer. And again, he wrote poetry:

> *Now no more I doubt or wait*
> *All my fears are vanished,*
> *Summer's coming dear, though late,*
> *Fogs and frosts are banished.*

In 1860, a merger between Marischal College and a neighboring college eliminated Maxwell's position. He then applied and was accepted to an open position at King's College in London. He and Katherine rented a house in the newly developed Kensington district, close to Hyde Park and Kensington Gardens where they enjoyed strolls and horseback riding. On nice days, Maxwell walked the four miles to work, taking him right by the Royal Institution that Faraday, though now retired, regularly visited.

Maxwell continued his experimental work at home, setting up a laboratory in the attic, which ran the length of the house. He perfected his color box, now nearly eight feet long. With it, he performed many series of experiments on color vision with Katherine as one of his most important subjects. He spent so much time working by the window with his color box that the neighbors thought him mad, staring for so long into a coffin.

In addition to his optical work, Maxwell performed experiments related to the kinetic theory of gases, a theory proposed by the Swiss physicist and mathematician Daniel Bernoulli. According to this theory, measurable quantities like temperature and pressure can be accounted for by assuming that a gas is made up of individual molecules that move about, only interacting when they collide with each other. It was thought that the temperature should depend on the speed of the molecules: the hotter the temperature, the faster the molecules, but it was impossible to go further than this without understanding how the speeds were distributed in the gas. Were they all moving at the same speed? This had been assumed in the work on kinetic theory by the German physicist Rudolf Clausius.

However, Maxwell demonstrated that even if all the molecules in the gas started at the same speed, their speeds would quickly be distributed in the shape of a lopsided bell where the most probable value depended on the temperature. He illustrated this with the help of an analogy between the distribution of speeds in a gas and the distribution of bullet marks on a target, assuming all shots are aimed at the center by the same marksman and the probability of any shot is inversely related to its error. He borrowed mathematics developed by Laplace and solved the problem in a few short paragraphs.

With this distribution, it was possible to solve all kinds of problems, connecting the hypothetical world of tiny, randomly moving particles to measurable quantities. He estimated that at ordinary temperatures and pressures, each molecule collides on average eight billion times per second. He also made a surprising prediction: the internal friction of the gas should be independent of pressure.

This was a testable prediction that could either strengthen or destroy the kinetic theory of gases depending on the results. No one had yet performed such an experiment, and so, in his attic laboratory alongside his optical work, Maxwell did. He heated the attic with a large fire that Katherine stoked for hours at a time, keeping track of its pressure and temperature. Alternately, he cooled the attic with ice. Ultimately, the results of these experiments confirmed his prediction, giving strong support to the kinetic theory of gases.

After moving to London, Maxwell began attending the Friday Evening Discourses at the Royal Institution, and after a time, Faraday invited him to give one. He chose as his subject the three primary colors of light and color vision. He wanted to demonstrate color mixing, but his color top was too small for the large lecture hall. With a colleague who was an expert on black and white photography, he devised a way to incorporate color into the process. On May 7, 1861, before the audience at the Royal Institution, Maxwell projected the first color photograph.

By now, Maxwell and Faraday's interactions extended beyond a mere exchange of letters. They had dined together on the occasion of Maxwell's lecture. Another time, while Maxwell was struggling to get out of the crowded lecture theater at the Royal Institution, Faraday called out to him, "Ho, Maxwell, cannot you get out? If any man can find his way through a crowd it should be you." No doubt this proximity to Faraday inspired Maxwell; during this period, he returned to his work on electricity and magnetism.

In his earlier work, Maxwell was only able to apply his fluid analogy to *static* electricity and magnetism. In these cases, the action-at-a-distance theories and Maxwell's equations based on Faraday's lines of force were indistinguishable. Consequently, he couldn't test Faraday's idea that electric and magnetic action should take time. He also couldn't account for Faraday's explanation of magnetic induction, that a current is induced in a loop of wire when the loop cuts across the magnetic lines of force. Maxwell needed a new analogy.

He first added vortices to his fluid model, allowing the fluid to rotate about the lines of magnetic force. He then replaced the fluid vortices with small, solid, spherical, spinning cells that fill the space as the fluid had in his original model. In his final analogy, he likened the universe to a gigantic machine filled with cogs and rollers. The rollers represent moving particles of electricity, while the cogs spin out vortices of magnetic power. Furthermore, Maxwell endowed the rollers and cogs with elasticity. Accordingly, the displacement of a cog or roller triggers a similar displacement in an adjacent one.

With this new analogy and appropriate assumptions, Maxwell could account for magnetic induction, the relationship between optical and electrical properties of matter, and the magnetic rotation of a plane of polarization in a beam of light. He also made two new predictions. First, something equivalent to an electric current should be able to exist in a perfect insulator due to the displacement of neighboring charges; he called this a "displacement current." Second, he predicted that electromagnetic disturbances should spread out like waves that travel through space.

While in Glenlair during the summer of 1861, Maxwell used his machine-like model to derive a mathematical formula for the speed of an electromagnetic wave. However, he didn't have the necessary constants to calculate its numerical value. When he returned to London, he obtained the required constants and inserted them into his equation, obtaining a value of about 310 million meters per second, which lay within the accepted range for the speed of light. On October 19, he dashed off a note to Faraday, "I think we have new strong reason to believe, whether my theory is fact or not, that the luminiferous and the electromagnetic mediums are one."

Maxwell wrote up his theory in a paper published in 1862, "On the Physical Lines of Force," its title a tribute to a similarly named paper written by Faraday ten years earlier. Even as this paper was coming off the printing press, he was working on another version of the theory, which he wrote up in a paper, "A Dynamical Theory of the Electromagnetic Field." In this theory, Maxwell abandoned the use of an analogy altogether. Faraday's lines of force were replaced by continuous electric and magnetic fields that carry energy through space. The entire theory was a set of equations: Faraday's theory written in a form that he could never understand.

After spending five years in London, Maxwell decided to resign from his position and retire to Glenlair. He had new ideas on kinetic theory and heat that he wanted to pursue. He wanted to write a book that included his own theory of electromagnetism along with all the rest of the important work in the field in a way that would be accessible to other scientists. Furthermore, he wanted to devote more time to his estate and complete a project to expand the Glenlair house according to plans his father had drawn up.

Maxwell settled back into country life, doing work around the estate and riding horses with Katherine. He developed great skill at the horse, "riding the ring" for the amusement of children — throwing up a whip and catching it, leaping over bars, and so on. He spent evenings with Chaucer, Spenser, Milton, or Shakespeare, which he would read aloud to Katherine. He also kept up with scientific and personal correspondences, bringing letters to and from the rustic post office with his dogs running alongside him.

Maxwell had plenty of time now to pursue his scientific research. He worked on his book on electromagnetism, which, when completed, would be almost one thousand pages long. He wrote another book on the theory of heat. He wrote seventeen papers on a variety of topics, each in some way breaking new ground. In addition, he worked with a colleague in England on an experiment to determine the ratio of the electric and magnetic constants, which according to his theory determined the speed of an electromagnetic wave. With this new method, they obtained a value for the wave speed that again agreed with the currently accepted value of the speed of light.

In 1871, Cambridge University created its first Chair of Experimental Physics. The University had been given a large sum of money from its chancellor, the Duke of Devonshire, to construct and furnish a laboratory, and they needed someone to lead the project. Their first choice was William Thomson, who had built up a laboratory at the University of Glasgow. Their second choice was Hermann Helmholtz, a German experimentalist. When neither Thomson nor Helmholtz would stand for the position, they looked to Maxwell.

Since his retirement in Glenlair, Maxwell had been making yearly trips to Cambridge to act as an examiner for the Mathematics Tripos. He had done much during this time to make the questions more interesting and relevant to practical application. In fact, it was in part his influence that led to the creation of a laboratory. In addition, as he was told in a letter imploring him to consider the position, there was no one else in the least fit for the job. This was a rare opportunity for Maxwell, but he wasn't sure that he had the required experience to take on such a task. In the end, he accepted the position with the condition that he could leave after a year.

And so, with characteristic dedication, Maxwell led the construction of the Cambridge Laboratory. Although he drew inspiration from the laboratories in Glasgow and Oxford, many of the best features of the laboratory were of his own invention. The result was a great success, both practically and architecturally. He then supplied it with the highest quality fixtures and equipment, donating some of his own, buying what he could, and having the rest custom-made.

The laboratory was completed in the spring of 1874 and given the name "Cavendish Laboratory," both in tribute to its benefactor, the Duke of Devonshire, who shared that family name, and the duke's great-uncle, Henry Cavendish. Henry Cavendish, a social and scientific recluse, had published only two papers related to electricity during his lifetime but had prepared, in addition, twenty or so packets of manuscripts on both mathematical and experimental electricity. After his death, these papers were passed on to the duke who, upon the completion of the laboratory, handed them over to Maxwell so that they could be compiled and edited.

In these papers, Maxwell found a treasury of experimental work. Cavendish had demonstrated the inverse-square law of electricity more effectively and earlier than Charles Coulomb, for whom the law was named. He also arrived at something equivalent to Ohm's law, a law that quantitatively relates voltage, current, and electrical resistance, fifty years before Ohm. He did so even without the use of a battery or galvanometer, neither of which had been invented. Rather, he connected wires to the oppositely charged parts of a Leyden jar and then held the wires in his hand. He did this with various arrangements, determining the strength of the current by measuring how far up his arm he could feel the shock.

Maxwell could have delegated this task of editing and compiling Cavendish's papers, which took valuable time away from his own research. However, he felt that Cavendish's work was an important part of history that should be properly preserved, so he chose to do it himself, "walking the plank" with them, as he wrote to William Thomson. He took pains to write an interesting and informative narrative, even checking small details like whether the Royal Society had a garden in 1770. He also measured current the way Cavendish did, as an American scientist visiting the laboratory was horrified to discover — Maxwell and students sitting around with their sleeves rolled up, sending shocks up their arms!

Maxwell's work editing Cavendish's papers wasn't his only scientific work during this time. He finished his two treatises, wrote another book and several articles, and acted as co-editor for the ninth edition of the *Encyclopedia Britannica*. But it was his most important work, the humble efforts to preserve the legacy of another great scientist. The final version was published in October of 1879: *An Account of the Electrical Researches of the Honourable Henry Cavendish, F.R.S.*

This was the last work published by Maxwell during his lifetime. In the spring of 1879, his friends at Cambridge started to notice a decline in his appearance, the absence of spring in his step and sparkle in his eyes. He had been experiencing stomach pains for two years, which he managed by drinking a solution of carbonate of soda. In April, he finally wrote to the doctor about his concerns. He continued to visit the laboratory but only stayed for short periods. In June, he returned to Glenlair with Katherine for summer break. He wrote cheerful, humorous letters to his friends and family about everyone and everything except himself. On November 5, at the age of forty-eight, James Clerk Maxwell passed away peacefully, at the same age and from the same disease that his mother had died of forty years before.

# 7

# Two German Physicists

A new scientific truth does not triumph by
convincing its opponents and making them see the light,
but rather because its opponents eventually die,
and a new generation grows up that is familiar with it.

— Max Planck

At the end of the University of Berlin's summer term in 1879, Hermann von Helmholtz set a prize question for his students: to experimentally determine an upper limit on the mass of electrical current. Helmholtz had been studying Maxwell's theory of electromagnetism in which electric and magnetic influences are mediated by fields that travel at finite speeds. He had also studied another electromagnetic theory developed by the German physicist Wilhelm Weber that assumed action at a distance. Weber had overcome some difficulties in his theory by assuming that electrical current has inertia similar to massive bodies. As Helmholtz considered the two theories, he became more and more inclined to adopt Maxwell's approach. Nonetheless, he insisted that the ultimate decision between the theories should be made by experiment.

When Helmholtz posed the question, he had a particular student in mind — Heinrich Hertz. Hertz had been working in his laboratory at the Physical Institute in Berlin since the previous fall as a graduate student. He had shown himself to be an exceptional student, both theoretically and experimentally. He had great practical experience; he had alternated between engineering and architecture before finally settling on physics.

Hertz rose to the challenge of Helmholz's question. In a very high-precision experiment that he designed himself, personally modifying available equipment to minimize error and improve sensitivity, he determined that the upper limit on the inertia of electricity was too small to reasonably account for Weber's hypothesis. He won the prize and was awarded a gold medal.

The same year, through Helmholtz's recommendation, the Berlin Academy of Sciences offered a prize to anyone who could experimentally establish a connection between electromagnetic forces and the dielectric polarization of insulators. This could indicate the existence of the displacement currents predicted by Maxwell's theory. Helmholtz brought this problem to the attention of Hertz and promised that he would have the assistance of the Institute should he choose to pursue it. Hertz looked into the problem and quickly determined that the effect would be too small to give a decisive answer with the available equipment.

Instead, Hertz spent three months on a project studying electromagnetic induction, which became the core of his doctoral dissertation. He graduated magna cum laude in the spring of 1880, after which he remained at the University of Berlin for three years as Helmholtz's assistant. During this time, Hertz published fifteen articles in academic journals on various topics in the field of electricity and magnetism. He was one of the most gifted researchers in the most actively studied areas of physics.

In 1883, Hertz took the position of lecturer of mathematical physics at the University of Kiel. During this time, he expanded his theoretical knowledge of electromagnetic theory. He compared Maxwell's theory with competing theories, all of which assumed action at a distance, and reworked Maxwell's equations into a more convenient form. As he studied, he became more and more interested in Maxwell's theory, which he considered to be the most elegant of all the electromagnetic theories, although it had not yet been supported by decisive experimental evidence.

Desperate to return to experimental work, Hertz left Kiel to teach at the Technical High School in Karlsruhe where he would have access to its well-equipped laboratory for independent research. One day during a lecture hall demonstration of Faraday's principle of induction, he observed a spark in a side circuit. He had studied electromagnetism enough to recognize its significance. He speculated and soon verified that the spark was the result of electromagnetic waves — standing waves similar to sound waves in a pipe organ.

With this result, Hertz was now confident that it was possible to solve the question posed by the Berlin Academy. Even so, with various experimental setups over several months, he was unable to obtain any satisfactory results. Eventually, he began to approach the problem more broadly.

The motivation behind the Academy's question was to test Maxwell's electromagnetic theory experimentally. The specific question dealt with displacement currents in insulators; however, another prediction of the theory was the existence of electromagnetic waves that travel through space at a finite speed.

Hertz began to focus on this new problem instead — attempting to detect electromagnetic waves in the air — which avoided the complicated analysis of forces in dense matter. In addition, the detection of electromagnetic waves in air would have special significance for Faraday's and therefore Maxwell's view. According to this view, the electromagnetic forces can disentangle themselves from material bodies and continue to exist as conditions or modifications in air or even in empty space. Furthermore, according to Maxwell's theory, any electromagnetic influence travels at a finite speed through a vacuum or the air, which matches the speed of light almost perfectly, making the hypothesis that light is an electrical phenomenon highly probable. This would be an extraordinary result if verified.

Accordingly, Hertz set out to detect electromagnetic waves in air. His initial results, however, seemed to indicate that electromagnetic waves, if they exist, travel with infinite speed. Disheartened, he gave up. A few weeks later, realizing that a negative result that disproved Maxwell's theory would be just as important as a positive result, he resumed his studies.

Hertz was more careful in his second attempt, especially in his efforts to minimize any chance reflections off of walls and nearby objects. This time, he obtained clear results indicating a finite wave speed, although greater than in a wire. Then, in a series of experiments conducted over many months, he explored the properties of these waves. He found that they are reflected by any metal but can pass through thick wood unimpeded. They refract and can be polarized, like light. They also travel at a speed nearly equal to the speed of light.

The journal where Hertz had published his papers, "unable to comply with the numerous applications made for copies of Professor Hertz's researches," invited him to compile his papers into a single volume so that they would be accessible to the general public. These papers, as William Thomson praised in the preface to the English edition, were a "splendid consummation" of the theory developed over the last half-century, beginning with Faraday's curved lines of force, in which there exists a single ether for light, heat, electricity, and magnetism.

In this same collection, Hertz included a paper inspired by a peculiar observation he had made early in his research. At the time, he had been exploring the effect of electrical resonance. The spark produced by the discharge of an induction coil, called the active spark, induced another spark in the circuit, the passive spark. When he shielded the passive spark from outside light in order to measure it more precisely, he noticed that its spark length decreased dramatically.

Hertz attempted to understand the reason for this, why shielding the spark would have such an effect when, theoretically, the electricity causing the spark ran through the wires. He determined that it could not be an electrostatic or electromagnetic effect because both conductors and perfect non-conductors such as glass, paraffin, and ebonite act as effective shields.

Additional experiments showed that the passive spark was augmented due to light from the active spark; shielding the passive spark eliminated this augmentation and therefore decreased its intensity. Furthermore, the active spark was only one such source that could cause the augmentation. For example, burning magnesium would work equally well to increase the strength of the passive spark. Burning hydrogen, however, *did not* produce the effect.

Finally, Hertz discovered that the light responsible for the augmentation of the spark could not be visible light. He knew this because glass and mica, both transparent to visible light, acted as effective shields. Using a prism, he was able to determine that the light must be ultraviolet light; the rays that caused an increase in spark length were situated outside the visible colors, beyond the violet end of the spectrum. Satisfied that this phenomenon had no effect on his research on electromagnetic waves, Hertz abandoned this line of inquiry, encouraging others to study it under simpler circumstances. Others did continue to study this phenomenon — the excitation of electricity due to ultraviolet radiation. It became known as the "photoelectric effect."

In 1892, the same year he published his electrical papers, Hertz became sick and underwent a series of operations. He died at the age of thirty-six, on New Year's Day in 1894. That February, a young German physicist named Max Planck gave a memorial address in his honor before the Physical Society in Berlin.

Planck was almost the same age as Hertz, born in 1858. Hertz was born just a year earlier in 1857. They both studied physics at the Universities in Munich and Berlin, just missing each other multiple times. They earned their doctoral degrees within a year of each other and shared many of the same teachers.

Planck liked neatness and order. He was a careful lecturer and proofreader. He was also a concert-level pianist. He ultimately decided against a career in music because he didn't think his compositions were good enough. His attention to detail eventually earned him the position of editor-in-chief at the German research journal *Annalen der Physik*, which he held for almost four decades.

From an early age, Planck was attracted to the abstract and

theoretical aspects of physics. It came to him like a revelation that the universe is something absolute and independent of man while also knowable through reason. He was particularly captivated by the idea of energy conservation, that there exists something in nature that can change forms but never be created nor destroyed. It appeared to him that the search for the laws of nature was "the most sublime scientific pursuit in life."

In college, Planck told one of his professors that he wanted to study physics. His professor tried to dissuade him, claiming that all the important discoveries had already been made. Newton had figured out gravity, Maxwell had figured out electromagnetism, and the newly created field of thermodynamics was complete. There was nothing left to do but fill in little holes. Planck said that he would happily do just that.

Planck did find a hole, an underdeveloped concept in the field of thermodynamics — entropy, a measure of disorder. The concept of entropy was introduced in the early 1850s by the German physicist Rudolf Clausius as a way to keep track of energy waste in a heat engine. Planck studied Clausius's writings and appreciated his clear formulation of the first and second laws of thermodynamics. Here, in these laws, Planck found something that he could develop. Clausius's statement of the second law of thermodynamics, that heat does not spontaneously flow from a cold to a hot body, had to be supplemented by the additional assumption that heat can only flow in this direction at the cost of a corresponding compensation in nature. Planck found another way to state this using the concept of entropy: the sum of the entropies of all the bodies involved in a natural process tends to increase with time.

Planck's restatement of the second law of thermodynamics in terms of entropy formed the central point of his doctoral dissertation, which he completed in 1879. None of his professors expressed any interest in his work. Gustav Kirchhoff was critical of its contents, and it is likely that Helmholtz didn't even read it. Clausius, who of all people should have appreciated it, did not respond when Planck reached out to him by letter. And when Planck traveled to Bonn to see him in person, Clausius was nowhere to be found.

Despite its poor reception, Planck's doctoral thesis was accepted, most likely on account of his other work. He subsequently gained employment as an instructor in Munich. There, he continued his study of entropy, which he believed was, besides energy, the most important physical quantity used to describe a physical system.

In the spring of 1885, Planck was offered an associate professorship in theoretical physics at the University of Kiel in succession to Hertz. Planck suspected that he had gotten this position because of some intervention by his father, who was a close friend of the physics professor at the University, but that didn't bother him. He would make the most of this great opportunity. He moved to Kiel and soon finished a paper he had been working on containing his answer to a prize question offered by the University of Göttingen on the nature of energy. He submitted it and was awarded second place, although neither of the other two entries won any prize. He then returned to his work on entropy.

Planck remained in Kiel for four years, after which he was invited to take Kirchhoff's place as professor of physics at the University of Berlin following his death. Soon after arriving in Berlin, Planck was given an assignment. The Institute for Theoretical Physics had just received the gift of a large harmonium with untempered tuning. Planck was asked to use the instrument in a study of the untempered scale. After a careful investigation, he determined that in all cases the tempered scale was more pleasing to the ear.

After this side project, Planck again returned to the study of entropy. Using the concept of entropy, he put forth the argument that the flow of heat from hot to cold is fundamentally different from the falling of a body from high to low. His view was almost universally opposed as most scientists believed that the two were equivalent. In accordance with this, they suggested, there should not exist any absolute zero temperature just as there is no absolute zero position. Eventually, Planck's view prevailed due to the work of Ludwig Boltzmann, who proved the key difference between these two processes using atomic theory. Planck's own work on the subject, however, became irrelevant.

Planck soon shifted his focus to another problem. New experiments at the German Physico-Technical Institute drew attention to the subject of the thermal spectrum, which led Planck to Kirchhoff's studies in the area. Kirchhoff had determined that in time, an evacuated cavity surrounded by totally reflecting walls will reach thermal equilibrium. The radiation emitted from the cavity during this process depended solely on the body's temperature. The simplicity of this result had great appeal for Planck, who began to tackle the problem theoretically.

He first approached the problem using Maxwell's electrodynamic theory of light. He assumed the cavity was filled with simple oscillators with different vibrational frequencies subject to small damping forces. With this model, Planck obtained the remarkable result that in thermal equilibrium, the radiation energy is independent of the damping constant of the oscillators. Therefore, he could replace the energy of the oscillator with the energy of radiation, exchanging a complicated structure for a simple one.

This was a success, but something was still missing. Although Planck could explain Kirchhoff's result that the radiation emitted from a cavity in thermal equilibrium depends only on temperature, he could not derive the specific shape of the spectral energy distribution with this model. Undeterred, he turned his attention back to his favorite subject, entropy.

Planck had long believed that entropy was one of the most important properties of physical systems because its maximum value indicates a state of equilibrium. He suspected that a crucial relationship lay in the dependence of entropy on energy. As was typical, Planck's colleagues were indifferent to his idea, but their indifference was now a blessing. While a multitude of outstanding physicists worked on the problem, focusing their attention exclusively on the dependence of the radiation spectrum on temperature, Planck could work out his calculations at leisure, with absolute thoroughness and without fear of competition.

Eventually, he established two simple limits on the relationship between entropy and energy, one valid for small energies and the other for large energies. Using these together with relevant empirical data, Planck obtained a new radiation formula valid for the full range of energies. He presented his results at a meeting of the Berlin Physical Society on October 19, 1900.

The next morning, one of Planck's colleagues greeted him with the news that he had checked the formula against his own measurements and, at every point, found satisfactory agreement. Other colleagues initially believed that they had discovered discrepancies, but after more careful analysis, they realized that this was due to their own calculational errors. Subsequent measurements confirmed Planck's formula again and again; the more precise the measurements, the better the agreement.

Planck had discovered a formula, but he also wanted to find physical meaning in the equation. He was confident in his thermodynamic treatment; now he probed the problem at a microscopic level, returning to an electrodynamic model in which the cavity is filled with radiating oscillators, each with a single frequency of vibration. He combined the two approaches by introducing a relationship between entropy and the probability that an oscillator has a certain vibrational energy.

With this, Planck was able to derive the equation that had so successfully described the distribution of radiation energies. However, in order to do this, he had to make one more assumption: each oscillator has an energy equal to a fixed constant times its vibrational frequency. This constant, he claimed, was a new constant of nature. He labeled it with the symbol "h" and called it the "quantum of action." He didn't yet understand how this new constant would figure into physical processes. He made multiple attempts to fit it into the framework of existing theories, but none were successful.

# 8

# A Quantum of Light

It is undeniable that there is an extended body of facts
pertaining to radiation which indicate that light
has certain inherent qualities that put its comprehension
far from either the Newton emission theory of light
or the view of wave theory. Hence it is my opinion that the next
phase of the growth of theoretical physics will bring us
a theory of light which will reveal itself as
a kind of mixture of wave and emission theory.

— Albert Einstein

In the spring of 1918, the scientific community of Berlin was preparing a banquet to celebrate Planck's sixtieth birthday. As his friend and colleague, Einstein was chosen to be the master of ceremonies. In his speech, Einstein described the reasons that most people choose a life of science, usually either to display their talents or in the hope of profit. "Should an angel of God descend and drive from the Temple of Science all those who belong to the categories I have mentioned," he continued, "I fear the temple would be nearly emptied. But a few worshipers would still remain — some from former times and some from ours. To these latter belongs our Planck. And that is why we love him."

Einstein and Planck worked together at the University of Berlin from 1914 until Einstein left Germany in 1933. They built on each other's theories. They played music together, Planck on the piano and Einstein on the violin. They respectfully disagreed about politics, each faithful to his own version of idealism.

They became colleagues in Berlin, but their relationship had begun many years before, while Einstein was still an unknown, working as a patent clerk. It was Planck, as editor of *Annalen der Physik*, who read Einstein's early papers and approved them for publication. Very few physicists, if any, would have appreciated these early papers as much as Planck must have appreciated them. They all included as a central concept his favorite subject: entropy.

In March of 1905, Einstein submitted an article for publication that must have aroused particular interest in Planck, for it built on his own work — specifically papers he had published in 1900 and 1901 proposing a new formula to describe the distribution of radiation energy emitted from a perfect absorber and emitter of radiation, called a blackbody.

Einstein began the paper establishing a "profound formal distinction" between the way current theory viewed a dilute gas and low density radiation. A gas is modeled as a system of tiny, discrete, randomly moving particles, while light is assumed to be continuous radiation. Although this view of light had in many ways been successful, it leads to the absurd prediction that the energy radiated in certain systems is infinite.

After justifying this last statement mathematically, Einstein rederived two of Planck's results: his energy equation and calculation of the mass of a hydrogen atom. Einstein's methods were very similar to Planck's, utilizing the concept of entropy together with thermodynamic principles. However, through these derivations Einstein demonstrated that the results do not depend on specific assumptions regarding the emission and transmission of radiation. This allowed him to retain Planck's mathematical results while making a conceptual shift.

Planck had assumed that the quantization of radiation energy was due to the presence of discrete electrical oscillators, each with a characteristic frequency, which fill the cavity of the blackbody. Einstein suggested, rather, that the quantization of radiation stems from light itself being quantized. In particular, he proposed that a light ray consists of a finite number of energy quanta localized in space that can only be produced and absorbed as complete units.

Einstein then applied this quantum model of light to three different phenomena: fluorescence, the photoelectric effect, and the ionization of gas by ultraviolet light. In all of these situations, he assumed that a single quantum of light is absorbed fully by the matter upon which it is incident. In the case of fluorescence, this leads to the emission of another quantum of light. In the photoelectric effect, it leads to the release of a single electron. If the light is absorbed by a gas, an individual gas molecule is ionized.

Einstein analyzed these systems making use of the principle of energy conservation. In all three cases, he used whatever experimental data was available to check his conclusions and in every case found them satisfactory. Furthermore, in each case he made at least one testable prediction. He predicted that there should be no lower limit on the intensity of the light capable of producing fluorescence. He also determined a cutoff frequency below which the photoelectric effect would not be observed. Lastly, he derived an equation relating the amount of light absorbed by a gas to the amount of gas that could be ionized.

Einstein was proposing that light is made of particles, but he was not advocating for an abandonment of the wave theory of light; reflection, refraction, diffraction, and other optical phenomena had been well explained by the wave theory. Planck argued that these two views are incompatible. Light cannot be a particle since it is a wave. If we were to adopt this quantum view, he claimed, the theory of light would be set back by centuries, "into the age when Christiaan

Huygens dared to fight against the mighty emission theory of Isaac Newton."

In September of 1905, Einstein published a paper addressing an asymmetry in Maxwell's electrodynamic theory. According to the theory, when a magnet moves toward a stationary conductor, an electric field is produced that does not exist when the conductor moves toward the magnet. This is a violation of the principle of relativity since the two situations are identical except for which of the two, the magnet or the conductor, is moving. Einstein suggested that the distinction between these two situations is merely a formal one, unrelated to the actual phenomena, since the measurable current produced in each case is the same.

Einstein proposed that this anomaly could be eliminated by first elevating the principle of relativity to a postulate. Therefore, the laws of electrodynamics should be valid in the same reference frames in which the laws of mechanics are valid. Second, the speed of light in a vacuum should take the same value independent of the motion of the emitting body.

With these two postulates, Einstein was able to correct the asymmetry in electrodynamic theory. By adopting appropriate equations relating space and time in one reference frame to another, a magnet moving toward a stationary conductor yields the same fields and forces as when the conductor moves toward the magnet. His assumption of the constancy of the speed of light was also consistent with experimentalists inability to measure its motion with respect to the ether. In fact, this assumption makes the existence of this hypothetical medium superfluous.

In the same month, Einstein published another paper extending these ideas, demonstrating that when a body radiates a certain amount of energy, its mass decreases by an amount equal to the energy radiated divided by the speed of light squared. These two papers taken together constituted the introduction to what became known as the special theory of relativity.

Einstein's papers on relativity, as Planck wrote to him, were much appreciated. Planck and others soon began to apply and extend the theory. In September of 1908, the theory of relativity received its first public acknowledgement at the meeting of German Natural Scientists in Cologne. Einstein's former mathematics professor, Hermann Minkowski, who had reformulated the theory in four-dimensional spacetime, spoke to its importance saying, "Henceforth space on its own and time on its own will decline into mere shadows." No one paid any attention to Einstein's light quantum.


In 1909, Einstein wrote another paper putting forth more strongly the idea that light sometimes acts like a particle and other times a wave. That light is a wave, he began, could not be doubted after it was observed that it experiences interference and diffraction. Maxwell's discovery that light can be understood as an electromagnetic wave added strength and clarity to the wave theory of light. Moreover, since it can travel through a vacuum, it was believed that light must travel through some special medium, referred to as the luminous ether. Although the ether had never been detected, it seemed obvious that this substance must exist. As evidence of this, Einstein quoted a textbook that had come out in 1902, which claimed that the probability of the existence of ether "borders extraordinarily closely on certainty."

Einstein, however, believed that the ether hypothesis was fundamentally flawed. The American physicists Michelson and Morley had performed high precision experiments over many years attempting to determine the influence of the ether on the speed of light but were unable to detect any effect. The most reasonable conclusion from these experiments, Einstein suggested, was that the ether does not exist.

If light is not a disturbance in a medium, it must be an independent entity. Furthermore, according to his theory of relativity, when a body radiates energy, the mass of that body decreases. Therefore, matter can be converted into energy in the form of light. Light, he concluded, is "something that matter itself consists of."

Einstein then reviewed a series of questions that the wave theory of light could not answer: Why does only the color of light, not its intensity, determine whether or not a photochemical reaction will occur? Why are high frequency light rays more effective at exciting chemical activity than low frequency rays? Why does the intensity of the light not affect the kinetic energy of the emitted electrons when light shines on a metal surface? And why are high temperatures required for the emission of high frequency light?

Einstein next turned to Planck's radiation formula, demonstrating that the equation implies the existence of two independent sources of fluctuation in the radiation pressure. The first term assumes random interference between uncorrelated waves; it dominates when the energy density of the radiation is large. The second term can be accounted for by assuming a random distribution of particles colliding, similar to that of a low density gas — this term dominates at low energy density. As a result, it is possible for light to sometimes act like a wave and other times like a particle within a single theory.

Einstein then proposed that perhaps the electromagnetic fields predicted by Maxwell's equations are bound to singular points. He suggested that we can visualize light as a collection of point particles surrounded by wave-like fields whose strength decreases as the distance from their center increases. If the fields overlap, light acts like a wave; otherwise, it acts like a collection of particles. Einstein's conclusion was that the wave-particle aspects of radiation "should not be considered to be non-unifiable." He ended the paper inviting others to work with him toward this goal: "Who would have enough imagination to construct a theory of vibration from this foundation?"

# 9

# Bohr's Atom

We, arrogant youngsters, we ventured to doubt
This thesis of Bohr and we wished to find out
If really a deep psychological facet
Of criminal law does make virtue an asset.

So the three of us went to the center of town
And there at a gunshop spent many a crown
On pistols and lead, and now Bohr had to prove
That in fact the defendant is quickest to move.

Bohr accepted the challenge without ever a frown;
He drew when we drew... and shot each of us down.

— Henrik Casimir

Christian Bohr was a passionate advocate of equality and independence for women. In this spirit, he taught courses at Copenhagen University for adult women seeking degrees, through which he met his wife Ellen Adler. Ellen was a strong, sincere woman who cast a warm glow over everything, "for lovableness was her being." Ellen's sister founded Hanna Adler's School, the first grade school in Denmark where girls and boys studied together. This is the family that nurtured the Danish physicist Niels Bohr.

Bohr was particularly close to his brother, Harald. About a year younger, Harald was generally thought to be the "quicker" of the two. In fact, Harald became an internationally renowned mathematician. Bohr was not a mathematician, but he had something else. According to Harald, "Niels possessed a superhuman, intuitive insight into the secrets of nature, and could even comprehend truth without needing to translate it into ordinary language, to which mathematics belongs."

Of course, Bohr was not born with extraordinary intuition. He gained physical intuition by doing experiments. And he gained mathematical intuition by doing calculations. Once, after one of his sons witnessed a seemingly miraculous display of his intuition at a scientific meeting, he explained, "If you had any idea how hard I worked in my early years, how I calculated and recalculated, then you would understand that later one can easily arrive at a result by intuition."

Bohr did his undergraduate work at the University of Copenhagen, studying under Christian Christiansen, a friend of his father and the only physicist at the University. He earned his master's and doctorate there too, also under the supervision of Christiansen. For his doctoral thesis, he demonstrated that the electron theory could not explain known properties of metals. His defense drew a large crowd but few questions. Christiansen, Bohr's only examiner, said that there was probably no one in all of Denmark who knew enough to judge his thesis. In conclusion, he regretted that the paper was not written in a foreign language.

This was exactly what Bohr wanted to hear, and he had already begun to translate the paper into English. He completed the translation along with some minor revisions and applied for a scholarship to study abroad. The application was successful, and with "ignorance and reckless courage," he headed to England, hoping to discuss his ideas about electron theory with J.J. Thomson.

Joseph John Thomson entered Trinity College as a sizar in 1876. He became one of the top students at Cambridge in his graduating class, earning the title of Second Wrangler on the Mathematics Tripos and second place in the Smith Prize. After graduating, he became a fellow at Trinity. During his fellowship, he developed a mathematical model of the atom visualized as a smoke ring; this work earned him a master's degree as well as the Adams Prize, an award chosen by Cambridge's mathematics faculty for distinguished work in mathematical science. In 1884, he was chosen to be the Chair of Physics at the Cavendish Laboratory.

Some were surprised by this appointment, for Thomson had not yet done any significant experimental work, but his appointment was certainly not a disappointment. Thomson became interested in cathode rays, glowing rays that appear when electricity is applied to a glass container mostly emptied of air. Cathode rays had been discovered in the 1700s, but only slow progress had been made in understanding them. It was known that they appear at the positive end of a battery and don't spread out like ordinary light rays; instead, they tend to form a focused beam. They also bend in the presence of a magnetic field.

There was no consensus as to the nature of these rays. Some speculated that they are a disturbance in the ether. Others believed that the rays mark the paths of electrically charged particles. This second view had the advantage of having testable predictions; it was this line of research that Thomson pursued.

Thomson produced cathode rays in a glass tube and observed them under different conditions. He recorded the path of the ray by taking photographs. Through several experiments, he determined that the cathode rays respond to electric and magnetic fields just as if they were negatively charged particles. In summary, he wrote, "I can see no escape from the conclusion that they are charges of negative electricity carried by particles of matter."

Thomson then measured their mass-to-charge ratio in different circumstances. He varied the type of gas used, the pressure, and the electrodes used to produce the rays. In all of these experiments, he found that the mass-to-charge ratio remained constant. He concluded, therefore, that the cathode rays produced by different metals are all made up of the same negatively charged particles.

The existence of these negatively charged subatomic particles, which became known as "electrons," prompted Thomson to speculate on the structure of atoms. He developed a model in which an atom is made up of negatively charged electrons evenly distributed throughout a positively charged substance, like plums in plum pudding. The electrons repel each other and are attracted to the center of the charge distribution; equilibrium is reached when the forces of repulsion and attraction are balanced. Thomson and his students began performing calculations with this model, attempting to connect it with known properties of elements.

For his work on the electron, both experimental and theoretical, Thomson received the 1906 Nobel Prize in Physics. Two years later, he was knighted. It is no wonder that Bohr was filled with excitement as he left Denmark for England with the intention of discussing with Thomson the flaws that he had found in his theory. He was still excited when he left his meeting with Thomson, who had greeted him warmly and said that he would be interested in reading Bohr's thesis.

While Bohr waited to see what Thomson would make of his "many criticisms," he immersed himself in the English culture and language, reading such classic works as *David Copperfield*, looking up every word he didn't know in a dictionary. Weeks passed and then months, and he still hadn't heard back from Thomson. He visited the Cavendish Laboratory in hopes of talking with Thomson but only managed to speak with him for a few minutes on a few small points.

Bohr was anxious but patient. When he returned for a second term, he decided not to go to the laboratory anymore but only to attend lectures and "read, read, read." He would also think and calculate a bit. He was so fond of Thomson, but unfortunately the great man had little time to spare. Bohr went to his lectures, though. He raved to his brother about one on the motion of a golf ball. How entertaining and illuminating! And Thomson had such splendid experiments and a "flashing, sparkling humor."

Thomson never did read Bohr's dissertation; he passed it on to the editor of the *Cambridge Philosophical Magazine*, who thought it was too long to publish. Bohr had been at Cambridge for two terms and still had not gotten the chance to discuss electron theory with Thomson. Finally, in the spring of 1912, another opportunity came his way. The annual Cavendish banquet brought together scientists from various universities. It was there, through a friend of his father, that he was introduced to an experimentalist working in Manchester, Ernest Rutherford.

Ernest Rutherford was born in New Zealand, one of twelve children. His father was a miller and his mother a school teacher who emphasized the importance of education. With little money and many children, inventiveness was part of their life. Rutherford was given his first science book at the age of ten and built a cannon. To the surprise of his family, it exploded.

With the help of scholarships, Rutherford attended private secondary school and then went on to university, where he excelled at experimental science. In 1895, at the age of twenty-four, he went to Cambridge as the first research student at the Cavendish Laboratory to work with J. J. Thomson. While Thomson worked mainly with cathode rays, Rutherford became interested in alpha particles, positively charged particles that spontaneously come out of elements such as uranium. Although their research went in different directions, they both had the same goal: to understand the structure of the atom.

Thomson had his plum pudding model, but Rutherford chose to investigate matter experimentally. To this end, he set up what was essentially a ballistics experiment. He used thin, gold foil as the target, while alpha particles played the role of the bullets. To detect where these particles hit, he set up a phosphorescent screen behind the foil. Green sparks appeared when the alpha particles impacted the screen, which his assistant observed with a magnifier. The results were surprising — most of the particles went through undeflected, while a few bounced almost directly back.

Conceptually, these results indicated the presence of a hard, dense, positively charged center of the atom, which Rutherford called the "nucleus." Most of the rest of the atom seemed to be empty space, allowing an alpha particle to pass straight through. Rutherford then performed quantitative analysis using classical physics to verify this model and determine the charge of the nucleus.

When Bohr and Rutherford met, they bonded immediately, sharing a sense of humor and love of soccer. But more importantly, by means of completely different paths, they had arrived at the same conclusion: Thomson's model of the atom was incorrect. And so, with rekindled hope, Bohr followed Rutherford to Manchester.

In Manchester, Rutherford and his team continued to probe the atom, always coming to the same conclusion that all of the positive charge is focused at a tiny point in the center of the atom. Meanwhile, Bohr worked alongside them, thinking and calculating. He stayed on with Rutherford through April, May, and June. In July, he returned home... to get married!

In Denmark, Bohr continued to think and calculate. In February of 1913, he finally fit the pieces together. He quickly wrote it all down in a paper that he sent to Rutherford. Rutherford replied that the paper was too long. Rather than continuing the dialogue through letters, Bohr traveled back to Manchester so that he and Rutherford could "fight matters out" in person. He stayed in Manchester for four weeks, during which time, with patience and passion, he explained every last detail of his paper to Rutherford. In the end, Rutherford agreed with Bohr; it all had to be kept.

The first part of Bohr's paper appeared in the July edition of the *Cambridge Philosophical Magazine*. It started with Rutherford's model of the atom, a positively charged nucleus surrounded by electrons that are attracted to the nucleus because of their negative charge. Bohr assumed that electrons orbit the nucleus of an atom like planets orbit the Sun. This, however, contradicted electromagnetic theory; an accelerating electron should radiate and quickly lose all of its energy within a fraction of a second.

Others had gotten to this point and abandoned the approach because of its inconsistency with electrodynamic theory, but Bohr pushed forward. He hypothesized that electrons might exist only in discrete energy levels that he called stationary states. An electron can jump from one state to another, but it can't exist in an in-between state. Bohr then connected this idea of stationary states to atomic spectra.

When light from a burning gas is shined through a prism, distinct bands of color can be seen. For example, light from burning hydrogen is separated into four distinct bands of color: red, blue-green, blue, and violet. The spectrum of each element is unique and can be used as a fingerprint to identify the composition of unknown gases. Analysis of the spectra of gases had been used since the 1860s to identify elements within compounds and also to discover new elements.

The spectrum of hydrogen was particularly well known, and in 1885, the Swiss mathematician Johann Balmer discovered an equation that accounted for all the observed lines and also predicted others. It was generalized three years later by the Swedish physicist Johannes Rydberg. These equations led to the discovery of many new lines in the hydrogen spectrum, all of which were accurately predicted by Rydberg's equation.

Despite the progress made in spectral analysis, scientists had no theory to explain the existence of these lines. Even more surprisingly, there had hardly been an attempt to understand them. According to Bohr, the lines were viewed like the "beautiful patterns on the wings of summer birds, whose beauty one can marvel at, but which no one presumes will display fundamental biological regularities." Now, Bohr could explain them.

The starting point of his explanation was his model of the atom consisting of electrons in stationary states orbiting the nucleus. Next, Bohr made three assumptions. First, the stationary states are described by ordinary mechanics, with the usual equations for centripetal force and energy. Bohr assumed circular orbits for simplicity, claiming that elliptical orbits would yield the same results.

Bohr's second assumption was that when an electron makes a transition between stationary states, it emits or absorbs light according to Planck's energy equation. Specifically, the frequency of the light is equal to the energy difference between levels divided by Planck's constant. If each element has unique energy levels, it would also produce a unique spectrum when the electrons transition from one level to another, consistent with experiment.

Lastly, Bohr assumed that the work required to remove an electron from its orbit equals half of the Planck energy. He later derived this equation by requiring that for very large, slow orbits, the frequency of the light emitted as the electron moves between energy states is equal to the orbital frequency of the electron as determined using ordinary mechanics. This was the beginning of what would later be called the "correspondence principle," the requirement that the new physics must connect in some way with the old.

With these three assumptions, Bohr derived Rydberg's spectral equation for hydrogen in terms of the charge and mass of the electron and Planck's constant, obtaining satisfactory numerical agreement. Furthermore, he explained the existence of certain spectral lines found in stars, called Pickering lines, which are *not* predicted by the Rydberg equation. He showed that these are lines of singly ionized helium.

Bohr's brother, Harald, was a mathematics professor at Göttingen University when the paper came out. Amused, he shared with Niels the reaction of the German physicists, who said that "the whole thing was some awful nonsense, bordering on fraud." Einstein's initial response was also negative; he claimed that if Bohr were right, physics was at an end! But then one of England's physicists, E. J. Evans, tested Bohr's claim about the Pickering lines in the laboratory and confirmed that the lines, in fact, did come from ionized helium as predicted by Bohr's model. When Einstein heard this, his big eyes got even bigger, and he said, "then this is one of the greatest discoveries." And indeed it was.

All of a sudden, following Bohr's lead, scientists throughout Europe began to pay attention to the atomic spectra of elements. In Manchester and Oxford, researchers focused on x-ray spectra, invisible electromagnetic radiation whose waves have thousands of times the frequency of visible light. They found that the spectral lines moved regularly from element to element, indicating that some fixed quantity was changing. This was consistent with Bohr's model, in which the quantity changing is the charge of the atomic nucleus.

Physicists working independently in Germany and Italy discovered that when an electric field is applied to the light from an element, each spectral line splits into several lines. This was a very similar effect to one that was discovered just before the turn of the century, the Zeeman effect, in which each spectral line of an element splits into several when subjected to a magnetic field. Hearing about the new experiments, Rutherford wrote to Bohr, reminding him of the Zeeman effect. Could he apply his theory to these observations?

Bohr received Rutherford's letter in December, and he immediately began to work on these new problems. The following March, he published a paper with his investigations. He also built on his original model, adding a relativistic correction to the circular orbits he had assumed. Others, too, began to work on the theory. The German physicist Arnold Sommerfeld, following Bohr's ideas, applied relativity theory to the non-periodic movement of electrons following elliptical orbits. In doing so, he was able to account for an observation made several decades before by the American physicist Albert Michelson — each of hydrogen's spectral lines when observed closely is actually two lines.

In the spring of 1920, Bohr traveled to Germany to give an address before the Physical Society of Berlin. The world war had isolated German scientists, limiting their travel and communication with much of the rest of Europe. Now, the war was over and some kind of normalcy had returned; the German scientists wanted to catch up on what they had missed.

Bohr spoke to the Germans in German. He was in Berlin, the hometown of Planck and Einstein and could assume that his audience was well acquainted with the work of both. He knew that they were familiar with Planck's work on blackbody radiation, in which charged oscillators in a cavity can only emit radiation in discrete amounts, called energy quanta. He also knew that they had read about Einstein's successful interpretation of the photoelectric effect using this idea of an energy quantum. Finally, he knew that they understood Einstein's relativity theory, and there were probably several in the audience who had even helped develop it.

Undoubtedly, the Germans were also well aware of the work of their countrymen in the area of atomic spectra — the discovery of the splitting of hydrogen's lines by Johannes Stark, which they called the "Stark effect," Arnold Somerfeld's relativistic interpretation of hydrogen's fine structure, and so on. But there was so much more to share, and Bohr began his speech with his regrets that he didn't have more time. It would be impossible, he claimed, to give a comprehensive survey of even only the most important results obtained with his theory of atomic spectra in a single address.

Bohr started at the beginning, even though he could not take the time to delve into any of the individual histories of the many parts that he was attempting to put together. He began with Rutherford's model of the atom, electrons surrounding a dense, positively charged nucleus. It was now believed that the number of electrons in a neutral atom of each element equals its position on the periodic table, its "atomic number." It should therefore be possible to explain the chemical and physical properties of elements from their atomic number.

Bohr then turned to atomic spectra, reviewing the important points from his 1913 paper. He explained that it is possible to account for the existence of the sharp lines in atomic spectra if it is assumed that electrons exist in stable, stationary states. It is possible to account for the existence of these stable orbits by assuming that an atom can only emit energy in discrete quanta according to Planck's energy equation.

This model had been applied to hydrogen's spectrum with great success. In the simple case in which only one electron orbits the nucleus, the motion of the electron is periodic and for large, slow orbital states, the difference in frequency between adjacent states corresponds to the frequency of the light emitted during that transition. This model not only applied to the well known visible lines of hydrogen, but also to its x-ray spectrum, the splitting of each line into doublets, and the modification of the lines when hydrogen is subject to electric and magnetic fields.

However, when even one additional electron is added to the model, everything becomes much more difficult. The resulting motions of the electrons, no longer periodic, are exceedingly more complicated, even in the case of helium, which has only two electrons. Compounding this problem, the atomic spectra do not give direct information about the energy levels of the stationary states, only the energy changes when an electron moves from one state to another.

Nevertheless, much progress had been made in this area too. By observing patterns in the spectral lines of some elements, it was possible to deduce the stationary states from the energy differences indicated by the lines. Bohr used the lines of sodium as an example. Certain series had been identified in its spectrum: "sharp series" (S), "principle series" (P), "diffuse series" (D), etc. These had been pieced together to create energy levels, which he illustrated with a diagram — the first energy level contained one sublevel from the S-series, the second energy level contained S and P sublevels, the third energy level contained S, P, and D-sublevels, and so on.

In addition, many theoretical advances had been made in determining the atomic energy levels of elements. Several physicists took part in this work — Sommerfeld, the Austrian physicist Paul Ehrenfest, and even Planck himself. Originally, Bohr's orbits were two-dimensional, lying in only one plane. This had been generalized to a more realistic model in three dimensions. Physicists had been able to determine the periodic components of non-periodic motion in some cases with a mathematical tool known as separation of variables. With this new method, it was apparent that each state must be represented by three integers, rather than the one integer in Bohr's original model — an integer for each independent way an electron can move in three-dimensional space.

Another development was made by considering angular momentum. Angular momentum conservation along with Bohr's correspondence principle suggested that, like energy, any change in angular momentum should only occur in discrete amounts, resulting in an equation analogous to Planck's energy equation. Furthermore, the radiation emitted by the atom when an electron undergoes a transition must exhibit circular polarization, a prediction important for radiation theory. Finally, these considerations led to the exclusion of certain transitions between stationary states as well as predicting that some transitions were more likely than others, facilitating the calculation of the relative intensity of some of the spectral lines.

So much had been done, but there was still so much left to do. In closing, Bohr emphasized that some special assumptions that had been made might have to be abandoned. Despite great progress, except for its application to hydrogen, the theory still lacked quantitative agreement with experiment.

One night in June, Bohr lay in bed awake all night — his thoughts and the white ceiling, bright in the Danish summer night, would not let him sleep. It was 1921, and he and his wife, Margrethe, had rented an old peasant cottage in Tibirke Bakker in the wooded hills of North Zealand for the summer. In the morning, he called the owner with a request: "You see, it's like this, it's so incredibly light at night, and I'm lying in bed with so many thoughts going round in my head, so if it's possible, I'd really be terribly happy if I could be allowed to paint the ceiling in the bedroom dark blue."

Bohr kept thinking and thinking, and when he was invited back to Germany, this time by faculty at the University of Göttingen for a series of lectures, he had something more to share: he had figured out the periodic table.

In 1869, the Russian chemist Dmitri Mendeleev discovered that when the known elements were arranged in order of increasing atomic weight, a repeating pattern of physical and chemical properties emerged. He also observed gaps in these patterns, which led to the prediction of new elements that were then found with the properties predicted by the table. This periodic table was an enormous achievement, the culmination of Lavoisier's theory of elements, an end in itself. Although there had been some attempts to understand the reason behind the periodicity of the elements, primarily by J.J. Thomson with his electron theory, no progress had yet been made.

Bohr could now account for the periodic trends simply by assuming that the electrons in each atom are organized in groups, each of which has a fixed number of electrons. If this was correct, he concluded, the periodic table should be organized *not* in order of increasing atomic weight, but of increasing nuclear charge. In most cases, the two methods of organizing the elements coincided. However, in a few cases Bohr's method predicted a different order, an order more consistent with periodic trends.

Bohr's theory also gave a different prediction for the properties of a missing element, which would be the seventy-second element on the periodic table. According to current theory, this element should belong to the rare earth group; it had been searched for repeatedly without success. However, Bohr's theory indicated that element seventy-two should be like its neighbor on the other side, zirconium. With this view in mind, two physicists working with Bohr searched for the missing element by analyzing the x-ray spectrum of zircon minerals. And here they found it, element seventy-two, which they named "hafnium" after the Latin name for Copenhagen.

# 10

# Uncertainty

Everything is determined,
the beginning as well as the end,
by forces over which we have no control.
It is determined for the
insect as well as for the star.
Human beings, vegetables or cosmic dust,
we all dance to a mysterious tune,
intoned in the distance by an invisible player.

— Albert Einstein


Harmony allows itself only to be sensed, never grasped,
and if we attempt to grasp it, it slips through our fingers...

— Niels Bohr

Einstein did not play a leading role in the development of quantum theory, but he was not uninterested in it. He was also not uninterested in Bohr. The two met for the first time in the spring of 1920 in Berlin, on the occasion of Bohr's address to the Berlin Physical Society. They talked, and afterwards Einstein followed up with a letter. "Rarely in my life has a man given me such joy by his mere presence as you have," he wrote. Bohr quickly responded, "For me one of the greatest experiences ever was to meet you and speak with you."

Einstein and Bohr were both awarded the Physics Nobel Prize in the year 1922, although Einstein's was actually for the previous year. This time Bohr took the initiative. He wrote a letter to congratulate Einstein, declaring that it was the greatest honor and joy to receive the prize at the same time as him. He also expressed his relief that Einstein had been awarded the prize first. Einstein responded that without exaggeration Bohr's letter had pleased him as much as the award. He then added, "I find especially charming your fear that you might have received the prize before me — this is truly Bohrian."

Einstein did not go to the Nobel Prize ceremony that December to receive his award; he was on a tour through Asia at the time. It was not until the following summer that he traveled to Stockholm to receive his prize and give his speech. On the way home, he stopped in Denmark to visit Bohr. Bohr met Einstein at the train station in Copenhagen and then took him on a streetcar to go to his mother's house in Bredgade. They began discussing the theory of cognition and, absorbed in conversation, missed their stop and rode all the way to the end of the line. On the way back, they again missed their stop. Later, as they made their way to Bohr's Institute, they missed their stop once more.

Bohr had spent four years working toward the creation of a research institution — no doubt, he was excited to be able to share this with Einstein. He had fundraised with the help of one of his friends to purchase a building site in Copenhagen. When they finally received approval to start the construction, the world war had just ended, devaluing the kroner and inflating prices throughout Europe. They received additional donations, enough to complete the building and furnish it with a lattice spectrograph. On March 3, 1921, the "University of Copenhagen Institute of Theoretical Physics" officially opened.

This was Bohr's dream: to have a center devoted to the study of fundamental science. No scientist works in isolation; every new idea builds on other ideas. Bohr felt that physics was on the brink of a new discovery. His atomic model had been a big step, but no one, and certainly not Bohr himself, thought it was the final step. He wanted to create an environment in which experimentalists and theorists with different perspectives could collaborate and collectively figure it all out.

Furthermore, Bohr wanted to create a space that fostered international collaboration. He believed this was especially important in light of the recent war — he hoped that scientists could lead the way toward a better understanding among nations. He also wanted to include scientists with new ideas, unprejudiced by the old physics. To these ends, he actively recruited young scientists from around the world to his institution, and many others were sent by recommendation.

One of Bohr's early recruits was a twenty-year-old German named Werner Heisenberg. Heisenberg was one of Arnold Sommerfeld's students and very interested in the new quantum theory. He accompanied Sommerfeld to hear Bohr speak at the University of Göttingen. After one of the lectures, Heisenberg put forward an objection to Bohr's theory. Bohr suggested that they go on a walk that afternoon so that they could dive more deeply into the problem. They walked the length and breadth of Hainberg's tree-covered ridges. Toward the end of the walk, Bohr invited Heisenberg to come to Copenhagen.

In the spring of 1924, funded by a fellowship, Heisenberg made his way to Copenhagen to research with Bohr. He knew no Danish, but as soon as it was understood that he would be working with Bohr, all the doors opened for him. Even so, his first days at the Institute were not easy. He met brilliant scientists from all over the world who were greatly superior to him in their ability to speak many languages, familiarity with the culture and poetry of many countries, and even in their knowledge of physics. They were also exceptional musicians. But, as they quickly found out, Heisenberg could play music too; he was an excellent piano player. Soon, he began to make music with the other scientists, Bohr beating time with his powerful hands.

Everyone made Heisenberg feel welcome, and he must have been especially cheered to see the Austrian, Wolfgang Pauli. Heisenberg and Pauli had worked together in Munich and again in Göttingen. They were good friends despite their very different habits of life. Heisenberg loved the daylight — walking in the mountains, swimming, and cooking a simple meal next to a lake. Pauli preferred the night. He enjoyed spending his evenings in bars and cafes; he would even do his best work at night and then come to school late the next day, missing morning lectures. This resulted in a little ribbing, but it never got in the way of their friendship.

Pauli was logical and critical. Heisenberg appreciated these qualities, as they were a perfect contrast to his own strong imagination. It was Pauli who offered guidance when Heisenberg was trying to decide between a career in pure mathematics and one in theoretical physics. His talks with Pauli were also the most valuable part of Sommerfeld's seminar. Unfortunately, their contrasting ways of life made it difficult to find time for long conversations. One time, though, Pauli dared to enter Heisenberg's world, going on a bike tour with him and another mutual friend. The conversations that began on this trip were to have a lasting influence on them both.

Early on, Pauli had been interested in Einstein's theory of relativity. At the age of eighteen, he made his scientific debut with a paper on general relativity. He was later asked to write a review article on the subject for an encyclopedia of mathematics, an over two-hundred-page paper that Einstein praised. In the end, though, Pauli believed that special relativity was a closed subject and something that one just had to learn. Although general relativity was still wide open, each experiment was accompanied by a hundred pages of mathematical derivations, making it almost impossible to tell if anything was really correct.

At the University of Munich, Pauli turned toward quantum theory and shortly after graduating spent a year in Copenhagen. Although he returned to Germany afterward, he continued to collaborate with Bohr and made two valuable contributions to his model. First, he explained the splitting of spectral lines into doublets, called fine structure, by the introduction of an additional quantum number. This was eventually called "spin," a name chosen because the quantity was visualized as the spinning of an electron.

Second, Pauli proposed the "exclusion principle," that no two electrons in an atom can share the same set of quantum numbers. With this principle, it was possible to create electron models of all the elements, from hydrogen to uranium. Now, they understood everything — the existence of all of the observed atomic spectral lines and the structure of the entire periodic table.

At the same time, they also understood nothing. They had a complete model of the atom and the periodic table that classically made absolutely no sense. Electrons orbited the nucleus of an atom in a way that all their energy should be radiated away in a fraction of a second. Spin, if interpreted classically, violates the theory of relativity. Pauli's exclusion principle, which restricts the number of electrons in an energy shell, makes no sense in a model where electrons are point-like particles. Finally, Bohr's model didn't lead to correct quantitative predictions for the spectral lines of any atoms except for those with only one electron.

Pauli announced his exclusion principle in January of 1925. That spring, Heisenberg returned to Göttingen where he had been a *privatdozent* since the previous summer. Toward the end of May, he became so sick with seasonal allergies that he took a two-week leave of absence and left immediately for the rocky island of Heligoland, far from blossoms and meadows. He took daily walks and long swims. With little to distract him, he returned to the problem of quantum theory.

Heisenberg was skeptical of the reality of the orbits in Bohr's model, a skepticism that others, including Bohr, shared. Bohr had suggested that language can be used for the atom only as words are used in poetry. With this kind of encouragement, Heisenberg attempted an entirely different approach to the atom, discarding altogether the picture of electrons in physical orbits around the nucleus. Instead, he focused on the transitions between pairs of states, transitions that could be directly observed through the light that is emitted or absorbed when an electron jumps from one state to another. He also held fast to Bohr's correspondence principle: the new must in some way connect to the old.

Heisenberg had the idea, but now he had to test whether it could be used to obtain consistent results. Specifically, he had to find out whether the energy principle would still apply. The first few terms seemed to work; he got excited and then started to make careless mistakes in his calculations. It was almost three o'clock in the morning before he was confident in his results. The energy principle held for all the terms. Too excited to sleep, he headed out to the southern tip of the island, climbed a rock that jutted out into the sea, and waited for the Sun to rise.

223

Heisenberg wrote up his results and, upon returning to Göttingen, sent a copy to Pauli. In an accompanying letter, he wrote, "Everything is still vague and unclear to me, but it seems as if the electrons will no more move in orbits." Pauli, usually his most severe critic, replied with warm encouragement. Heisenberg submitted his initial results toward the end of July. He continued working out the mathematics of the theory with two colleagues from Göttingen, Max Born and Pascual Jordan, creating what became known as the "matrix mechanics" formulation of quantum mechanics.

In 1917, Einstein noticed something in his relativity theory: the relativistic equation for energy coincides with Planck's energy equation when the mass is set equal to zero. Therefore, light can be viewed as a *massless particle.* Importantly, this implies that a light particle should have momentum, and the equation for its momentum comes straight from relativity theory.

About five years later, the American experimentalist Arthur Compton demonstrated that the scattering of light by electrons could be explained by assuming that particles of light collide elastically with electrons, conserving energy and momentum. Compton's experiment was conducted using x-rays, the same kind of radiation used to demonstrate the interference of light when it is passed through crystals. It had finally become impossible to ignore the conclusion that, as Einstein had suggested in 1905, light sometimes acts like a particle and other times, a wave.

In 1924, the French nobleman Louis de Broglie wrote a doctoral thesis in which he suggested that the same applies to matter: matter sometimes acts like a particle and other times like a wave. The core of de Broglie's thesis was that wave-particle duality as a general principle was already embedded in our physical theories. To support this claim, he demonstrated the mathematical equivalence between Fermat's least time principle — a principle that governs the motion of waves and can be used to derive such optical phenomena as reflection and refraction — and Maupertuis' principle, which governs the motion of a massive particle.

Therefore, de Broglie concluded, the wave-particle duality that had been applied to light should also apply to matter. He then derived Bohr's quantization condition for the energy levels by assuming that the electron is a wave. With this, he gave quantum theory a conceptual base. Light quanta are a result of the wave-particle duality of light, while quantized energy levels in atoms are a result of the wave-particle duality of matter.

De Broglie's thesis was primarily theoretical, and during his defense a committee member asked whether the matter waves assumed by his theory could be experimentally detected. De Broglie was ready with a response. Electrons should diffract when passed through crystals in a manner similar to the diffraction of light through a regular array of slits. His committee was skeptical, believing that these matter waves were just a mathematical trick or model; his doctoral advisor sent a copy of the thesis to Einstein asking for his opinion.

Einstein was impressed and inclined to believe, along with de Broglie, in the physical existence of matter waves. He shared de Broglie's work with Max Born at the University of Göttingen, who passed it along to the head of the experimental physics department, James Franck, who informed him that two Americans had already observed the predicted effect!

Early in 1925, Einstein published a paper commenting on de Broglie's work. "I believe that it involves more than merely an analogy," he wrote. Inspired by Einstein's paper, Erwin Schrödinger, a physics professor working in Zürich, began working on the problem. Using wave mechanics, he calculated the stationary states of the hydrogen atom. He was disappointed at first when he obtained numbers that did not agree with atomic spectra and abandoned the project for several months. In reality, he was doing everything correctly except for not incorporating "spin," which he had not yet heard about.

Schrödinger returned to the problem after he was asked to give a talk explaining de Broglie's work. He found that if he ignored relativistic effects, which he had included in his first attempt, he obtained satisfactory agreement with experiment. Furthermore, although his assumptions and physical picture were fundamentally different from Heisenberg's theory, Schrödinger was able to demonstrate that the equations resulting from the two theories are identical.

In July of 1926, Schrödinger was invited to Munich to present his wave theory of the atom. At first, the audience was enthusiastic, but then, Schrödinger began to interpret the theory in a way that contradicted all previous results. He described the "electron waves" as material waves, like sound waves in a pipe organ. Heisenberg, who was in the audience, put forward several objections. Wilhelm Wien, the head of experimental physics at the University of Munich, interceded on Schrödinger's behalf, expressing criticism of Heisenberg. Heisenberg reported the incident to Bohr, who promptly invited Schrödinger to Copenhagen.

The showdown began at the Copenhagen railway station. Bohr and Schrödinger were engaged in heated discussion for several days, from early in the morning until late at night. Bohr would not give up. Schrödinger's insistence on the physical existence of electron matter waves was not consistent with experimental evidence; it could not even explain Planck's law. Schrödinger continued to raise objections. Bohr was relentless, meeting every objection with a point-by-point refutation. Before long, Schrödinger became ill, presumably from exhaustion.

Schrödinger stayed at Bohr's home throughout his illness with Bohr's wife, Margrethe, patiently nursing him. Bohr hardly left his side and could be heard saying, "But Schrödinger, you must still admit, that..." Finally, Schrödinger declared, "If we still have to go on with this confounded quantum leaping, then I am sorry I ever had anything to do with atomic theory," to which Bohr replied, "But the rest of us are so thankful that you did."


In May of 1926, Heisenberg returned to Copenhagen as a university lecturer and assistant to Bohr. Toward the end of February in 1927, Bohr went on a skiing holiday in Norway, and Heisenberg could give his thoughts free rein. He had been in Copenhagen when Schrödinger visited Bohr and witnessed the struggle between the two men. He believed that their difficulty in communicating was due to a conflict of concepts, specifically because of the complicated nature of quantities such as position and velocity when applied to the tiny world of the atom.

In developing quantum mechanics, Heisenberg had assumed an infinite number of "virtual" vibrators with frequencies equal to all the possible frequencies that the photons emitted by a given atom could have, representing them with a matrix. Since the frequencies were represented with a matrix, he decided to also represent quantities such as position and momentum with matrices.

Matrices don't necessarily follow the commutative property of multiplication, meaning that if A and B are matrices, A x B does not in general equal B x A. Heisenberg assumed that the matrices for position and momentum *don't* commute and set MOMENTUM x

POSITION - POSITION x MOMENTUM to a finite number. When he set this number to a value proportional to Planck's constant, he obtained a system of equations that led to values of frequencies and relative intensities of spectral lines consistent with experiment.

Heisenberg felt comfortable representing the atom with abstract mathematics. He believed that the atom itself was abstract, not visualizable in any ordinary way. Mathematics was as good of a language as any to describe it. Bohr, in contrast, had always emphasized the physical picture; he feared that the formal mathematical structure would obscure the physical core of the problem. Heisenberg now attempted to connect the mathematical formalism of quantum mechanics with a physical picture.

Heisenberg decided to make the non-commutability between certain observables, like momentum and position, the cornerstone of his interpretation. Expressed another way, the non-commutability of position and momentum implies that the uncertainty in an object's position times the uncertainty in its momentum is always greater than a finite value. Therefore, it is impossible to know both where an electron is and how fast it's moving; there is inherent uncertainty in nature, regardless of the precision of our measuring instruments.

Heisenberg then came up with a thought experiment to illustrate the concept of uncertainty with an imaginary gamma ray microscope. A microscope can only detect objects larger than the wavelength of the light used. If high energy, short wavelength radiation is used, the position of an electron can be detected with extremely high precision. However, the high energy of such rays would disturb the electron's velocity and, therefore, its momentum. Although it would be possible to decrease the disturbance to the electron's momentum by using lower energy light, this would increase the uncertainty in measuring its position.

Heisenberg wrote up his analysis in a paper and sent it to Pauli, who gave a favorable reply. Meanwhile, in Norway, Bohr was thinking about the same problem and decided to make wave-particle duality central to his interpretation — that the wave and particle theories of light and matter are not mutually exclusive, but rather complement each other. When Bohr returned, he and Heisenberg struggled through weeks of discussions about their competing interpretations, discussions that were "not without tension." In the end, though, they came to the conclusion that Heisenberg's uncertainty relation was a more general expression of Bohr's complementarity principle. They agreed.

The achievement of Bohr and Heisenberg in their interpretation of quantum theory was celebrated in Copenhagen, but elsewhere, it provoked debate. Notably, Einstein objected to the assumption, crucial to the Copenhagen interpretation of quantum mechanics, that uncertainty was an inherent part of nature. Already in 1924, Einstein had expressed objections to probability concepts that had by that time begun to creep into quantum theory: "I find the idea quite intolerable that an electron exposed to radiation should choose of its own free will, not only its moment to jump off, but also its direction. In that case, I would rather be a cobbler, or even an employee in a gaming-house, than a physicist."

At that point, Einstein believed that a fully developed theory would get rid of any kind of probability or indeterminism. But now scientists had a fully developed theory that had excellent agreement with experiment, and both Bohr and Heisenberg were insisting not only that uncertainty was crucial to the theory but that it is an indispensable part of nature. Einstein set out to prove that they were wrong.

His first challenge came at the Solvay conference in October of 1927. Einstein approached Bohr during breakfast in the hotel with a thought experiment that he believed proved that it was possible to determine the position of a particle without interference from an instrument. Imagine that electrons pass first through a single slit, then through a double slit, and finally are detected by a photographic plate. The combined effect of individual particles hitting the plate should produce an interference pattern of alternating bright and dark spots, similar to what would be produced using light or any other wave. In this way, the electrons would be acting as both particles and waves.

Up to that point, Bohr agreed. Einstein then suggested that it would be possible to introduce a mechanism to control the momentum transmitted to the screen. In this way, one could determine which of the two slits the electron had passed through; it would thereby be possible to determine both the position and momentum of a particle with unlimited precision.

Heisenberg and Pauli were with Bohr when Einstein described his thought experiment, and the three continued to discuss it on and off throughout the day. By late afternoon, Bohr had his answer ready, which he offered to Einstein during dinner at the hotel. If it were possible to know precisely the electron's velocity and position before going through the slits, it would also be possible to know which slit it should go through and where it would hit on the screen. It then should make no difference if the other slit is covered.

However, if one of the slits is covered, the pattern of electrons detected on the screen changes. Since it would be absurd to imagine that an electron passing through one slit knows that the other slit is covered, Bohr's analysis demonstrated that it is impossible to ignore in this case the influence of the instruments on the electron. Einstein reluctantly gave in: "Yes, but do you believe that Almighty God plays dice?"

Einstein met Bohr at the next Solvay Conference with another thought experiment. Imagine a box containing a single photon and a clock that can open a shutter over a hole in the box. A photographic plate is placed 300 million meters from the hole so that if the shutter opens at noon, the photon produces a spot on the plate one second after noon. Using the equivalence between mass and energy from special relativity, it is possible to determine the energy of the photon by means of a scale. Therefore, Einstein claimed, it is possible to measure energy and time — two complementary quantities — with unlimited precision.

Einstein had challenged Bohr by making use of special relativity. Bohr then refuted him using general relativity: there would be uncertainty in the measurement of time due to the uncertainty in *where* the photon is located in the gravitational field during the weighing.

Einstein had fled Europe by the time of the next Solvay Conference in 1933, but that same year Bohr traveled to see him in the United States. Einstein offered another challenge, which Bohr again refuted. And when Einstein teamed up with two other scientists, Boris Podolsky and Nathan Rosen, to come up with another challenge, Bohr handily refuted this as well. Einstein could not accept that quantum theory, with uncertainty at its core, was complete. And Bohr wouldn't be satisfied until he could convince Einstein that it was.

Einstein and Bohr met for the last time in 1948. Einstein had offered Bohr the use of his large room at Princeton; he would use the small room. Bohr called in his assistant — he had something important to dictate. He spoke and circled around a large oval table in the room and then broke off, muttering, "Einstein...Einstein...Einstein." He walked faster around the room, almost running, and then made his way over to a window, still muttering, "Einstein...Einstein..."

While Bohr stood by the window, Einstein crept into the room, motioning to the assistant to keep quiet, and made for Bohr's tobacco jar. Suddenly, Bohr turned around with a firm "Einstein," and Einstein was right there, as if he had summoned him! Einstein then explained his mission, and they all burst out laughing.

# 11

# Legacy

They laugh at me,
the man of dynamite as a man of peace.
But, since men don't listen to reason,
it is necessary to invent an instrument
of death which, through fear,
will make Humanity move to peace.

— Alfred Nobel

Alfred Nobel was an inventor and entrepreneur with hundreds of patents to his credit, his most famous being dynamite. He created one of the first multinational companies, with ninety factories in twenty countries around the world. Nobel was a citizen of the world, with houses in six different countries. "My home is where I am found working," he wrote, "and I work anywhere." By the end of his life, Nobel had amassed a huge fortune.

In November of 1895, at the age of sixty-two, Nobel wrote the third and final version of his will. In his will, he specified that the majority of his fortune be used to establish annual awards in the following categories: Physics, Chemistry, Medicine, Literature, and Peace. The awards would be given to those who had "conferred the greatest benefit to mankind" during the previous year.

The idea of merit-based awards was not new; it began during the French Revolution as an alternative to what had been, essentially, birth-based awards in the form of money and titles. Several of these sprang up during the 19th century, first created by governments and later by private organizations. What was so unique about Nobel's prizes is that they were *international*. Nobel was Swedish by birth and chose primarily Swedish groups to award the prizes. However, he specifically requested that in selecting the recipients, "no consideration be given to the nationality of the candidates, but that the most worthy shall receive the prize."

The 19th century saw the growth of railroad systems, telegraph technology, the first World Exposition, and the first modern Olympics, held in 1896. That same year on December 10, Alfred Nobel died. Now, every year on Nobel's deathday, people from all over the world gather together in Stockholm, Sweden to celebrate the accomplishments of scientists, writers, and peace-makers.

The first Nobel Prizes were given out in 1901, five years after Alfred Nobel's death. That year, the physics prize was awarded to the German scientist Wilhelm Röntgen for his discovery of x-rays. Röntgen had made this discovery in 1895, while studying cathode rays. He had darkened his laboratory and shielded the apparatus with black cardboard in order to see the rays more clearly. When he turned the apparatus on, although no visible light escaped, he noticed that something began to glow *outside* the shielded area. He looked around and found that it came from a fluorescent screen resting on a bench.

Röntgen began to experiment with these new rays. He tried different shields: more cardboard, books, various metals. Only lead could completely block them. What were these penetrating rays? He had no idea. When he told his wife about them, she thought he was crazy. He brought her into the laboratory and focused a beam of the rays on her hand, which was placed on a photographic plate. When he developed the photograph, she could see the bones in her hand and declared, "I have seen my death!"

Röntgen announced his discovery at the end of December 1895. He called the rays "X-rays," where "X" stood for "unknown," declining to call them a new kind of light or electricity. Others called them "Röntgen rays." Scientists immediately began experimenting with the rays. Newspapers began publishing popular articles with ghostly photographs of animal skeletons. Medical professionals started using x-rays to aid in diagnosis, and soon a medical student discovered that they could also be used to treat tumors. In February, the first radiation therapy facility was founded.

Thomas Edison, catching the excitement, began to develop an x-ray light bulb. In May, a prototype was demonstrated at New York's Electrical Exhibit, allowing the public to look at their own bones. Soon after, Edison's head researcher, Clarence Dally, developed lesions all over his body and his hair fell out. Edison immediately terminated the project, but Dally continued. Eventually, his burns became cancerous, and both of his arms were amputated in an attempt to save his life. He died in 1898 at the age of thirty-nine.

When Röntgen won the 1901 Nobel Prize, he still didn't know what the rays were. The Nobel committee referred to them as "the remarkable rays subsequently named after him." It was speculated that they might be high energy electromagnetic waves. This was confirmed by the German physicist Max von Laue with crystal diffraction, who won the 1914 Nobel Prize for this work. In 1927, Arthur Compton was awarded the Nobel Prize for demonstrating that x-rays are also particles.

In the same year that Röntgen's rays were made famous, another kind of ray was discovered by a French physicist working in Paris, Henri Becquerel. Becquerel was experimenting with phosphorescent substances, following in the footsteps of his father and grandfather.

Soon, quantum theory would explain these glowing substances in terms of energy levels, but in 1896, they were still a scientific mystery.

Inspired by Röntgen's use of photographic plates, Becquerel tried similar techniques with phosphorescent materials. At the time, he was experimenting with uranyl potassium sulfate, exposing it to sunlight and then photographing its emission. However, one day he developed a plate that had been lying next to uranium salt in a dark drawer. To his surprise, the picture showed that the salt had been radiating. He left the uranium salt in the dark for months, but still it continued to radiate! The invisible rays seemed to come from *within* the matter. Not observing this effect with any other salts, Becquerel assumed that the rays came from uranium and called them "uranic rays."

With so much excitement surrounding electromagnetism and the recently discovered x-rays, few paid any attention to Becquerel's uranic rays. However, a Polish graduate student working in Paris, choosing to avoid the lengthy background research that would be involved in studying either of these fields, decided to probe this new area. Her name was Marie Curie.

Curie was born Marja Skłodowska in Warsaw in 1867, which at that time was part of the Russian Empire. Three years earlier, Polish nationalists had attacked the Russian government in Poland and were defeated. The Russians, vengeful, began a process of "Russification" that targeted the Polish education system. One consequence of this was that women were no longer allowed to enter universities.

Polish women did not accept this subjugation and banded together to create an education collective, where women taught each other what they knew best. It was called the "Floating University" because they continually changed the location of classes to avoid getting caught. This is where Marja Skłodowska went to university, supplementing her "formal" education with extensive reading. She also took an advanced mathematics course by mail with the help of her father, and she persuaded a factory chemist to teach her chemistry.

In 1891, Skłodowska went to Paris, filling out her registration card at the University of Paris as "Marie" Skłodowska. Three years later, she had filled in the many gaps in her education, earned a masters degree in physics and mathematics, and gained fluency in French. After completing her second degree, she returned to Poland, hoping to earn her doctoral degree in her homeland. Unfortunately, the situation in Poland made this impossible, and soon after, she returned to Paris.

While in Poland, Skłodowska was courted by the French scientist Pierre Curie, whom she had been introduced to in Paris when looking for lab space. She resisted, saying that she wanted to stay in Poland to help her country. He wrote to her in simple Polish, offering to move to Poland and learn her language. He offered her a shared life devoted to science. Soon after returning to Paris, Marie and Pierre married. Marie dressed practically, in a dark suit over a blouse, so she could go work in the laboratory right after the wedding.

Toward the end of 1897, after giving birth to their first child, Marie Skłodowska Curie decided on the topic for her dissertation, the study of Becquerel's uranic rays. Using a piezo electrometer that Pierre helped to invent, Marie could quantify the radiation coming out of the uranium. She confirmed Becquerel's results that heating, cooling, wetting, drying, illuminating, compressing, and pulverizing didn't change the radiation; the amount of radiation emitted only depended on the quantity of uranium. She went further in her interpretation of this evidence than Becquerel, who had assumed that the radiation came from inside the matter. She speculated that the radiation did not come from chemical reactions, but from some change within the atom. If this were true, she concluded, we would be forced to abandon the assumption of the immutability of atoms.

Experimenting with other minerals, Curie found that torbernite was twice as radiant as uranium, and pitchblende, four times as radiant. She concluded that there must be another more powerful element within these substances. Pierre, fascinated, abandoned his own research on crystals and joined her. They purchased seven tons of pitchblende from Bohemia and constructed a primitive laboratory from a hut with a glass roof that had formerly been used by medical students for cadaver dissections. With no ventilation in the hut, they did most of their work in the adjacent courtyard, carting the materials and equipment back and forth.

In July 1898, the Curies announced a new element found in pitchblende, which they named "polonium," the Latin name for Poland. They also coined a word for elements that spontaneously radiate, calling them "radio-active" elements. By the end of the year, they had isolated and identified another radioactive element, which they called "radium." They continued this work for another three years out of their little laboratory hut, Pierre investigating their properties and Marie experimenting with large-scale ways to isolate the new elements. In 1902, using about 40 tons of chemicals, they finally succeeded in obtaining a tenth of a gram of pure radium chloride, enough to determine radium's atomic weight.

In 1903, the French Academy of Sciences nominated Henri Becquerel and Pierre Curie for the Nobel Prize in Physics, with no mention of Marie. The Swedish mathematician and women's rights activist Gösta Mittag-Leffler noticed this omission. He made inquiries, pulled some strings, and in the end, Henri Becquerel shared the prize with both Marie and Pierre Curie. This made international news, and both the Curies and the Nobel Prize became famous. And Marie Curie became an inspiration for women throughout the world.

Thus began the public's love affair with the Curies and their precious radium. *The Cosmopolitan*, an international women's fashion magazine, called the element "life, energy, immortal warmth." After someone noticed the invigorating effect of radium on mice, all kinds of radium-laced products appeared: bath salts, chocolate, bottled water, and so on. Radium was also believed to be a cure-all and was used to treat such ailments as diabetes, epilepsy, heart disease, hysteria, infection, rheumatism, and senility. The recklessness with which the public embraced radium could only be rivaled by its previous infatuation with x-rays, and the story had a similar ending.

In 1900, Pierre Curie, hearing of reports from a dentist on the biological effects of radium, performed similar experiments on himself. He was excited to discover that radium caused a burn-like lesion on the skin that developed progressively and took months to heal. Therefore, like x-rays, radium could be used to treat cancerous tumors with the advantage that it didn't require any sophisticated electrical equipment: just insert a tiny bit of radium into a tumor, and the radiation would kill the cancerous cells. Because of the short range of its radiation, it would cause minimal damage to healthy cells.

The technique proved to be very successful. However, Pierre soon became sick from an unknown cause; the doctors thought he had rheumatism. Many years later it was discovered that the energizing effect of radium that had been observed was actually due to an increase in red blood cells trying to defend against the poisonous element, which both cured and caused cancer. Because of his illness, Pierre was too sick to go to Stockholm in 1903 to receive the Nobel Prize. Marie, too, declined to go to the ceremony. Pierre never recovered and died in 1906 in a street accident with a carriage.

Marie Curie would eventually die from prolonged exposure

to radiation, but she was alive and well to receive a second Nobel Prize in 1911, this time in chemistry. It was a solo award, although it was for the work she had done with Pierre. Consistent with the spirit of the award — to ensure that those who have made great contributions have the financial independence to continue their work — the Nobel Prize could not be awarded posthumously. Marie traveled to Stockholm to receive the prize. Afterwards, with the help of the prize money, she continued her research, focusing almost exclusively on the use of radium in cancer treatment.

Curie's eldest child, Irène, accompanied her to Stockholm in 1911. In 1935, Irène won her own Nobel Prize — a joint award with her husband Jean Frédéric Joliot for the artificial production of radioactive elements. Although Marie was not alive to celebrate this award with her daughter, having died the year before, she was alive to celebrate the achievement. After synthesizing radioactive phosphorus, Irène and Frédéric brought it to her in a tube. Her face lit up. She checked the radiation levels with a radiation counter, heard it clicking, and declared to her daughter, "We have returned to the glorious days of the old laboratory!"


In 1906, J.J. Thomson won the Nobel Prize in Physics for his discovery that cathode rays are made of particles called electrons. About three decades later, his son George Thomson shared the Physics Prize with the American physicist Clinton Davisson for demonstrating that an electron is also a wave.

The 1908 Nobel Prize in Chemistry was awarded to Ernest Rutherford for his research on the disintegration of elements and the chemistry of radioactive substances, work that he began in 1897 while in Cambridge with J.J. Thomson. Rutherford, interested in radiation, had followed up on Becquerel's early experiments with uranium. He discovered two different kinds of radiation, which he called alpha and beta radiation. Later, Becquerel determined that beta radiation was made of electrons, and Rutherford found that alpha radiation was composed of helium nuclei.

Rutherford also discovered that each radioactive substance has a characteristic half-life, the time it takes for half of a sample to decay. In this way, he introduced probability into the theory of radioactivity. Furthermore, he determined that alpha particles don't have enough kinetic energy to escape the nucleus. Quantum theory would soon clarify this and connect it with the concept of half-life: if all matter acts like particles and waves, the particles that make up the nucleus don't have definite positions or velocities. So, although classically the alpha particles *cannot* escape the nucleus, the uncertainty inherent in nature makes it possible.

Rutherford was also one of the first to speculate on the nature of radioactivity, proposing that it is the process of transmutation of elements. For example, if a nucleus ejects an alpha particle that has a charge of +2, it should turn into an element two steps down on the periodic table. In this way, Rutherford discovered decay series, one radioactive substance decaying into another until the products of the decay are stable. In 1919, he produced the first artificial nuclear reaction, bombarding nitrogen with alpha particles to create hydrogen and oxygen.

Rutherford had several research partners throughout his years in Manchester and afterwards in Cambridge, where he took over J.J. Thomson's position as director of the Cavendish Laboratory. Although none of these collaborators shared his 1908 Nobel Prize, two of them received their own award. First, Frederick Soddy received the 1921 Nobel Prize in Chemistry for his discovery that atoms of the same element can have different masses. He called these *isotopes* from the Greek word meaning "same place," because the isotopes are at the same place on the periodic table. Another of Rutherford's collaborators, James Chadwick, explained the existence of isotopes with his discovery of the neutron, a massive but electrically neutral particle in the nucleus of an atom. Isotopes of a given element have the same charge but different numbers of neutrons and, therefore, different masses. Chadwick received the 1935 Nobel Prize in Physics for this discovery.

Many Nobel Prizes in Physics, starting in 1902 and continuing for decades, were awarded for research related to the development of the quantum theory of the atom. In 1902, Hendrik Lorentz and Pieter Zeeman won the prize for their discovery of the Zeeman effect: the splitting of spectral lines in a magnetic field. In 1919, Johannes Stark won the Prize for a similar discovery, the splitting of spectral lines due to an electric field, called the Stark effect.

In 1911, Wilhelm Wien won the Nobel Prize for his discovery of the laws governing blackbody radiation, and in 1918, Max Planck won the award for his discovery of energy quanta. Einstein won the 1921 Prize for his discovery of the "law of the photoelectric effect," although at this point, his conceptual explanation of the photoelectric effect in terms of particles of light was not yet accepted. The award was also "for his services to Theoretical Physics," which was likely a nod to his relativity theory. In 1923, Robert Millikan won the Prize for his experimental work on the photoelectric effect, which confirmed Einstein's predictions even though Millikan did not believe Einstein's quantum hypothesis that light can sometimes be viewed as a particle.

Bohr won the 1922 Physics Prize for his quantum model of the atom, and de Broglie won it seven years later for discovering the wave nature of matter, giving a conceptual base to Bohr's model. Heisenberg was awarded the Prize in 1932 for the creation of quantum mechanics. The next year, Schrödinger and Paul Dirac shared the Nobel Prize in Physics, the former for his quantum theory of the atom and the latter for his relativistic version of quantum mechanics that included the spin of the electron.

Enrico Fermi was born at the turn of the century in Rome and grew up in an unheated apartment near a train station. He was almost entirely self-taught, learning through books and his own exploration. He built motors and drafted detailed technical drawings of airplanes. He scavenged for math and physics books at the flea market in Campo de' Fioris, where three centuries earlier Giordano Bruno had been burned at the stake.

Fermi received a full scholarship to the University of Pisa, where there was hardly a physics program. In a way, this was an advantage — Fermi and the other two physics students were allowed full use of the research laboratories, were given keys to the library and instrument cabinets, and received permission to try any experiment that they could conceive of and had the means to perform. Fermi already had abundant knowledge, which he shared with the other students. He even taught one interested professor Einstein's theory of relativity.

Fermi studied x-rays, improvising much of the equipment needed for his experiments. He graduated with a laurea at the young age of twenty. The Italian Ministry of Education offered one fellowship a year for postdoctoral studies, which he won. With this, he studied for seven months in Göttingen with Max Born alongside

Heisenberg and Pauli. When Fermi returned to Italy, he became a lecturer of mathematical physics and mechanics at the University of Florence.

In 1927, Fermi was recruited to join the faculty at the University of Rome by its dean of physics, Orso Mario Corbino, who was determined to begin a world-class program in Italy. Franco Rasetti, one of Fermi's friends from the University of Pisa, joined them, as did several talented students. The group immediately made such an explosion in international publishing that a physics conference held in Italy that year attracted such acclaimed physicists as Rutherford, Planck, Bohr, and Heisenberg. The Italians were then honored with various international invitations, especially from Germany.

The department was housed in a monastery on a small hill in the middle of Rome, surrounded by palm trees and bamboo thickets. The group grew so close that the members developed their own accent, which evolved from college friends Fermi and Rasetti playfully imitating each other. They also had nicknames. Fermi, the first professor in Rome to teach quantum mechanics, was "the Pope" since it took profound faith to believe that matter and light are both particles *and* waves and that mass can become energy and energy can become mass. Rasetti was "the Cardinal," and their colleague Emilio Segrè, known for his temper, was "the Basilisk."

In 1932, Chadwick discovered the neutron, and two years later, Irène and Frédéric Joliot-Curie created the first artificially radioactive elements by bombarding atoms with alpha particles. These discoveries gave Fermi an idea. What if he bombarded atoms with *neutrons* instead of alpha particles? Neutrons, being electrically neutral, should be able to "sneak up" on a nucleus more easily than alpha particles, which, because of their positive charge, are strongly repelled by the positive nucleus. The Joliot-Curies could only produce radioactive nuclei from lighter atoms like boron, magnesium, and aluminum. With a neutron, it should be possible to penetrate even the largest atoms.

In order to test this, Fermi and his team gathered samples of elements from the largest chemical supply store in Rome: aluminum, carbon, platinum, gold, and many others — about sixty different elements in total. One by one, they tested each element. They wrapped the sample around a glass tube that contained the source of the neutrons, made from beryllium and radon, and enclosed it all in a lead box. They placed the bombarded sample in one room and, to avoid contamination from the radioactive source, put their homemade radiation counter in another room. Since excited nuclei often decay in less than a minute, the scientists would then race through the adjoining hallway to take data, Fermi as fast as any of them.

They started with the lightest elements, which all failed to become radioactive until they got to the ninth element, fluorine. From then on, they had astounding success — out of the sixty elements tested, forty could be transformed into radioactive isotopes. They marched up the periodic table, ending with the ninety-second element, uranium. They bombarded it, and preliminary results suggested that they had created an element with ninety-three protons, unlike any element found on Earth.

Fermi wasn't sure of this result and kept turning things over in his mind. He thought he should test the effect of inserting a piece of lead between the neutron source and the target. He carefully manufactured a piece of lead but kept postponing putting it into place. Finally, on a whim, he took a piece of paraffin wax and placed it where the lead should have been. The radioactivity levels soared.

Fermi made his discovery in the morning, and by lunchtime, he had shared with his group what he believed to be the reason for the morning's results: the hydrogen in the paraffin slowed down the neutrons, making them more likely to be absorbed by neighboring nuclei. After their siesta, the group tested Fermi's idea by using the most hydrogen-rich source they could find — they plunged the neutron source and the target into a goldfish pond. Again, the radioactivity levels soared!

In 1938, Fermi was awarded the Nobel Prize in Physics for his work on radioactivity. Although the Prizes were usually kept secret, he knew that he had won even before receiving the call from Stockholm; Bohr had told him. A few months earlier, following in the footsteps of Germany, the Italian dictator Benito Mussolini had launched a campaign against Italy's Jews. Although Fermi was Catholic and his children were raised Catholic, his wife was Jewish. With Bohr's forewarning, they had time to plan their escape. They traveled to Stockholm with a nursemaid for their two young children so that Fermi could accept his Prize. With the prize money, they booked passage on an ocean liner headed for New York City, where they began a new life.

In January of 1937, Hitler announced a law prohibiting any German national from receiving a Nobel Prize. He had been provoked to this decision by the awarding of the Peace Prize to the journalist Carl von Ossietzky, a pacifist who openly spoke out against the Nazi regime. In 1939, three Germans won Nobel Prizes — two in chemistry and one in medicine. The three, under various threats, were forced to refuse the awards. All efforts from Stockholm to persuade Hitler to reconsider his position regarding the Prizes failed. In response to these refusals and the outbreak of world war, the 1939 Nobel award ceremony was canceled.

The Nobel Prizes were either canceled or reserved until the end of the war in 1945, but the Nobel committee continued their work. They sent invitations to nominators, although many did not reply. They also continued to evaluate and rank candidates. Moreover, during this time, scientists continued to do science, and many also worked toward peace.

Bohr, with his wide network of friends and colleagues around the world, had been helping to protect targeted scientists since Hitler's rise to power in 1933. He sought out scientists who might be dismissed because of new "racial laws" and found them employment elsewhere, often in newly created positions. Many of them went to England or the United States. Others went to Sweden or took a position at Bohr's Institute in Copenhagen.

All records relating to these activities were burned in the early hours of April 9, 1940, when Germany occupied Denmark. Suspecting his involvement, the German authorities repeatedly tried to trap Bohr. The Resistance movement urged him to leave again and again. The United States tried to lure him to their country, offering him the use of the best research facilities and assistance with travel arrangements. In early 1943, he received a secret message from James Chadwick, delivered on a tiny microfilm inside a key, with a plea to come to England. And yet, he stayed, believing he could still be of help in his country and fearful lest his departure cause difficulties for others once the Germans discovered that he had fled.

In September of 1943, Bohr received a message from the Swedish ambassador in Copenhagen that he and his brother Harald were in danger of immediate arrest by the Germans in connection with the deportation of Jews from Denmark. The Danish Resistance arranged for Bohr and his family to escape by boat to Sweden that very night. Then, riding in the bomb compartment of a little airplane, they passed over German-occupied Norway to London. From London, they eventually made their way to the United States.


One of Bohr's earliest rescue missions was to a young Austrian-born Jewish physicist working in Hamburg named Otto Frisch. During a trip to Germany in 1933, Bohr sought him out in his laboratory, approached him, took him by his waistcoat buttons, and invited him to come and work at the Institute for a while. That evening, Frisch wrote to his mother, "You are not to worry about me any longer. Our Lord Himself has taken me by the waistcoat buttons, and He smiled at me."

Frisch was the nephew of Lise Meitner, another physicist working in Germany. Meitner had come to Berlin in 1907 after completing her doctoral degree in Vienna. By this time, she had already published original scientific research, which she had sent to Max Planck at the University of Berlin and Heinrich Rubens, the head of the University's Department of Experimental Physics. She requested that she be allowed to attend lectures and set up a workspace for the continued study of radioactive processes.

Although Rubens could not offer her a private workspace, she was welcome to work in his laboratory under his supervision. Meitner, however, did not want to work *under* Rubens or any other senior scientist. Fortunately, Rubens suggested another option — the chemist Otto Hahn would be interested in working with her. He was the same age as Meitner and had a good reputation in the field of radioactivity.

At that time, Hahn was working across town at the Berlin Institute with Emil Fischer. Fischer agreed that he and Meitner could use a small room that had been designed as a carpenter's work area for their experiments. Hahn soon had it adapted for use as a laboratory. As soon as the room was ready, they set up their equipment — three electroscopes for counting alpha, beta, and gamma rays. Here, Meitner and Hahn conducted their own experiments, working in the evenings after they had finished up at their respective institutions.

They worked well together, Meitner's mathematical skill and creative thinking complementing Hahn's patience and attention to detail. They published, both individually and jointly, and gained international recognition for their work. They were also able to expand. By 1910, they had three rooms in the Institute at their disposal. In

1912, Planck offered them both positions in the newly founded Kaiser Wilhelm Institute of Chemistry, making Hahn the head of the radioactivity laboratory and Meitner a "guest physicist." The following year, Meitner became a paid "Scientific Associate," an equal to Hahn, and the radioactivity section became theirs. In 1918, Meitner became the head of her own radiophysics department, while Hahn became the head of the radiochemistry laboratory.

At the same time that Meitner was growing her research space, she was solidifying a place for herself at the University of Berlin. Just arrived from Vienna, Planck welcomed her and introduced her to other scientists, most importantly to his assistant Max von Laue, another scientist about her age. She made additional acquaintances during the informal Wednesday physics colloquiums, initially led by Rubens. In 1909, Planck invited her along with Hahn and von Laue to attend the Congress of Natural Sciences in Salzburg, Austria, where Einstein was the guest of honor.

In 1912, Planck formally gave Meitner a position as his assistant, succeeding von Laue — she was now in Planck's inner circle of scientists, soon to be joined by Einstein when he moved to Berlin two years later. In 1922, she became a lecturer at the University, and four years after that, an assistant professor, the first female physics professor in Germany.

On April 7, 1933, an anti-Semitic "Law for the Restoration of the Career Civil Service" was passed. Within the next few weeks, hundreds of professors, scientific researchers, and assistants throughout Germany were dismissed, and many others left on principle. On May 10, students and professors from the University of Berlin, dressed in brown shirts, threw twenty thousand books deemed "un-German" from the university library into a bonfire — books by Einstein, Karl Marx, Kafka, Proust, Thomas Mann, Helen Keller, and many others.

In mid-May, Planck arranged a visit with the new Chancellor of Germany, Adolf Hitler, trying to bring him to reason. Planck argued that many of the greats of German culture along with many of the oldest families were Jewish. Moreover, expelling Jewish scientists would not only be a loss to Germany, but also a gain for other countries and potential enemies. This sent Hitler into a rage: "If the dismissal of Jewish scientists means the annihilation of contemporary German science, then we shall do without science for a few years!" As Hitler talked faster and faster, Planck withdrew into silence.

Although Jewish, Meitner initially kept her position at the

University of Berlin; as an Austrian, she was exempt from the law, which only applied to German citizens. However, in September, she was abruptly dismissed from the University and banned from presenting her scientific research in Germany. Planck and Hahn both wrote letters in protest: She should be considered as having tenure at the University. She was a veteran; she served on the front line in the world war as an x-ray technician. She was a leading radium researcher both in Germany and internationally. Despite these efforts, the authorities would not reverse their decision.

The swastika was hung outside the Kaiser Wilhelm Institute, and most of the chemistry department was staffed by party members devoted to weapons research and development, but Meitner kept her position there and buried herself in work. In 1934, new inspiration came when she heard about a discovery made in Rome.

Enrico Fermi and his group had bombarded uranium with neutrons to create artificial radioactivity. With this method, they had been able to induce radioactivity in atoms even with the most positively charged nuclei, including uranium. Furthermore, it was possible that in the bombardment of uranium they had created transuranics, elements with atomic numbers larger than uranium. However, they had not yet carried out a careful search for such particles.

This was a challenge, one that Meitner eagerly accepted. She urged Hahn to join her. They hadn't published a paper together in ten years, each busy in their respective laboratories. But now, she would need his chemical expertise to help analyze the radioactive products. Hahn agreed, reasoning that any new radioactive elements they discover might eventually have a medical use, which would please their industry sponsors.

Meitner and Hahn immediately got to work. They first confirmed the results obtained in Rome, but then more careful analysis confused rather than clarified the problem. They found substances with three distinct half-lives. They couldn't figure out how they were related; they didn't fit into any known decay series. Fermi's team found the same thing — decay products with three distinct half-lives.

There were several new theoretical models attempting to account for the observations. Von Weizsäcker, who worked with Meitner and Hahn, suggested that the nucleus is made of stacked alpha particles arranged in a non-spherical lump like a "nuclear sausage." Bohr and a couple of colleagues proposed a more sophisticated, compound nucleus, where the space inside the potential barrier created by the electric charge of the nucleus was a sort of "mush."

The German chemist Ida Noddack had a completely differ-ent idea. First, she challenged Fermi's assumption that the products they were finding were transuranic — to make that claim, one would first have to eliminate the isotopes of all known elements, not just elements down to lead as Fermi had done. She then suggested that when heavy nuclei are bombarded with neutrons, the nuclei should break apart into several large fragments that are isotopes of known elements but not neighbors of the irradiated elements as all were assuming.

Meitner and Hahn found this last suggestion ridiculous and continued their work searching for transuranics, publishing eight pa-pers on the subject between 1935 and 1936. Then everything became even more confusing. In Paris, Irène Curie and her collaborator Paul Savitch reported another product emitted when uranium was irra-diated with neutrons, a short-lived particle with a 3.5-hour half-life resembling an isotope of thorium.

How could they have missed thorium, two steps down on the periodic table from uranium, an element that would result from a simple alpha decay of uranium? Fritz Strassmann, the principal assistant who had been working with Hahn and Meitner on this project, got to work reanalyzing the data but again failed to find any radioactive thorium. In January of 1938, they wrote to Curie, noting the details of their own experiments and politely suggesting that she had made a huge mistake. They advised her to make a public retrac-tion so they would not have to publish their criticism.

Then, much to their surprise, researchers from the Universi-ty of Michigan, with a powerful neutron source generated with the help of a cyclotron, confirmed the Paris group's discovery of a 3.5-hour half-life product and found many others that had never before been reported. The Vienna Radium Institute began looking for alpha particles, which should have been released in the transmutation of uranium into thorium, but found none. At Berkeley, researchers also failed to find alpha particles and were perplexed by the "unorthodox" decay schemes for the transuranic elements, which were completely unlike the decay patterns of other heavy elements. They began look-ing at x-ray spectra as a way to unequivocally identify the numerous radioactive products, and researchers at Cambridge soon followed suit.

Another article came out from the Paris group, claiming that the 3.5-hour half-life substance actually resembled actinium. Then, in the fall of 1938, they retracted that conclusion as well, suggesting that the substance was really lanthanum, an element nearly halfway down the period table from uranium.

On September 15, 1935, the Nuremberg Laws were passed, depriving all Jews of German citizenship. These laws were enforced with rage and terror — business owners, schoolchildren, and prominent men and women across Germany were arrested, detained, and brutalized for no reason except their background or political views. Throughout, Meitner carried on with her work at the Institute and even traveled abroad to give a lecture in Switzerland, protected by her Austrian citizenship.

Then in March of 1938, Austria was annexed by Germany. Almost immediately, friends and colleagues from around the world reached out to help Meitner. She was offered a lecture position in Zürich, with all expenses paid for her relocation. She was sent a pledge of financial support from a colleague at the University of Chicago, the first step toward an application to emigrate to the United States. She also received a letter from Bohr, friendly but serious, inviting her to Copenhagen.

Meitner decided to go to Copenhagen, but then the Danish consulate declared her Austrian passport invalid and refused her entry. Carl Bosch, who had recently succeeded Planck as the president of the Kaiser Wilhelm Society, wrote to the Reich Minister of Education on her behalf, requesting that she be allowed to travel to a neutral foreign country. To support his request, he cited her international scientific reputation. Soon after, he received the following response: "It is considered undesirable that renowned Jews [underlined!] should leave Germany for abroad to act there against the interests of Germany according to their inner persuasion."

Meitner continued day after day, going to the Institute and spending long hours in the laboratory with Hahn and Strassmann, where they continued their work on transuranics. On July 4, Bosch notified Hahn that the policy prohibiting Jewish scientists from leaving Germany was being strictly enforced and warned that Meitner was in danger. Although Bohr, by this time, had found her a safe haven and job in Sweden, no visa or entry permit had yet arrived. And so, in an international collaborative effort, Meitner was smuggled out of Germany into safety, through Holland and Denmark, arriving in Sweden in late August.

Meitner and Hahn wrote letters back and forth. He updated her on Institute affairs and his research; she expressed frustration at not having the proper equipment to conduct her research. Meitner was now employed at the new Research Institute for Physics in Stockholm, where they were constructing a building that would eventually be equipped with a cyclotron and many large x-ray and spectroscopic apparatus. At the time, though, the building was still a long way from completion, and in the meantime, she only had a small space without basic equipment or a suitable collaborator.

In November, Meitner and Hahn met briefly at Bohr's Institute where Hahn was giving a lecture, sharing his most recent experimental results: the observation of radium-like products during the irradiation of uranium. Afterwards, he had to hurry home to take care of personal matters. In December, just before heading to the coast of Sweden for the Christmas holiday, Meitner received a letter from Hahn that threatened to invalidate their last four years of work with transuranics.

He began the letter, "The thing is: there is something so ODD about the 'radium isotopes' that for the moment we don't want to tell anyone but you." Hahn then described the results of experiments that he and Strassmann had performed, prompted by Meitner. More and more, they were coming to the "awful" conclusion that the radium-like isotopes behave not like radium, but rather like barium — an element halfway down the periodic table from uranium! Perhaps Meitner might be able to suggest some "fantastic" explanation. They themselves realized that uranium cannot just burst into barium.

Meitner took this as a challenge. She continued to think about it when she arrived in Kungälv, a village on the coast of Sweden, where she was spending the holidays with friends and her nephew, Otto Frisch. Preoccupied, she invited Frisch for a walk in the woods; he skied through the snow while she walked briskly beside him. Meitner quickly dismissed the idea that they had made a mistake. Hahn was too good of a chemist. But how could barium be formed from uranium?

According to current models, the nucleus was not a brittle solid that could be cut up or broken. Bohr had worked on a model in which the nucleus is more like a liquid drop. Could the droplet burst apart? They began to speculate — perhaps a droplet could divide into two gradually, first elongating and then eventually breaking in two. The key, Meitner suggested, was in its surface tension. Electrical charge was known to decrease the surface tension of a nucleus. Would the charge of a uranium nucleus be enough to allow it to divide?

At this point, they stopped walking and sat down on a snow-covered log. Meitner took out a pencil and a scrap of paper from her coat pocket, and she and Frisch began calculating. Yes, the charge of a uranium nucleus was large enough for it to overcome its surface tension almost completely, so a uranium nucleus might be like a very unstable drop, ready to split at the slightest impulse.

Assuming the nucleus did split apart, the products would experience a great repulsion due to their like charges, propelling them at high speeds. This should result in about 200 million electron volts of energy. Where could that energy come from? Meitner continued to calculate. The combined mass of the two resulting nuclei would be smaller than that of the original nucleus by about one fifth the mass of a proton. Using Einstein's mass-energy equivalence, the energy released would be about 200 million electron volts — everything fit!

After returning to Copenhagen, Frisch paid a visit to Bohr, who was getting ready for a trip to the United States. Frisch reminded Bohr of the liquid droplet model of the nucleus and shared the ideas that he and Meitner had discussed. Bohr agreed in every way and thought that they should publish immediately! Frisch called his aunt, and they decided to write an article for *Nature*. He quickly started a draft and, upon Bohr's request, rode out to his house so they could go over everything more carefully. Frisch went home and typed up the first two pages of the paper, which he handed to Bohr the next day at the railroad station just before he departed for the United States.

On January 16, the same day that Frisch mailed the completed paper off to *Nature*, Bohr's ship docked at the West 57th Street pier in New York City. He was met by twenty-seven-year-old John Wheeler, an American physicist who had spent a year in Copenhagen as a fellow and was now working at Princeton. The two had worked together on the compound model of the nucleus — Bohr excitedly described the new experiments and the new interpretation. He was also warmly met by Enrico and Laura Fermi who were just settling into their new life, grateful for Bohr's help in their recent escape from Italy.

Fermi and his new colleagues at Columbia University quickly got to work bombarding uranium with neutrons, verifying the existence of fast-moving fragments consistent with Meitner and Frisch's theory. Frisch had already performed such experiments in Copenhagen, and Meitner, once she was able to gather together the appropriate equipment, did the same in Stockholm. Additional confirmation came as the news spread like wildfire through the scientific community. The *Times* magazine blasted the news to the American public through a sensational announcement: two Germans had split open an atom.

But not everyone was so eager to have this news spread. Notably, the Hungarian physicist Leo Szilard worried about the implications of such a discovery. He had been a younger colleague of Meitner at the University of Berlin, dismissed at the same time. Upon departing Germany, Szilard moved to the United States. Although unable to find an academic position, he continued to follow Meitner and Hahn's nuclear research through scientific journals. When he read the article written by Meitner and Frisch in the mid-February issue of *Nature*, he took the opportunity to share his concerns.

Szilard thought that it was likely that neutrons are released when a uranium nucleus breaks apart. In this case, it should be possible to create a chain reaction — the neutrons released from one reaction could cause other reactions, and so on — which could lead to the construction of a powerful bomb. In the current era, dominated by fascist leaders bent on racial domination, the use of such weapons could change the course of history. In short, Szilard believed it was imperative to keep any information that might be used in the construction of such a bomb away from fascist countries.

Szilard first sought out Fermi, who had voiced similar ideas about the possibility of a chain reaction. Fermi, however, did not seem concerned; he estimated the probability of a sustained chain reaction at only around ten percent. Szilard, not finding comfort in this estimation, then reached out to Frédéric Joliot in Paris, suggesting that research surrounding the fission of nuclei should only be discussed privately among scientists from England, France, and the United States and that there should be no publications in the case that a sustained chain reaction might be possible. Joliot ignored this advice.

Meanwhile, the political situation in Europe was deteriorating. In March, Hitler's army invaded Czechoslovakia, forcing its president to accept German "protection." In April, Mussolini's army conquered Albania. With Europe on the brink of collapse, Szilard decided to take stronger action. He sought out a former colleague from

Berlin, Albert Einstein, who was vacationing on Long Island.

Einstein listened with interest to Szilard's concerns and then with his help drafted a letter to President Roosevelt. In the letter, Einstein informed Roosevelt of the likelihood that a chain reaction could be set up in a large mass of uranium, which could lead to the development of an extraordinarily powerful bomb. Furthermore, he wrote that Germany had stopped the sale of uranium from the recently acquired Czechoslovakian mines. Other sources of uranium were in Canada and, most importantly, in the Belgian Congo. Finally, he suggested that Roosevelt set up permanent communications between his administration and the scientists working on such research.

Einstein's letter to Roosevelt was dated August 2, 1939. On September 1, Germany invaded Poland. Two days later, Great Britain and France declared war on Germany. On September 17, the Soviet Union invaded Poland from the east. Ten days after that, Poland surrendered and was divided between the two conquering nations. At the end of October, Einstein received a reply. Roosevelt thanked him for the interesting and important letter. In response, he formed a committee, which included representatives of the Army and Navy, to thoroughly investigate Einstein's suggestions regarding uranium.

On March 7, 1940, Einstein sent another, more urgent letter that he hoped would reach Roosevelt. Since the outbreak of war, he wrote, interest in uranium in Germany had intensified. Furthermore, von Weizsäcker, the son of the German Undersecretary of State, had been collaborating on nuclear research involving uranium with scientists at Kaiser Wilhelm's Institute of Chemistry. His leadership had now been extended to the Institute of Physics, and their research was being conducted in great secrecy.

Otto Frisch, who had taken a position in England after the outbreak of war, and another Jewish émigré, Rudolf Peierls, began to consider the feasibility of an atomic bomb. They estimated the amount of highly fissionable uranium, the isotope U-235, required for a chain reaction. They performed calculations pertaining to the separation of U-235 from the more abundant isotope, U-238. They addressed other technical difficulties like whether fast as well as slow neutrons could initiate a fission reaction. Their conclusion was that the construction of a bomb was not only possible, but that it was possible within a reasonable time frame given large-scale government funding.

Frisch and Peierls sent a summary of their investigations to their university leader, Mark Oliphant. Impressed, Oliphant suggested that they write a more formal letter, which they did, and Oliphant passed it onto the Royal Air Defense Council. This resulted in the formation of a top-secret committee that was code-named "Tube Alloys." Slowly, word of the British initiative made its way across the Atlantic, and in October of 1941, following Britain's lead, Roosevelt authorized preliminary research on the construction of an atomic bomb. In December, after the United States was attacked by Japan, American efforts began in earnest.

Initially, the research was centered at the University of Chicago. Fermi, who had made a first attempt at a self-sustaining fission reactor while at Columbia, joined the group. Glenn Seaborg, a physicist from the University of Berkeley who had recently discovered a new highly fissionable element called plutonium, came shortly after. They worked together with their colleagues to create a fission reactor that would be used to create the highly fissionable plutonium, which could be used as an alternative to U-235 as the fuel for an atomic bomb.

They made progress, but they weren't fast enough. In September, a Corps of Engineers brigadier general named Leslie Groves was appointed to run the program, which soon became known as the "Manhattan Project." Groves estimated that the first bomb would have to be ready to use no later than the summer of 1945 to have any chance of beating the Germans. The scientists had been debating the best way to cool the nuclear reactor for months. Should they use water or helium? Groves gave them five days to decide. On December 2, 1942, Fermi and his colleagues, in a squash court inside the abandoned University of Chicago football stadium, succeeded in creating the first sustained nuclear chain reaction.

Around the same time, the U.S. State Department confirmed rumors of the mass extermination of Jewish men, women, and children in Nazi-occupied Europe. American soldiers were enlisting in droves, many of whom were being killed across Europe and in Japan. Moreover, news had come that German scientists, headed by Werner Heisenberg, had succeeded in producing a chain reaction possibly several months earlier. There was no time to waste.

By that point, most of the science necessary for the construction of an atomic bomb had been worked out. Now, it was time for the production to begin. The Americans had uranium. Within a couple weeks of his appointment, Groves had signed a four-hundred-ton-a-month contract with a company that had stockpiled uranium on Staten Island. The uranium had come from the Belgian Congo; just before the German army had taken over the mines, the entire on-hand stock of uranium had been shipped out.

The Americans needed industrial sized reactors to produce plutonium for a plutonium bomb, and they needed industrial sized machines to separate out the U-235 for a uranium bomb. Under Groves's direction, hundreds of thousands of acres were used to set up secret towns across the country devoted to production. Families were evacuated, barracks were erected, and trailer parks were set up to make room for more than a hundred thousand men and women who were involved in some way in these efforts. In two years, enough fissionable U-235 and plutonium was produced for eighteen bombs.

Meanwhile, the British continued their own research, and in mid-1943, they merged with the American program. This brought Frisch and Peierls to Los Alamos, New Mexico, the center of the American efforts. Niels Bohr had recently escaped from Denmark and was brought to England, where the British scientists brought him up to date on the project. They then sent him off to Los Alamos, where he arrived in December of 1943.

Bohr brought with him a picture drawn by Heisenberg on a visit to Copenhagen in 1941 after the Nazis had occupied Denmark. Heisenberg had reached out to Bohr, who reluctantly agreed to meet with him. During dinner in Copenhagen, Heisenberg made a sketch of what he was working on and gave it to Bohr — a cylinder with lines sticking out. Bohr was filled with dread, convinced that it was a bomb. He now showed it to colleagues at Los Alamos, who all agreed that, rather, it was a nuclear reactor and control rods.

Earlier that month, British forces had bombed the Institute of Theoretical Physics in Leipzig where Heisenberg was working, destroying his equipment and all of his working scientific papers. They then headed to Berlin, where they bombed Hahn's nuclear fission laboratory. At this time, British intelligence was becoming more and more convinced that the Germans were not working on an atomic bomb. After Bohr left, Groves received an intelligence report that suggested the same. He shared this information with Robert Oppenheimer, the scientific head of the project. Oppenheimer shrugged his shoulders; they would finish what they began.

Einstein was not part of the project, but he knew enough about it to write another letter to Roosevelt in the spring of 1945, attempting to prevent the use of an atomic bomb against civilian targets. Roosevelt died on April 12; Einstein's letter was found unopened on his desk the following day. On May 7, Germany surrendered. A group of scientists working on the project wrote a report to the Secretary of War with a plea that the bomb be revealed to the world in an uninhabited area. But no one was listening to the scientists now.

On the morning of August 6, the four-ton uranium bomb "Little Boy" was dropped over the Japanese city of Hiroshima, killing over a hundred thousand people immediately. Almost as many were left wounded, and tens of thousands died in the aftermath. On August 9, the slightly larger plutonium bomb "Fat Man" with nearly twice the explosive power decimated the city of Nagasaki. Shortly after, Japan surrendered.


The war was now over, and the awarding of Nobel Prizes resumed. In November of 1945, the committees made their usual calls to the winners of the Prizes. They could not find Otto Hahn to let him know that he had won the 1944 Prize in Chemistry, which had been deferred, for his discovery of nuclear fission. He and nine other German scientists, including Max von Laue, Heisenberg, and von Weizsäcker, had been captured by the Allies and brought to Farm Hill, a country estate near Cambridge, England where they were kept under surveillance. Hahn learned of the Prize first through a BBC broadcast; he and his German colleagues celebrated.

Still detained at Farm Hill, Hahn was unable to attend the Stockholm award ceremony that year. Due to special considerations, he was able to receive his Prize at the 1946 ceremony. Meitner was in the audience to congratulate him. She had been nominated with Hahn for the chemistry award for over a decade and many additional times alone for the physics award. For a multitude of reasons, Meitner did not share this Prize with him, nor did she ever receive a Nobel Prize. However, in 1949, she and Hahn together were recognized in Germany with its highest award, the Max Planck medal. And in 1966, Meitner, Hahn, and Strassmann were awarded the prestigious Enrico Fermi Award.

The awarding of the Nobel Prizes continued. Pauli won the 1945 Prize in Physics for his exclusion principle. The 1949 Prize went to Hideki Yukawa for his theoretical work predicting the existence of a short-lived mediating particle based on Heisenberg's uncertainty principle. The following year, Cecil Power won the Prize for the detection of such a particle. In 1954, Max Born won the award for his statistical interpretation of the quantum mechanical wavefunction.

And on and on and on, all the way up to today, scientists continue to be awarded Nobel Prizes for their work verifying, extending, and applying the quantum theory of matter and light developed in the first three decades of the 20th century.

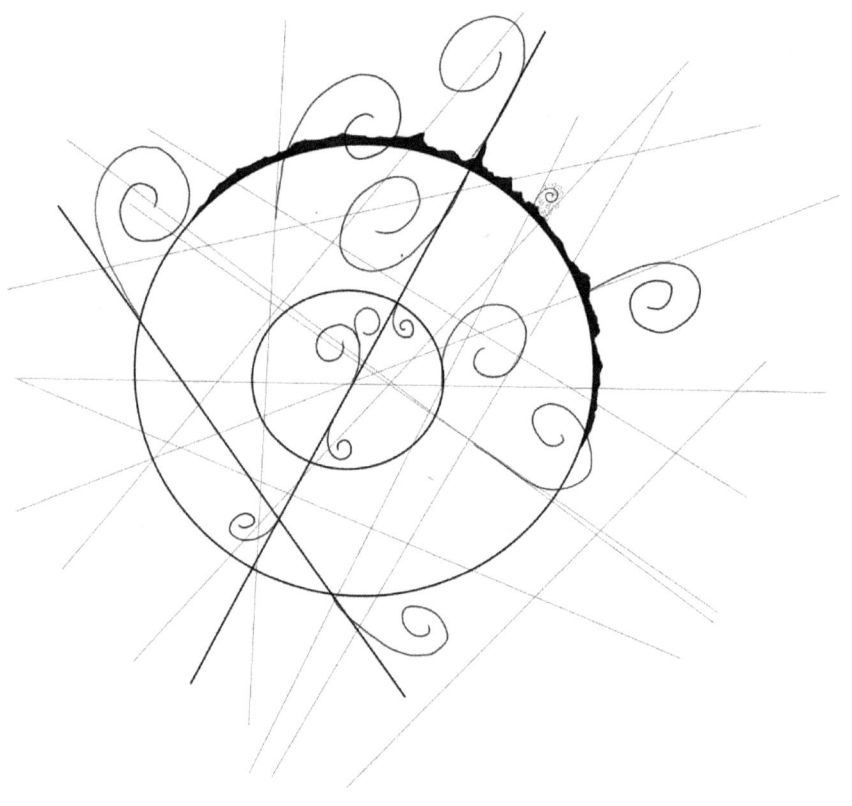

# BOOK III

# Frontiers
# of
# Physics

# 1

# A New Quantum

But it is human nature to believe that the
phenomena we know are the only ones that exist,
and whenever some chance discovery extends
the limits of our knowledge
we are filled with amazement.

— Marie Curie

Hideki Ogawa was born in Tokyo, but he grew up in Kyoto. It was here, in Kyoto, that he built his own Imperial City out of blocks. It was in Kyoto that he built miniature gardens, furnishing them with bridges, farmhouses, and shrines. It was in Kyoto that he studied the Chinese classics with his grandfather, laboring to memorize kanji even before he could understand them. And later in life, returning to Japan after spending many years in the United States, it was only when the train approached Kyoto that he felt like he was home.

Hideki's father was a geographer and geologist by profession, a professor at the Imperial University in Kyoto. Nevertheless, he had wide interests, and to support every interest, he would collect every possible book. His books would fill their library, spilling out into the other rooms until, eventually, he would decide that they must move to a larger home. In this way, the family, which included grandparents and eventually seven children, moved from one house to another, time and time again.

At home and in school, young Ogawa was quiet and withdrawn. He would respond to all complicated matters with *Iwan*, "I won't say." He kept his school desk with his notebooks, brushes, and inkwell exceptionally neat. He could not settle down to work at home unless his desk was oriented parallel to the lines of the tatami mats. He would hardly speak but was always first in his class; he wore a badge with a red ribbon for this honor.

Ogawa went to middle school with a liberal principal, teachers with a sense of humor, and interesting classmates. They gave each other nicknames in jest. Hideki's name was *Gonbei*, "no-name." He was sensitive and increasingly quiet, but he found joy in mathematics, especially geometry. The prospect of finally grasping the solution to a difficult problem gave him something to live for.

Near the end of the summer in 1922, the eleventh year of Taisho, the newspapers announced that Dr. Einstein would be visiting Japan. The trip had been arranged by the publisher of the magazine *Kaizo-sha*. A few days before his arrival, the magazine released its December edition, the "Einstein issue." Einstein's photo was on the cover, and within the magazine were fifteen articles about his life and work. Stories and pictures of Einstein continued to appear in Japanese newspapers throughout his tour of the country.

Einstein stayed one night in Kyoto at the beginning of his trip and returned toward the end to give a lecture on the principle of relativity. Ogawa did not attend; the event was only important in retrospect. At this time, during his last year of middle school, he was uninterested in anything outside of his head. He did not know who he was or where he was going; he felt like a boat without a rudder. When a classmate said to him, "You will become like Einstein," Ogawa had no idea what he was talking about. He had no thoughts of becoming a physicist, but for some reason, the comment made him happy. Recalling these events much later, he felt that they made a tiny crack in the sheet of ice that blocked his ship.

In high school, still without any clear goal, Ogawa filled the spaces in his day with reading — mostly literature and Taoist philosophy — and his thoughts centered on reading. Despite this, he became freer and more cheerful, even participating in interclass sports competitions and the cheering squad. During these years, his aspirations changed several times. Mathematics had been his favorite subject; he had felt excitement by the leaps of imagination required to solve difficult problems. Now note-taking in his mathematics class became a speed race, and he was required to present arguments not only with perfect reasoning but also that matched his teacher's proofs. As a consequence, his joy in mathematics waned.

Hideki was like a closed book to his father, so much so that he considered sending him to trade school after high school rather than university. His mother objected. His father then shared this idea with one of Hideki's teachers; the teacher offered to adopt the boy and pay for his university expenses himself. The father was satisfied that he should send Hideki to university. But what should he study?

One day, he handed Hideki a thousand-page college-level geology textbook written in English with many pictures and diagrams and said, "Read this book. If you find it interesting, then you should go into geology." Around this time, a questionnaire about college was passed around; Hideki wrote on the line that asked for the desired major: "geology." Meanwhile, his father continued to give him geology books. He quickly tired of them and left them unread. The next time a college questionnaire was passed around he indicated his intended major as "physics."

Ogawa took physics in high school with an American textbook that included sample problems at the end of each chapter. He began solving these problems ahead of the class, completing them as fast as he could to test his abilities. Soon classmates began asking him for help, more and more until it became very time-consuming. Then someone had an idea — Ogawa could give lectures. He agreed, and at seventeen years old, he began giving physics lectures to a classroom of twenty to thirty of his peers.

As Ogawa's interest in physics grew, so too did his dissatisfaction with the subject as it was taught at his school. He searched for books to extend his knowledge at the bookstore in Kyoto. One day, he found an English translation of a book written by the German physicist Fritz Reiche called *Quantum Theory*. In this book, Ogawa read about Planck's quantum and Bohr's atom.

The book was difficult to understand but interesting. Ogawa could feel that theoretical physics was in a state of confusion. At the conclusion of the book, Reiche wrote: "Over all these problems there hovers at the present time a mysterious obscurity. In spite of the enormous empirical and theoretical material which lies before us, the flame of thought which shall illumine the obscurity is still wanting. Let us hope that the day is not far distant when the mighty labours of our generation will be brought to a successful conclusion." Ogawa had never before received such stimulation or encouragement from a book.

In 1926, Ogawa entered Kyoto University to study physics. Soon after, he attended a lecture sponsored by the Society of Arts and Sciences titled "The Present and Past of Physics." The lecture was given by Hantaro Nagaoka, the most important scientist in Japan. Nagaoka spoke about the metamorphosis of physics that had occurred during the past two decades. After the talk, Ogawa bought a book on quantum theory written by Max Born. One of Born's students, Werner Heisenberg, had developed a new quantum theory. Born, recognizing its worth, worked with Heisenberg and another student to develop the theory, which he wrote up in a book. It was short, only a hundred and ten pages, but all of it was new.

Ogawa had never before wanted to be part of a team, but now he wanted to be part of this team. He wanted to contribute something to the new quantum theory. He began spending all of his spare time in the library. He attempted to read all of the journal articles on the new theory, but there were too many, and more and more continued to be written. Eventually, he decided to focus on the papers of Erwin Schrödinger, who had developed an alternative version of quantum theory that while equivalent to Heisenberg's was much easier to understand.

In his third year, Ogawa was joined in his studies of quantum mechanics by a classmate, Shinichiro Tomonaga. He now had a companion. They worked together to learn the theory, and Ogawa kept looking for ways he could extend it. He thought of combining quantum mechanics with relativity theory, but then he found that this had already been done by the British physicist Paul Dirac. The new theory kept moving forward at such a fast rate that he began to worry that his hopes of making a discovery in physics might amount to nothing. He briefly considered becoming a priest.

Ogawa graduated in the spring of 1929, after which he remained at the University as an unpaid assistant in the Physics Department. It was a time of great economic depression, and there were few jobs available for college graduates. Many of his classmates stayed on as well; the depression turned them all into scholars. Most let their hair grow long, but Ogawa kept his cropped short and continued to wear his old student uniform.

Ogawa worked with the Tamaki group, which he had joined in his final year of college. Tomonaga was in the group too, continuing his study of quantum theory. It was an eclectic group of researchers focusing on a variety of theoretical topics including hydrodynamics, acoustics, and relativity. Tamaki exerted no pressure on the students and gave them complete freedom in their studies. This suited Ogawa perfectly; now he had to decide what to work on.

One idea was to work on a theory of the nucleus. The structure of the nucleus was completely unknown and, when considered in the framework of current theories, completely incomprehensible. It was believed to be made up of protons and electrons, since these were the only known matter particles and each had been seen coming out of a nucleus. However, to have electrons in the nucleus contradicted quantum theory.

Before attempting to study electrons in the nucleus, Ogawa decided to study electrons outside the nucleus but interacting with it. Specifically, he decided to use Dirac's relativistic quantum theory to study the effect of the magnetic force from the spin of the proton and electron in a hydrogen atom on its atomic spectrum — the hyperfine structure. He solved the problem, wrote up the results, and then handed the paper to Tamaki, who put it in a safe, saying he would read it later. Soon after, Fermi published a paper that took the same topic even further.

Ogawa continued to read papers on quantum theory and

came across one written by Heisenberg and Pauli on quantum electrodynamics. In the paper, they set out to extend Dirac's relativistic quantum theory to create a fully quantum mechanical theory of the electromagnetic field. However, when they attempted to complete Dirac's calculation accounting for all the energy exchanges, they obtained infinity in one of the final lines. How could the infinity be eliminated?

Ogawa read the paper many times and considered the problem daily. Week after week, he would wake up with optimism only to end the day in despair. Finally, he decided to find an easier problem. He had studied English and German, but he had never gotten very far learning French. He had time now and decided he would learn French. He enrolled in the Japan-French school, taking classes two to three times a week. He went to see the French movie "Under the Roof of Paris." He learned the words to its theme song in French and sang along.

He also wrote compositions in French: "I do not wish to seek the strong stimulus of the city, and I am also too lazy to travel into the country, far from the city. Because the Imperial Palace is near my house, I often take a walk in its grounds. Autumn is the best time. The fallen leaves, which carpet the paths between the ancient trees of the Palace make tiny sounds under my sandals; these sounds remain in my heart as an unforgettable echo."

Ogawa continued to devote himself to theoretical physics. He was now almost twenty-four. By that age, Heisenberg, Pauli, Dirac, and Fermi had already distinguished themselves in the field. Ogawa spent entire days in the research room immersed in scientific journals. When he became tired of studying, he sometimes drew floor plans of a simple room only for himself: a chair, a bed, a bath, and a bookcase.

In the spring of 1931, Yoshio Nishina came to Kyoto to give lectures on Heisenberg's quantum theory. Nishina was a trained electrical engineer who was sent to study abroad a few years after graduating. He spent a year at the Cavendish Laboratory, a year in Göttingen, and six years at Bohr's Institute in Copenhagen. He returned to Japan carrying with him the "spirit of Copenhagen." When he returned to Japan in 1928, he built up the Nishina group in Tokyo that focused primarily on the atomic nucleus.

Nishina's lectures and the social interaction with him during his visit transformed Ogawa. Nishina's personality put him at ease

— the quiet, withdrawn young man could talk easily with the older scientist. During the visit, Ogawa and his companion, Tomonaga, had intense discussions with Nishina on nuclear physics. The next year, Tomonaga left Kyoto to work with Nishina. Ogawa, though solitary by nature, felt the loss of his friend.

By this time, though, another person had entered his life. The previous autumn, Ogawa had been surprised with a marriage proposal delivered to him by the Secretary of Kyoto University. The other party was Sumi Yukawa, the youngest daughter of the proprietor of the Yukawa Gastrointestinal Hospital. The match was satisfactory to all those concerned, and in April of 1932, they were married. Hideki went to live in the Yukawa home in the busy commercial seaport of Osaka. He was adopted into the family and took their name — he was now Hideki Yukawa.

That same year brought many changes to the subject of physics. It was discovered that, along with protons, neutral particles called "neutrons" reside in the nucleus of atoms. Soon after, new positively charged particles with the same mass as the electron were discovered, which were named "positrons." Lastly, scientists succeeded in breaking apart nuclei by bombarding them with high-speed protons. Suddenly, there was great interest in nuclear physics.

Heisenberg responded with a trio of papers published between 1932 and 1933. Yukawa translated these papers into Japanese, adding his own introduction to give context to Heisenberg's theory. This was his first publication. Soon after, Yukawa was offered a lectureship at Osaka University, and a year later he became the head of his own group.

During this time, Yukawa spent long days struggling to understand the nucleus. It was difficult but satisfying work, like that of a traveler carrying a heavy load up a steep slope. He enjoyed the liveliness of Osaka as well as the quiet and neatness of his new home, which gave relief to his strained mind. He allowed himself to be changed by the new environment, which awakened his initiative and activity.

One day, his father came by to visit and suggested to his father-in-law, "Why not send Hideki to a foreign country to study?" His father-in-law said that he would think about it. Hideki rejected the proposal. He did not want to go to a foreign country and talk with foreign scientists until he had something new that he could call his own. He didn't care how many times he failed in the attempt. He wanted to put everything into his work — knowledge, passion, and will.

Yukawa continued to think about the nucleus. How was the nuclear force related to the other forces? Quantum mechanics had united mechanical, chemical, electrical, and magnetic forces into a single force, the electromagnetic force. Gravity is another force. Were there only two fundamental forces, or was the nuclear force a third? Is the nuclear force fundamental? It seemed so. Gravity and the electromagnetic force are both too weak to account for nuclear binding.

If the nuclear force is a fundamental force, then according to quantum mechanics it should be accompanied by a particle mediator. In electromagnetism, charged particles "play catch" with the photon, which is a wave-like field. This applies to protons and electrons as well as positrons; attraction and repulsion are conveyed through a game of catch with photons. If the nuclear force can also be visualized as a game of catch, what is the nature of the ball? Heisenberg and others had suggested that it might be electrons.

In 1934, Fermi published a new paper on beta decay, a process by which a neutron becomes a proton while ejecting an electron. Similarly, a proton becomes a neutron with the ejection of a positron. However, experiments indicated that energy as well as other quantities are not conserved in these interactions. Bohr and Heisenberg were satisfied with this, believing that energy conservation must break down in the nuclear region. They welcomed this failure, anticipating that it might lead to radically new concepts of space and time.

Pauli objected, however, and proposed instead that a low-mass neutral particle must be emitted with the electron during beta decay so that energy and angular momentum can be conserved. Fermi used this idea, calling Pauli's particle a "neutrino," to explain beta decay: an electron is accompanied by an antineutrino, and a positron is accompanied by a neutrino. In this way, Fermi quantitatively demonstrated that energy and angular momentum are conserved in the nuclear region.

Yukawa wondered if the neutrons and protons could be playing catch with a pair of particles, perhaps with an electron and a neutrino. Others had already considered this and found that such an exchange could not account for the strength of the nuclear force. Then he had a thought. Perhaps the particle that mediates the nuclear force is not the electron, positron, or even the neutrino, but rather a *new* particle.

During the day, Yukawa became lost in equations. At night, interesting ideas grew, usually to disappear in the morning light. One night in October of 1934, shortly after the birth of his second child, an idea came to him: the mass of the particle that mediates the nuclear force should be inversely related to the force's range. An estimate of this mass could easily be calculated using quantum theory. He jotted this down in a notebook that he had placed beside his bed.

After waking up in the morning, Yukawa calculated. The effective range of the nuclear force is extremely small, about 0.02 trillionths of a centimeter. With this, he predicted that the mass of the particle mediating the nuclear field should be about two hundred times the mass of an electron. Its charge should be the same as the electron's, and both positive and negative versions should exist. Why had such a massive, charged particle never been detected before? He calculated again. It would require about 100 million electron-Volts to create such a particle, more energy than what was available in even the most powerful accelerators of the time.

Yukawa shared these results at his usual lunch with the Kikuchi group. Kikuchi was supportive and suggested that if such a particle exists, it could be detected in the Wilson cloud chamber. Yukawa agreed that the particle should be found in cosmic rays. Soon after, he presented his idea at a meeting of the Physico-Mathematical Society of Japan in Osaka, where Nishina was in attendance. Nishina was very interested in the theory and congratulated him.

Yukawa began to write up a paper. He proposed the existence of massive mediating particles, which he referred to as "heavy quanta" to distinguish them from photons. In the nucleus, protons and neutrons play catch with these quanta. A proton tosses a positive quantum and becomes a neutron; a neutron catches it and becomes a proton. This game of catch conceptually accounted for the force among protons and neutrons in the nucleus. Yukawa modeled the situation mathematically too, suggesting an equation for the nuclear potential in which it decays exponentially, accounting for the very small range of the nuclear force.

Similarly, he explained the interaction between a pair of nuclei during a collision. The only difference is that in this case, the game of catch is not only among protons and neutrons in the same nucleus, but also among those in adjacent, colliding nuclei. With his method, Yukawa was also able to account for the beta decay of a nucleus. In that case, after the heavy quantum is created, it decays into a pair of particles rather than being caught by a proton or neutron; a positive quantum decays into a positron and neutrino, and a negative quantum decays into an electron and antineutrino. Similarly, it was possible to explain the known decay of a neutron into a proton, electron, and antineutrino.

In this way, Yukawa was able to explain all the nuclear interactions — the binding of the nucleus, beta decay, collisions, and the decay of a free neutron — within a single theory that was consistent in all ways with known conservation laws, special relativity, and quantum theory. Furthermore, he made an important distinction. The interaction among protons and neutrons responsible for nuclear binding and collisions occurs with a large probability, while the interaction responsible for beta decay and the decay of a free neutron occurs with a small probability. He therefore used two separate coupling constants to account for two separate nuclear forces, one strong and the other weak. He also connected his theory to those of Heisenberg and Fermi and compared his results with experiments whenever possible. He found the correspondence satisfactory in all respects.

Yukawa wrote the paper quickly, urged on by his wife: "Please write the English paper and show it to the world." He completed it in a month, submitting it to the Society at the end of November. He now felt like a traveler "who rests himself at a small tea shop at the top of a mountain slope." Finally, he had made a discovery.

# 2

# Infinities

We are too powerless to make assumptions based only on reasoning. We must beg instruction from nature herself.

— Shinichiro Tomonaga

In 1928, the English physicist Paul Dirac created a relativistic version of quantum theory. In this version, he was able to mathematically incorporate the spin quantum number, which had been missing from the original theory, and thereby account for the splitting of some spectral lines that had not yet been explained. In addition, Dirac made a surprising prediction: the existence of anti-particles, particles with the same mass as ordinary particles but with opposite charge. This was soon confirmed when scientists studying cosmic rays detected antielectrons, positively charged particles with the same mass as electrons, which became known as positrons.

In 1947, the American physicist Willis Lamb found a discrepancy when comparing hydrogen's spectral lines to those predicted by Dirac's theory. With increasingly precise measurements of the lines, it became apparent that two of hydrogen's energy levels are slightly shifted as compared to the predicted values. This pushed theoretical physicists to revisit Dirac's theory.

Scientists soon realized that the reason for this discrepancy is that Dirac's calculations didn't fully incorporate the interaction between electrons and photons. Since his theory took into consideration the most probable states of the electron, its match to experiment was very close, but it didn't account for *all* the possibilities. For example, it didn't account for the possibility that the electron might spontaneously create and absorb a photon. It also didn't account for the possibility that the electron could create a particle-antiparticle pair, which would then self-annihilate. In fact, there was an infinite number of possibilities, and it was necessary to include *all* the possibilities to get the numbers right.

Physicists began calculating the effects of these infinite possibilities. Infinities were everywhere! They couldn't complete any calculation without getting absurd answers. This problem arose every time they tried to include interactions that extended to zero distance. If they cut off the calculations just short of zero, they got finite answers, but their answers depended on that cutoff. Therefore, all calculations yielded arbitrary answers.

Eventually, the German-American physicist Hans Bethe found a way to get rid of the infinities. In the original theory, with numerous photons and virtual particles spontaneously coming in and out of existence, all the quantities get very fuzzy and complicated, and the "mass" parameter in the theory doesn't actually represent the particle's measured mass. However, when Bethe forced the mass to match its measured value, all the infinities disappeared, and everything worked out perfectly. Other physicists, notably Richard Feynman and Julian Schwinger in the United States and Shinichiro Tomonaga in Japan, built on this idea. With different methods of calculation, they arrived at identical values for the observed shift as well as other measurable quantities that were in excellent agreement with experiment.

This process of getting rid of infinities was called "renormalization," and the resulting theory, a fully worked-out quantum theory of electricity and magnetism, was called "Quantum Electrodynamics." The theory was extraordinarily successful. With quantum electrodynamics, it was possible to derive various wave properties of light such as reflection, refraction, and interference. It could also be used to explain experiments that had previously been explained with a particle view of light, like blackbody radiation and the photoelectric effect. Even more impressive, it could explain experiments, such as the double-slit experiment, in which light sometimes acts like a particle and other times a wave.

Furthermore, quantum electrodynamics could explain the polarization of light, now identified with the photon's spin. It could be used to explain the stability of atoms, the periodic table, the bonding of atoms, and the minute details of atomic and molecular spectra. It also accounted for electric and magnetic fields, which Faraday had imagined as real entities filling the space between charged particles; in quantum electrodynamics, these fields became photons — particle-waves that continually blink in and out of existence mediating the electromagnetic force.

With technological advances, experimental values gained precision, and with advances in computational power, theoretical values gained precision. Still, with each additional digit of precision, the theoretical values continued to agree with the experimental values. Although the majority of the systems studied with this theory were small, atomic-sized systems, scientists also found experimental agreement when analyzing astronomical systems using the theory, systems a hundred times larger than the Earth.

The success of quantum electrodynamics was so great that scientists used it as a model for the strong nuclear force. This resulted in a theory called "Quantum Chromodynamics," which was subsequently supported by convincing experimental evidence. Next, scientists attempted to use quantum electrodynamics to model the weak nuclear force. They discovered that it acts so much like the electromagnetic force that instead of creating a new theory, they succeeded in incorporating these two forces into a *single* theory. They named the unification of the electromagnetic and weak nuclear forces the "electroweak force."

With such remarkable success quantizing the electromagnetic and nuclear forces, scientists then turned their attention to gravity. But when they attempted to model it analogously to electrodynamics, they were left with infinities that wouldn't go away. They therefore labeled gravity "non-renormalizable." This was the first attempt to create a quantum theory of gravity, but it would not be the last.

# 3

# The Theory of Everything

But what is equally important,
and sobering, is how often we fool ourselves.
And we fool ourselves not only individually but en masse.
The tendency of a group of human beings to
quickly come to believe something that its
individual members will later see as
obviously false is truly amazing.
Some of the worst tragedies of the last century
happened because well-meaning people fell for easy
solutions proposed by bad leaders.

— Lee Smolin

In the August 1936 edition of the *Physical Review*, American physicists Carl Anderson and Seth Neddermeyer published their photographs from the cloud chamber atop Pikes Peak — photographs of particle tracks made by cosmic rays. There were a few tracks that they could not clearly identify as either protons or electrons. In Japan, Hideki Yukawa studied these photographs and then submitted a letter to *Nature* suggesting that it was "not altogether impossible" that the ambiguous particles were the heavy quanta that he had predicted.

Yukawa's paper was rejected by the editor, though he did succeed in publishing the same note the following summer in a Japanese journal. By then, several groups had reported similar observations, and scientists had become interested in Yukawa's theory. The observed particles, which were named "muons," were not actually Yukawa's particles. However, his heavy quanta were found in cosmic rays about a decade later and named "pions."

While some scientists continued to search for new particles in cosmic rays, others used particle accelerators to smash protons together at higher and higher speeds. In this way, they were able to artificially produce numerous variations of muons, pions, and other particles, which they named "lambdas," "sigmas," "rhos," and so on through the Greek alphabet. They added numbers to the letters to distinguish among similar particles. There were so many new particles that they came to be collectively referred to as the "particle zoo." It was like the growing list of elements that had eventually come together to form the periodic table.

This collection of particles was named "hadrons," and in 1961, two physicists — the American Murray Gell-Mann and the Israeli Yuval Ne'eman — discovered a pattern when organizing the particles by mass, charge, and spin, called the "eightfold way." Based on the pattern, Gell-Mann predicted a new particle. In November of 1963, a large group of scientists at Brookhaven National Laboratory conducted a thorough search for this particle, taking over fifty thousand photographs. Finally, they found it, a particle with the mass, charge, and spin Gell-Mann had predicted.

This discovery won over most scientists to the eightfold way, but now they asked: What was the *reason* for these patterns? The symmetry of the theory seemed to indicate that hadrons might be made up of more basic particles. It was a big puzzle, and Gell-Mann solved it — again sharing the credit, this time with the Russian-American George Zweig. They assumed the existence of fundamental particles, which were named "quarks." According to the theory, each quark has a half-integer spin and ⅓ or ⅔ the charge of a proton or electron. Quarks come in three "flavors" — up, down, and strange, and they also have antiparticle counterparts. Furthermore, each quark has a "color charge"; they always join in such a way that the composite particles are color neutral. This model accounted for all of the known hadrons and predicted the existence of others, many of which were soon found.

In the hills behind Stanford University, a new instrument was built — a two-mile-long electron accelerator. Shortly after it became operational in 1967, scientists began to probe the inner structure of protons. By shooting a beam of electrons at protons, scientists were able to "see" inside a proton; the experiments indicated the existence of point-like charges within the proton, consistent with the quark model. This made practically every physicist a believer in the existence of quarks. However, despite all of their searching, scientists never discovered even a tiny track made by a quark. Their conclusion was that if quarks exist, they must be trapped within the larger particles.

One of the most successful theories to explain the apparent confinement of quarks within hadrons was the "Bag Model" proposed by Kenneth Johnson and his collaborators at MIT. In this model, hadrons are contained in confined spaces like a tiny bag with quarks inside. When hadrons collide, their bags overlap. During this time, quarks can rearrange and even multiply, creating new quark-anti-quark pairs from the energy of the collision. Using this model, many details about hadrons could be calculated that gave remarkable agreement with experiment. The success of this model inspired confidence in the hypothesis that quarks are permanently trapped inside hadrons.

Another theory that evolved alongside the bag model of quark confinement was a "String Model" of hadrons. The young Italian physicist Gabriele Veneziano had observed an interesting pattern in the scattering data between two hadrons and formulated an equation describing this pattern, which generated interest in Europe and the United States. Soon, a few were able to interpret Veneziano's equation using a physical picture. In this picture, the particles were modeled not as points, but as rubber bands or strings. The strings could stretch, contract, and vibrate. They could collide with each other, exchanging energy. The states of vibration of the strings corresponded to the various particles that were produced.

The bag and string theories worked well to explain different aspects of the observed behavior of hadrons. In 1973, however, the development of quantum chromodynamics replaced both of these models. According to quantum chromodynamics, the apparent confinement of quarks could be explained by the strong interactions among them, mediated by new particle-waves called "gluons." The theory had everything that was desired in a fundamental theory, with excellent experimental confirmation besides.

Nonetheless, before the introduction of quantum chromodynamics, much work had already gone into the development of the string theory of hadrons. Specifically, there had been much effort to turn this theory into a fundamental theory consistent with special relativity and quantum theory. The only way they were able to do this was by assuming twenty-five spatial dimensions and faster-than-light "tachyons." Another problem with the theory was that there were no half-integer-spin particles and therefore no quarks.

Soon after, two separate variations on the original string theory were proposed — both with "supersymmetry" among the particles — that eliminated the tachyons, reduced the spatial dimensions from twenty-five to nine, and incorporated half-integer-spin particles. Several continued to develop the theory, and in 1974, the two physicists Joël Scherk and John Schwarz found that one of the particles predicted by the theory might be a graviton, the massless particle that was assumed to mediate the force of gravity! Consequently, they proposed, string theory might actually be *the* theory that unifies all the forces and all the particles.

What's more, the entire theory arose from only a few simple assumptions. The movement of a one-dimensional string makes a two-dimensional surface in spacetime, with an area approximately equal to its length multiplied by time. The string moves in such a way to minimize this area; in this way, the motion of the strings is determined. Furthermore, with the breaking and joining of the strings, the existence of all the forces and particles could be explained. For all of this, the theory only had two fundamental constants — the tension of the string and a coupling constant indicating the probability of a string breaking in two, which then gives rise to a force. It was an exciting prospect: the unification of all of physics! Despite this potential, the theory initially generated little interest within the scientific community.

Ten years later, the tide turned. Schwarz and a new collaborator, Michael Green, performed a calculation that provided evidence that string theory was a finite and consistent theory, at least in nine spatial dimensions. The infinities that had plagued previous attempts at creating a quantum theory of gravity were absent. Maybe this was it! Maybe string theory was the way to unite the two pillars of modern physics.

The scientific community, at least some influential subset of it, took notice. When Princeton physicist Edward Witten heard rumors about the paper, he called Schwarz asking for a copy. Schwarz sent him a draft by FedEx, and then within a couple days all the theoretical physicists at Princeton and its Institute for Advanced Study began working on the problem. Interest in the theory grew from there. Soon there were string theory conferences. Furthermore, many of the major universities and research institutions began holding string theory seminars. Among these was Harvard, which named theirs the "Postmodern Physics seminar." It seemed to some that this was the beginning of a new era in physics.

Those who worked on it, the "string theorists," felt that they had in their possession *the* theory of everything. A kind of elitism arose in this group based on the assumption that the only reason a physicist wouldn't work in this area is that they *couldn't*. It was, in fact, an extremely difficult mathematical theory, and only the best mathematicians and physicists dared to enter the field.

Detractors of the theory were equally passionate. They criticized the theory for its disconnect with experiment. Besides, string theory requires *nine* spatial dimensions, while we live in a world with only three. Where are the other six?

This was the question — where are the other six dimensions? Those working on the theory found that what was so simple and elegant in nine-dimensional space became very complicated when they tried to fit it to our three-dimensional universe. There were many different ways to curl up the extra dimensions, and each way gave rise to a different theory. Soon they had hundreds of versions, then thousands. By the early 1990s, they had hundreds of thousands of distinct theories, each with many free constants and all with extra particles and extra forces. Among these theories, there was not a single testable prediction.

Many abandoned string theory to study solid-state physics, biology, computer science, and even finance. Meanwhile, others stayed the course, and in the spring of 1995 at a string theory conference in Los Angeles, Edward Witten gave a talk that breathed some life into the fading field. He proposed that all of the string theories that had been discovered were special cases of a more general theory. He went on to describe some of the features that this theory would have.

Although there were by now millions of variations of string theory obtained by wrapping up the extra spatial dimensions in different ways, string theorists were only working on five distinct theories in nine-dimensional space. Connections that had recently been discovered between certain pairs of theories gave hope that these five theories were actually a single theory expressed in different forms.

Until now this had been a conjecture, but Witten presented strong evidence that it must be so. The overarching theory, he proposed, would have an additional spatial dimension. No consistent string theory existed in ten dimensions, at least no supersymmetric version. However, in the early 1980s, a theory had been developed not with strings, but with two-dimensional surfaces called "membranes" that move in ten-dimensional space. The theory was largely ignored because it was not known whether it could be made consistent with quantum theory. Witten now took up this theory and discovered a remarkable result.

If one of the spatial dimensions is a circle, it is possible to wrap one of the two dimensions of a membrane around the circle, leaving the other dimension — which looks just like a string — to move around in the remaining nine dimensions. Moreover, Witten discovered that by wrapping one dimension of the membrane in different ways around the circle, it was possible to get all five of the consistent string theories, and *only* those five theories. In this way, the connections that had been observed between pairs of the five theories were now explained.

This generated an enormous amount of excitement, and soon Witten gave the unifying theory a name: M-theory, where "M" didn't stand for anything since it wasn't yet a theory but only an idea for one. He invited others to fill in the rest of the name by developing the theory.

It was soon discovered that in addition to strings, higher dimensional objects were needed: two-, three-, four-, and five-dimensional surfaces that were called "branes." With these objects, it was not only possible to wrap up the extra dimensions, but it was also possible to wrap branes around loops and surfaces in the background geometry. This meant that there were infinite possibilities for the background of string theories, which gave hope for the existence of a background-independent unifying theory that would give rise to all of the others.

Furthermore, M-theory was applied to the study of black holes, massive astronomical objects with such a strong gravitational pull that even light cannot escape. Specifically, it was discovered that with branes it was possible to describe a certain type of black hole, called "extremal," which had been studied for many years from the perspective of general relativity. It was found that extremal black holes have identical thermodynamic properties to a system in which gravity has been turned off and many branes are wrapped around the extra dimensions. Undoubtedly, M-theory has had some successes.

However, after over two decades of development with some of the best mathematical minds working on the theory, it still has no testable predictions. It cannot even reproduce the theories that have already been verified. And, having not yet become an actual theory, it still has no proper name.

# 4

# The Big Bang

The last two thousand million years are slow evolution:
they are ashes and smoke of bright but very rapid fireworks.

— Georges Lemaître

The problem of infinity was difficult even for Aristotle. As he admitted, "contradictions result whether we suppose it exists or not." Then he made a decision regarding the universe: it is finite in space and infinite in time. The celestial sphere encloses the entirety of the universe; nothing, not even a void, exists beyond. And this spatially finite celestial sphere has always rotated and will always rotate, marking the passage of time without beginning or end.

Copernicus did nothing to challenge these aspects of Aristotle's universe, but the Italian philosopher Giordano Bruno did. Bruno proposed that the universe is infinite, containing an infinite number of solar systems. He also suggested that the other planets scattered throughout the universe are populated with life forms — plants and animals, including human beings, troglodytes, demons, and nymphs. Bruno retained, however, Aristotle's assumption of the timelessness of the universe, that it has always existed and will exist indefinitely into the future.

Newton did not publicly take a stance regarding the structure and creation of the universe. When asked why he was silent about such matters he supposedly responded, "I do not deal in conjectures." Privately, though, he did speculate. Prompted by questions posed by Reverend Richard Bentley, an English classics scholar and theologian, he wrote some of his thoughts down.

Bentley asked Newton about the physical consequences of the theory of gravity if applied to the universe. Newton had thought about this before, but Bentley's probing pushed him to think more deeply; his ideas evolved as Bentley questioned him, challenged him, and offered his own ideas and analysis.

In his first letter, Newton suggested that in a finite universe where all the matter is scattered evenly throughout and every particle attracts every other particle according to his laws, the interplay of forces would create one great spherical mass. However, in an infinite universe, it would be possible to have local condensation, giving rise to an infinite number of planetary systems scattered at great distances from one another. He concluded that only an infinite universe could be stable, with attraction in one direction counteracted by attraction in the opposite direction regardless of location.

However, by the end of the exchange of letters with Bentley, Newton seemed to be assuming a finite universe, or at least that the problem of stability is no greater in a finite than an infinite universe.

Furthermore, he began to puzzle over how stars, planets, moons, and comets could have taken shape as matter came together from remote regions of the universe. How was it possible that the planets and moons all orbit in nearly circular orbits and all in the same direction? If the formation of the planetary system occurred naturally, wouldn't the celestial objects have a variety of motions with respect to the Sun and each other — eccentric ellipses, hyperbolic orbits, and parabolas? And why was the motion of comets so different from the motion of the planets and moons?

In the end, Newton conceded that the formation of the universe would require a Cause that compared and adjusted the various masses and distances of the Sun, planets, and moons. To do this for so many bodies, the Cause could not be blind, but rather "very well skilled in Mechanicks and Geometry." Furthermore, to prevent the fixed stars from falling into each other, this intelligent and powerful being must have placed them at immense distances from one another.

Throughout these letters, Newton implicitly assumed the eternal nature of the universe. He tested this too, the idea that the universe is unchanging with the stars distributed evenly throughout. By assuming that the intrinsic brightness of all the stars is the same as our Sun's and that their apparent brightness decreases according to an inverse-square law, he was able to estimate the distance to many stars. His results were surprising; the number of stars appeared to increase too rapidly with distance. This implied that the stars farther away from the Earth are closer together than the ones nearby. He did not take this result seriously, however. Rather, he modified his assumptions in order to obtain the expected result.

In 1720, Edmund Halley read papers before the Royal Society that challenged the view that the universe is both infinite in time and space, posing the question: Why is the sky dark? In an infinite universe with a uniform distribution of stars, assuming that apparent brightness decreases according to an inverse-square law, the entire sky should be ablaze with light. Newton did not comment on this idea.

In the early 1900s, American astronomer Henrietta Leavitt cataloged thousands of Cepheid variable stars, stars that periodically brighten and dim. Analyzing the data, she discovered that the period of these stars is proportional to their apparent brightness. Assuming that these stars, which were found in a single cluster, were all at approximately the same distance from the Earth, her discovery implied that the period of a Cepheid variable is proportional to its intrinsic brightness.

Leavitt published her results in 1912, and quickly their significance was recognized. Cepheid variable stars, although rare, exist throughout the observable universe. The relationship between their period and intrinsic brightness offered a way to estimate the distance to stars too far away to determine using parallax. However, at this time, not even one known Cepheid variable star was close enough to measure its distance. In order for Leavitt's equation to be useful, it had to be calibrated with a star of known distance. Another American astronomer, Harlow Shapley, undertook this task and within a few years had found eleven Cepheids close enough to determine their distance through parallax. Leavitt's equation could then be used like a measuring stick to determine the distance to faraway stars.

Soon, this story joined another that had begun the previous century. Two German scientists, Robert Bunsen and Gustav Kirchhoff, had found that the spectrum of each element is unique and can be used to identify unknown gases. Furthermore, they observed that the frequencies of the bright lines in the spectra of burning elements match observed dark lines in the solar spectrum, lines that had been observed and cataloged many years earlier by the German physicist Joseph von Fraunhofer. In this way, astronomers could perform chemical analysis of the Sun and stars by comparing their spectra to those of known elements.

In the early 20th century, astronomers began photographing the spectra of stars. They noticed that the spectral lines of some nearby clusters of stars were all shifted toward the blue end of the color spectrum, while those of other clusters were shifted toward the red end of the spectrum. According to the Doppler effect, the frequency of a wave appears higher as the wave approaches an observer due to the compression of the waves, and it appears lower as the wave recedes. If the shifts in the spectral lines are interpreted as Doppler shifts, the results show that some clusters of stars are moving toward us and others away from us.

Astronomers collected data from many star systems and found that most are red-shifted. They could reach a more specific conclusion with the help of Cepheid variable stars found within the star systems: all distant star groups are red-shifted. Then in 1929,

American astronomer Edwin Hubble published a remarkable result. He found that the radial velocity of star groups is proportional to their distance from us. In his paper, Hubble was careful to avoid coming to any radical conclusions about the implications of his results regarding the universe. Others, however, did and came to a startling conclusion. The universe seems to be *expanding*.

This was almost, but not completely, unexpected. Two years earlier, the Belgian astrophysicist and Catholic priest Georges Lemaître anticipated this result at an international conference. He had come to the conclusion that the universe is expanding with the help of general relativity. He shared this result with Einstein, who told him, "Your calculations are correct, but your physical insight is abominable." Lemaître must have been delighted when Hubble's work confirmed his own conclusion.

If the universe is expanding, it must have once been compressed. No one knew what this would look like. It was such a crazy idea that the term "Big Bang" was used sarcastically to describe the beginning of the universe. Lemaître described it as "bright but very rapid fireworks." It *was* a crazy idea, but as astronomers accumulated more and more data that supported Hubble's result, it soon became clear that, whether it was a big bang or some other kind of cosmic event, the universe had a beginning.

With this recognition, the last remaining cinders of Aristotle's universe were blown away. In its place grew an entirely new field of study: cosmology. Although the word "cosmology" existed before this time, it referred only to the structure of what was assumed to be a static universe. Astronomers had merely been trying to catalog all the things in the universe. Now, cosmologists began to try to understand its origin, evolution, and future.

One of the first questions asked was, how old is the universe? This was a surprisingly easy question to answer, at least in approximation. We know that speed equals distance divided by time. For any galaxy, we can use this equation together with its distance from us and radial speed to approximate how long ago its position coincided with our own. This time is about the same for all galaxies and directly gives us the age of the universe. Hubble's original data underestimated the distances to galaxies and, consequently, the age of the universe. Now, we believe it to be about fourteen billion years old.

Scientists then began to try to figure out how the universe came into existence. They started with general relativity, which had been used to predict its expansion, attempting to trace the development of the universe back to the beginning of time. Their calculations yielded an absurd result, that the universe began with all the matter compressed into an infinitesimally small point.

Before this point, however, quantum effects would have become important. As a consequence, for the first time, physicists attempted to use general relativity and quantum theory together. General relativity was able to describe the evolution of the universe back to a very early time and then stepped aside, letting quantum theory take it from there.

Crucial to the development of this new theory was George Gamow, who had experience in both general relativity and quantum theory. As a university student in Leningrad, he had worked with Alexander Friedmann, the first physicist to predict the expansion of the universe from general relativity. After graduating in 1926, Gamow studied quantum theory in Göttingen. He then worked with Bohr in Copenhagen and Rutherford at the Cavendish Laboratory. Upon returning to Leningrad in 1931, he was employed at the Radium Institute, where he helped design the first cyclotron. In 1934, he defected to the United States, where he obtained a position at George Washington University. There, he turned his attention to the nuclear physics of stars and the big bang theory.

In 1946, Gamow published a paper suggesting that the conditions necessary for rapid nuclear reactions in the early universe existed for such a short time that, contrary to previous assumptions, they could not occur in equilibrium at constant temperature. He based his result on general relativity. Although he admitted that one should be cautious in extrapolating back to such an early period, the expression that he had used was really nothing more than a statement of the law of energy conservation, where the inertial expansion of the universe was counteracted by the attractive pull of gravity.

Because of the very short time period of rapid nuclear reactions, any neutrons present would not have had time to decay, which would affect predictions regarding the chemical composition of the universe. Gamow and other scientists refined this model as new information about nuclear reactions emerged, and a more specific timeline for nuclear reactions in the early universe was proposed, starting at the first hundredth of a second after the big bang. Eventually, the model yielded predictions for the proportions of helium and other light elements that were consistent with the actual proportions of our universe.

According to this model, nuclear reactions would have stopped within a few hours of the beginning of the universe. After that, the universe expanded and cooled. A hundred thousand years later, the universe was cool enough for atoms to form, and these atoms began to form clouds of atomic matter, which then condensed into galaxies and stars. Regions with rotating matter formed flat, disk-like galaxies like the Milky Way, while regions without rotation formed ovular, elliptical galaxies — all consistent with the universe we observe.

To be able to explain what we observe using known theories and reasonable assumptions was a great success. An even greater success of the theory came in the form of a prediction. Soon after Gamow introduced the model, it was shown that the radiation produced in the early stages of the universe should still be present today. The radiation would have expanded at the speed of light, cooling in the process; it should now be a few degrees above absolute zero.

In 1964, such radiation was detected. The radiation looked the same in all directions, not coming from any single source but rather from the entire universe. Moreover, its distribution was consistent with Planck's blackbody radiation formula, and its temperature was consistent with theoretical predictions. The observation of what came to be known as the cosmic microwave background radiation gave strong support to the big bang model of the universe and convinced most scientists of its basic correctness.

In the 1960s, scientists discovered that a quantum theory of the weak nuclear force could be made consistent with relativity provided that it also included the electromagnetic force. The resulting theory of the electroweak force included three new mediating particles in addition to the photon — positive, negative, and neutral quanta. At high temperatures, all four mediating particles were assumed to be massless, and the electromagnetic and weak nuclear forces were expected to be essentially the same. At cooler temperatures, the symmetry between the forces was broken, giving three of the mediating particles large masses, allowing for the short range of the nuclear force.

In the next few years, the theory was developed mathematically and found to be renormalizable, enabling precise calculations of the masses of the new particles. Although the accelerators at this time couldn't reach the required energies to produce these particles, other predictions of the theory were confirmed over the next several years. Consequently, the electroweak theory gained widespread acceptance. In 1979, the creators of the theory — Sheldon Glashow, Abdus Salam, and Steven Weinberg — shared the Nobel Prize in Physics for their work. In 1983, the three mediating particles, called $W^+$, $W^-$, and $Z^0$, were detected in high-energy proton-antiproton collisions at CERN, with masses consistent with their predicted values.

The success of this theory led to several attempts at unifying the electroweak force with the strong nuclear force to create a "grand unified theory." The concept underlying these attempts was based on the observation that as the temperature increases, the nuclear force weakens while the electroweak force gets stronger. Therefore, at some very high critical temperature, both forces should have the same strength and might be different versions of the same force. Unfortunately, reaching such high energies would require an accelerator about the size of our solar system.

However, there were some low-energy consequences of this theory that could be tested. For example, it predicted that protons could spontaneously decay into positrons, although the probability of this decay at achievable temperatures was minutely small. Another prediction was that as the early universe cooled, when it reached some critical temperature, huge magnetic monopoles — magnets with only a north or south pole — should have formed. These had not been detected, casting doubt on the theory.

In 1980, the young American physicist Alan Guth came up with an explanation for why we don't observe these monopoles. He proposed that in the first moments of the universe, the temperature cooled *beyond* the critical temperature that should have broken the symmetry between the electroweak and nuclear forces, without the phase transition taking place. This supercooling would create an unstable state with more energy than if it had undergone the phase transition. Guth demonstrated that this extra energy would have a repulsive effect, causing the universe to expand exponentially at speeds far surpassing the speed of light.

The suppression of the phase transition that would break the symmetry between the electroweak and strong nuclear forces should also suppress the creation of superheavy magnetic monopoles. In addition, Guth's theory addressed two major problems that had arisen in big bang cosmology: the universe is too smooth and too flat.

In every direction, on a large enough scale, the universe looks the same. The cosmic microwave background radiation is the same. The stars, galaxies, and galaxy clusters look the same. It is as if all the parts of the observable universe had been in contact at some time. However, there are parts of the universe that are too far apart to have ever had causal contact, assuming light has always had the same speed. This problem is solved if we assume that space itself expanded faster than the speed of light; in this way, it is possible that even light rays coming from opposite sides of the universe had once been in contact. The resulting enormity of the universe due to the exponential expansion explains something else — why, even as we look farther and farther in every direction, we have never seen any evidence of an end to the universe.

The second problem that Guth's theory addressed was what he referred to as the "flatness problem": the energy density of our universe is very close to what would be required to just halt its expansion. In the standard big bang model, this would require very fine tuning of the initial conditions. However, if the universe is supercooled according to Guth's model, the energy density of the early universe determines the expansion rate, naturally creating the conditions for a flat universe.

Guth's inflationary theory of the universe was well received by theoretical physicists and cosmologists alike, who modified and extended the idea. Its solution to the monopole problem became irrelevant as efforts toward grand unification theories waned. However, its explanations for the flatness and large-scale homogeneity of the universe continued to be relevant as powerful telescopes were sent aboard spacecrafts, peering deeper and deeper into space and verifying these features with greater and greater precision.

Subsequent development of the theory offered an experimental prediction. Quantum effects would have produced regular variations in the temperature and density of the early universe, creating the patterns of stars, galaxies, and galaxy clusters that we see today. The remnants of the early quantum fluctuations should also be observable in the cosmic microwave background radiation. These fluctuations were searched for and found, verifying not only the general features of these fluctuations, but even the detailed predictions of some versions of inflation theory.

The inflationary model of the universe has now been a standard part of big bang cosmology for almost four decades — the idea that soon after the big bang, the universe expanded exponentially, increasing its size by over twenty-five orders of magnitude in a fraction of a second and in doing so producing just the right conditions for a flat universe. Inflation stretched and smoothed out the universe to such an extent that billions of years later, in all directions, we see a uniformity that gives no evidence of any edge.

If we assume that the inflation theory is correct, the natural question is: What *caused* the rapid expansion of the early universe? Hundreds of proposals have been made as to what this exotic energy might be with no satisfactory conclusion so far. Furthermore, how and why did the rapid inflation end? Again, several theories have been proposed, none of which is completely satisfactory. One of the most popular of these is the multiverse theory, which claims that quantum fluctuations continually give rise to new inflationary periods that create new universes, a prediction that some critics refer to as a "multimess." Discouraged, these detractors suggest that we consider other paradigms — perhaps what we thought was a big bang was actually a transition from some preceding cosmological phase, a big bounce...

In 1928, Dirac determined an equation that the electron's quantum mechanical wave function must obey for it to be consistent with relativity. He noticed that the equation actually has two solutions, one for the electron and another for an identical particle with a positive charge. Rather than ignore this solution, he deduced that there should exist a positively charged counterpart to the electron. He went further than this, suggesting that the symmetry between positive and negative charges in quantum theory implies that there should be an oppositely charged "antiparticle" for every particle.

Three years after Dirac's prediction, an antielectron, also called a positron, was detected in cosmic rays. In the mid-fifties, antiprotons and antineutrons were artificially produced with the help of high-energy accelerators. In 1995, the first antiatoms were created — positrons bound to antiprotons, forming antihydrogen. And in 2018, spectral analysis of antihydrogen directly confirmed the symmetry between matter and antimatter; all of the observed spectral lines matched the spectral lines of hydrogen precisely, indicating identical energy levels.

This symmetry between matter and antimatter, which has been established both theoretically and experimentally, invites the question: Why is the universe, as far as we have observed, made almost entirely of matter? Matter and antimatter are always paired when created out of pure energy, and when they collide, they annihilate each other completely. Wouldn't the big bang have produced equal amounts of matter and antimatter? In fact, models of the early universe assume that at some point, the only particles that existed were electrons, positrons, neutrinos, antineutrinos, and photons. If so, where did all the antimatter go?

In answer to this question, scientists have suggested two general scenarios. First, there may have been slightly more matter than antimatter emerging from the big bang, and this excess is the source of all the matter in our universe today. Another idea is that the very early universe had the same amount of matter and antimatter, but there are reactions that treat matter and antimatter differently so that the two types are created and destroyed at different rates. A third possibility is that antimatter is still out there.

Since light emitted by matter is identical to light emitted by antimatter, as far as we know, distant stars and even entire galaxies could be made up of antimatter. Although the light is indistinguishable, a collision between matter and antimatter would create a powerful explosion of high-energy radiation. For example, a gram of matter colliding with a gram of antimatter would produce an explosion more powerful than the atomic bombs dropped on Japan during World War II.

Astronomers have searched far and wide for evidence of such a large release of energy. They have searched throughout our solar system. They have searched interstellar regions within our galaxy. They have searched in the space between adjacent galaxies, where matter from both galaxies might meet. They have never found any decisive evidence of the kind of explosion that would result from matter colliding with antimatter.

Astronomers have, however, observed events that at least allow for speculation about such collisions. In the 1960s, military satellites searching for prohibited nuclear testing detected high-energy bursts of radiation. These were extremely short events, usually less than ten seconds. They didn't repeat or leave any observable trace, which made it difficult to obtain any information about their origin. Eventually, enough data accumulated that scientists were able to determine that they occur uniformly in all directions. Therefore, these bursts were likely a result of very distant cosmic events, well beyond the local structure of our galaxy cluster.

By the late 1990s, astronomers were able to confirm that these gamma ray bursts originated very far away indeed — billions of light-years away. They identified two different kinds, which they called "long soft bursts" and "short hard bursts." The long soft bursts, lasting around ten seconds, were found to be jets of radiation associated with regions of active star formation. Evidence suggested that they somehow accompany the formation or development of a massive astronomical object, either a neutron star or black hole.

Astronomers have had more difficulty trying to determine the origin of the short hard bursts. These bursts last as little as one-tenth of a second and have higher energies, although they are usually not as intense. They are found far away from regions of active star formation, typically billions of light-years from their birth place. Various hypotheses have been proposed to account for these bursts, but none of them have been proven quantitatively. As a result, it seems that there is a *possibility* that at least some of these short bursts might arise from the presence of antimatter in our universe.

However, there is also a very real possibility that antimatter is not and never will be directly observed through any kind of collision with matter, though it may still be present in our universe. The enormous space between galaxy clusters opens up the possibility that entire clusters are made of antimatter and that matter and antimatter separated completely in the early moments of our universe. In that case, what would have caused them to separate? Wouldn't all the matter and antimatter just annihilate each other, leaving a universe filled only with radiation?

It would...unless matter and antimatter repel. Our theory of gravity assumes that they attract, that all particles experience a gravitational field in exactly the same way. Our cosmological models assume they attract. Our particle theories assume that photons are their own antiparticles, which would imply that light emitted from antimatter bends due to gravity in the same way as light emitted from matter. However, *experimentally*, this question has not yet been resolved. Gravity is so incredibly weak compared with every other force that, so far, it has been completely ignored in all experiments with antimatter, which means that we still don't know.

However, we will soon find out. With unprecedented effort, three independent, competing experiments are underway at CERN that are attempting to determine the acceleration of antihydrogen atoms due to the Earth's gravitational field. In addition, experiments are in development at University College London to study the gravitational free-fall of positronium atoms, unstable exotic atoms composed of an electron and a positron. Similarly, researchers at the Illinois Institute of Technology are attempting to measure the gravity experienced by muonium atoms, atoms composed of an antimuon and an electron.

They'll *probably* discover exactly what most people expect, that antimatter falls in a gravitational field just like matter. But what if they don't?

# 5

# Dark Matter, Dark Energy, and the End of the Universe

The effort to understand the universe is one of
the very few things that lifts human life a
little above the level of farce, and gives it some
of the grace of tragedy.

— Steven Weinberg

In August of 1916, Einstein spent three weeks in Leyden visiting his friends Hendrik Lorentz and Paul Ehrenfest, during which they spent much time in conversation about general relativity. The Dutch astronomer Willem de Sitter also joined in the discussions. The visit was, as Einstein wrote to his friend Michele Besso, "stimulating" and "re-invigorating." He told Ehrenfest that the visit had cheered and refreshed him. "Solitude can only be tolerated up to a certain limit," he wrote.

Einstein continued to think, calculate, and exchange ideas with Ehrenfest, Lorentz, de Sitter, and others. By February, he had developed a relativistic model of the universe. He wrote to de Sitter about it: "I am curious to see what you will say about the rather outlandish conception I have now set my sights on." He wrote to Ehrenfest, telling him that he had created something in gravitational theory "which exposes me a bit to the danger of being committed to a madhouse." He presented his cosmological model to the Prussian Academy of Sciences on February 8, 1917, publishing a paper on the subject the following week.

In the paper, Einstein brought the readers with him on his journey — "a rough and windy road" — in the hopes that this might make them more interested in his result. He began by considering the same finite universe that Newton had considered. However, while Newton was concerned with the collapse of the universe into one great mass, Einstein was concerned with the opposite, that a finite universe would evaporate. Radiation carrying energy would travel outward from stars into infinite space, never to return again. The same fate would come to individual stars that have enough energy to overcome the gravitational attraction of the other bodies. Statistically, this must happen from time to time. And so, like water in a pot without a lid, the universe would eventually evaporate away.

Einstein then posed the question: How can we use general relativity to create a cosmological model? One possibility would be to apply the same boundary condition at infinity as he had when calculating the perihelion precession of the planets, in which he had assumed that far away from the system, space is geometrically flat. However, applying such a boundary condition presupposes a preferred reference frame for the universe, contradicting the principle of relativity. In addition, this solution doesn't resolve the evaporation problem.

Another possibility would be to abandon the search for suitable boundary conditions altogether, which he was not yet ready to do since in his opinion this would be equivalent to giving up. Rejecting these two options, he turned to a third, radical possibility — modifying the equations of general relativity.

Einstein then made a number of assumptions about the universe. He assumed that the universe is spatially finite. In this way, boundary conditions at infinity arose naturally from the finiteness of the physical universe. He also assumed that at the largest scales, the distribution of matter in the universe, and therefore the curvature of space, is constant. Finally, he assumed that there exists a reference frame in which all matter is at rest, which he claimed was reasonable given the very low velocity of stars. In this reference frame, the universe can be approximated as a sphere.

With these assumptions, Einstein derived the energy-momentum tensor for this matter distribution as well as the metric, which describes the curvature of spacetime. However, he found that these don't satisfy the field equations of general relativity. To resolve this, he modified the equations by adding an extra cosmological term. This new formulation was now compatible with his model of the universe, consistent with the principle of relativity, and respected the conservation of energy and momentum.

The new term, proportional to a new universal constant, didn't affect predictions of the motion of the planets as long as the constant was very small. Furthermore, by applying the theory to the universe, it was possible to estimate the density of matter, total mass, and the radius of the universe in terms of the constant. In addition, with the new cosmological term it was possible to model a universe in equilibrium, one that would neither collapse in on itself nor evaporate away. Einstein ended the paper expressing satisfaction that he was able to present a consistent theory. However, he acknowledged that it was still uncertain whether the new theory would be supported by astronomical evidence.

In his paper, Einstein did not include any calculations based on his theory using known astronomical data, but he did do this in several correspondences during this period — with Ehrenfest, Besso, de Sitter, and others. Specifically, using an estimate of the density of the universe, he calculated the cosmic radius to be around ten million light-years. He expressed doubt in the correctness of this result since astronomers estimated the farthest stars to be at a distance of only ten thousand light-years.

Calculations were one thing, but another question that arose after Einstein's proposed modification of general relativity was, what did it mean? Soon after Einstein published his paper, Schrödinger demonstrated that a static, spatially finite universe could be obtained directly from the original field equations if a negative pressure term was added to the source tensor. Einstein responded by showing that Schrödinger's formulation was equivalent to his own modified theory if the negative pressure term was constant. In the same paper, Einstein gave his first physical interpretation of the cosmological term: it serves the purpose of distributing "gravitating negative masses" all over interstellar space. Less than a year later, he came up with an alternate interpretation, that the constant that appeared in the cosmological term was simply a mathematical constant of integration, not a universal constant associated with cosmology.

Einstein never appeared completely satisfied with his modification of general relativity, believing that it marred the beauty of the original theory. He had considered it necessary given his belief in a static universe. But, as he wrote to de Sitter, "Conviction is a good mainspring, but a bad judge!" In 1929, with Hubble's data demonstrating the expansion of the universe, observation judged his theory harshly. Einstein conceded, "The redshift of the distant nebulae has smashed my old construction like a hammer blow."

Close on the heels of Einstein's 1917 paper, de Sitter proposed an alternate cosmological model based on the same modified field equations, replacing Einstein's matter-filled three-dimensional spherical universe with a matter-free four-dimensional closed universe. Einstein disregarded this solution, claiming that it had no relationship to the physical world since it did not include matter. However, de Sitter's model, which predicted that light would be red-shifted, attracted much attention, especially that of other astronomers when new data emerged revealing the redshift of distant stars.

A few years later, the Russian physicist and mathematician Alexander Friedmann developed another cosmological model. In order to solve the field equations, he made two simplifying assumptions: the universe looks the same in every direction and the Earth doesn't occupy a special place within it. Importantly, he did *not* assume a static universe.

In his model, the universe had a beginning. It started from a point and expanded to its present size in what he referred to as the "time since the creation of the world." Einstein read the paper and publicly suggested that Friedmann had made a mathematical error. In fact, the error was his own, and so his criticism was quickly followed by a retraction. In an unpublished version of the retraction, Einstein declared that he did not believe that Friedmann's solution was realistic.

Although Friedmann published his paper in a prestigious German physics journal, it received almost no attention. This allowed another physicist, the Belgian priest Georges Lemaître, to rediscover his equations. However, while Friedmann obtained these equations as a mathematical result, Lemaître's view of the universe came directly from astronomical observations. He had worked with Arthur Eddington at the Cambridge Observatory, where he became interested in Einstein's and de Sitter's cosmological models. He had also spent time in the United States, visiting the Mount Washington Observatory and Lowell Observatory where they had observed the redshift of distant galaxies. Realizing the importance of this evidence, that it implied that the universe is expanding, he set out to create a realistic model of the universe.

In this way, Lemaître came up with a set of equations equivalent to Friedmann's that represented the expansion of a matter-filled universe. Assuming that all matter in the universe was once at the same place, the model made possible the calculation of the age of the universe — billions of years, large enough to account for stellar evolution. Furthermore, his model predicted that the velocity of galaxies should be proportional to their distance from us. He tested this formula with nearly a hundred measurements from Hubble and a colleague and in every case found satisfactory agreement. Lemaître arrived at the proportional relationship between redshift and distance two years before Hubble published the same relationship as an empirical result.

Lemaître sought out Einstein at the Solvay Conference in 1927 to share his results with him. This was the second time Einstein was faced with an expanding universe model as a consequence of his own theory, and this time he was prepared to accept the mathematics, if not yet the physical interpretation. Lemaître had the distinct impression when talking with Einstein about the redshifts and speeds of distant galaxies that Einstein had no awareness of the astronomical facts.

Lemaître's work, which was published in an obscure Belgian journal, received little attention. A few years later, Eddington came across it and, realizing its importance, published an English translation of the paper in the astronomical journal of London's Royal Society. Nevertheless, Lemaître's work still received such little attention that the equations he and Friedmann had developed were rediscovered again, this time by the American physicist Howard Robertson and the English physicist Arthur Walker. Credit was given to all four in naming the metric that summarizes the model — it is called the Friedmann-Lemaître-Robertson-Walker metric. The equations were named, in honor of the original discoverer, the Friedmann equations. For decades these equations remained in the background of physics, largely viewed as a mathematical curiosity.

In 1933, the Swiss-born astronomer Fritz Zwicky, who had been observing a group of galaxies known as the Coma cluster, found that the galaxies were moving too fast to be held in by the gravity of the cluster. Considering only the visible mass, they should disperse in a short time period. He came to the surprising conclusion that most of the mass of the galaxy cluster must be in the form of some kind of invisible, dark matter. Three years later, a similar discovery was made by astronomers studying the Virgo cluster.

In the late 1950s, astronomers Franz Kahn and Lodewijk Woltjer observed that the galaxy M31 was on a collision course with our own Milky Way. They concluded that the two galaxies must have completed almost a full orbit around each other since their inception in the early universe. With this information, they arrived at an estimate for the mass of our galaxy cluster that was considerably greater than the total visible mass. They suggested that most of the mass must be some kind of invisible matter, possibly in the form of a hot non-radiating gas. Apparently, they came to this conclusion without knowledge of the earlier papers that had hypothesized such non-luminous matter in the Coma and Virgo clusters.

Others studying galaxy clusters obtained similar results, that the mass consistent with a cluster's dynamics was much greater than the visible mass when assuming stable configurations. The Soviet astrophysicist Viktor Ambartsumian challenged their assumptions, suggesting rather that the galaxy clusters are unstable. In 1961, a conference was held in Santa Barbara, California to examine Ambartsumian's hypothesis, resulting in a lively debate. Both positions were problematic. Assuming that there is no invisible matter, most galaxy clusters should have already dissipated by now, contradicting observations. Alternatively, our theories are based on less than one percent of the matter that is actually there!

While scientists continued to debate the existence of dark matter, new astronomical objects were discovered in the sky: quasi-stellar radio sources, or quasars. The first of these powerful radio-emitters was discovered in 1963 — it was a hundred times brighter than any known radio source and receding from us at the astonishing speed of one-sixth the speed of light. In response, astronomers quickly put together a conference dedicated to the discovery: the "Texas Symposia on Relativistic Astrophysics." Astronomers suddenly became interested in the subject of general relativity, which for decades had been relegated to the quiet world of mathematics.

At the conference, the attending scientists celebrated the union between the two subjects. The relativists were suddenly "experts in a field they hardly knew existed," and the astrophysicists had "enlarged their domain, their empire, by the annexation of another subject." This union was strengthened when astronomers discovered that the number of quasars increased with increasing redshift, which brought up new questions regarding the formation and evolution of galaxies. It was around this same time that astronomers first observed cosmic microwave background radiation and neutron stars. The field of cosmology blossomed.

One of the key cosmological questions during this time became, as one astronomer expressed it, "a search for two numbers" — the Hubble constant and the deceleration parameter. The Hubble constant indicates the current expansion rate of the universe. The deceleration parameter determines its evolution, the rate at which gravity slows down the expansion, which in turn determines the universe's future. If Einstein's cosmological constant is taken to be zero, as many assumed, the deceleration parameter and therefore the fate of the universe should be determined entirely by its mass density.

As a consequence, estimating the density of the universe became one of the most important questions in the field of cosmology. Will the universe expand forever into infinite space? Or will the universe collapse in on itself due to the strength of its gravity? Or rather, is the density of the universe such that it will eventually reach a stable equilibrium? This last option had much philosophical and aesthetic appeal. However, all of the estimates of the density of the universe were too small for it to be true.

In 1974, two cosmological papers were published at about the same time, one by a group at Princeton and another by a group in Estonia. They were both the result of interdisciplinary efforts by physicists and astronomers, and they both came to the same conclusion: the mass density of the universe may have been underestimated by a factor of ten or more; its actual density seems to be close to the critical density that would produce a stable universe. Their conclusion was based in part on the "missing mass" hypothesized in galaxy clusters. This was connected for the first time to the similar observation that the motion of stars around the galactic center indicates large amounts of invisible matter. The papers concluded that most of the matter in the universe could be in the form of non-radiating, dark matter.

Since the publication of these papers, the search for dark matter has moved to the forefront of astronomical research. Scientists have discovered new ways to infer the existence of dark matter, notably through observations of the bending of light around galaxies, the extent of which can only be explained using current theories that assume the existence of invisible matter. Recently, new clues have emerged regarding the nature of this mysterious, invisible substance.

In 2006, astronomers viewed two colliding galaxies within the Bullet Cluster with optical and x-ray telescopes. They could see evidence of big explosions as the clouds of dust and gas collided. Close to the collision site, they detected two giant clusters of dark matter, observable through the bending of the light from galaxies behind them. Scientists deduced that while the dust and gas slowed down because of the electromagnetic resistance among the particles, the clumps of dark matter just continued along their paths, only acted upon by gravity. In this way, the dark matter became separated from the ordinary matter during the collision.

Multiple experiments are being conducted around the world to detect dark matter directly, although none have yet succeeded. Scientists are fairly certain it exists, has mass, and is present in galaxies. It does not seem to interact, or it only interacts very weakly, with the electromagnetic and nuclear forces; it therefore cannot be a group of planets, interstellar gas, or other regular matter. Scientists also do not think it could be neutrinos or black holes, at least the kinds that have been studied. Some speculate that dark matter might be made of new weakly interacting massive particles. Others hypothesize that there is no dark matter, but instead our theory of gravity is incorrect on large scales. Alternate theories have been proposed that make testable predictions, but none have yet had the predictive power of general relativity.

Will gravity halt the expansion of the universe and cause it to collapse? Or was the explosion that began our universe so great that matter will escape and the universe will expand forever? Or, by some chance or miracle, will our universe reach equilibrium? Theoretically, we can determine the fate of the universe if we know its density, but we don't. As an alternative, astronomers attempted to find ways to determine the deceleration parameter directly by looking into the past — billions of years into the past.

But they needed more than to just *see* billions of years into the past. They also needed a way to determine the distance to these astronomical objects, which is very, very difficult. The light from such objects must have passed by or through countless galaxies. It would have encountered an enormous amount of interstellar dust during the time it took to reach us. It is incredible to think that this light could retain any information after such a long trip for us to even identify an image, let alone allow us to determine the distance at which it was emitted. It seems like an impossible task, to measure the distance to objects billions of light-years away. Nevertheless, astronomers tried.

Advances in telescopes, as well as in almost every area of technology, were tremendous during the 20th century. However, even with the best telescopes it is impossible to distinguish one star from another when viewing stars billions of light-years away. Therefore, standard methods using Cepheid variable stars to determine distances fail. These stars are simply too dim. The only thing we can see at such distances is the faint glimmer produced by a whole galaxy of stars. What astronomers needed was a new astronomical tool to help determine distance. And it needed to be very bright.

Supernovas, for a short amount of time, can outshine an entire galaxy. They appear suddenly in the sky and then disappear. It was the observation of such stars that shook the confidence of Tycho Brahe and Galileo in the immutability of the heavens. With quantum theory, we now understand them. They are not new stars, as their name might suggest, but rather old stars that are collapsing. If a star is massive enough, its gravitational collapse triggers a nuclear chain reaction that causes the star to explode. This exploding star is a supernova. Supernovas are so bright that they can be seen billions of light-years away, and with the help of powerful telescopes and specially designed filters, they can be distinguished from the faint glimmer of the galaxies in which they reside.

However, for a star to be used to determine distance, astronomers must have a way to determine its intrinsic brightness. If they know a star's intrinsic brightness, they can use its apparent brightness together with the inverse-square law to determine its distance, as is done with Cepheid variable stars. Astronomers systematically studied supernovas for decades. They innovated computer detection systems and refined methods to filter and analyze the light. Finally, astronomers devised a way to use supernovas to determine the distance to objects billions of light-years away.

The key was a certain type of supernova star, called a Type 1a supernova, which always has the same mass when it collapses and therefore explodes with the same intensity, burning with the same intrinsic brightness. The reason for this consistency with regard to mass is that they grow from stars with masses below the amount required for them to collapse, called the Chandrasekhar limit — 1.4 times the mass of the Sun. They continue to gain mass from some companion star, and then just after they reach the Chandrasekhar limit, they explode. They are the brightest kind of supernova and their brightness was calibrated with the help of a Cepheid variable. Equipped with a tool to measure distances on the order of billions of light-years, two research teams set out to determine the fate of our universe.

The teams working on this were the Supernova Cosmology Project at Berkeley and the High-z Supernova Search Team in Australia. They were hoping to find out how quickly the expansion of the universe is slowing down; the greater the deceleration of the universe, the more likely it will end in collapse. By comparing the redshift of the supernovas to their distance and fitting the data to theoretical models, the groups believed that they could distinguish among the three different types of Friedmann equations, and therefore among the three possible fates of the universe. The teams arrived at the same conclusion, announcing their results in early 1998 within weeks of each other: the expansion of the universe isn't slowing down at all, it's speeding up.

Theorists began calculating. They started with the Friedmann equations with the cosmological term, which they soon named "dark energy," a mysterious energy that somehow causes matter to move away from other matter at an increasing rate, despite the mutual gravitational attraction of matter bodies. They set the cosmological constant to a value determined by the expansion data and then ran simulations with different proportions of regular matter, dark matter, and dark energy. To account for the patterns we see in the cosmic microwave background radiation, the universe should be composed of approximately 5% regular matter, 27% dark matter, and 68% dark energy. The same ratios also accounted for the observed galaxy distribution and the relative abundances of the most common elements in stars: hydrogen, helium, and lithium. The theory built on the existence of dark energy has been so successful it is now called the "standard model of cosmology."

The cosmological constant was introduced in the attempt to create a gravitational theory based on the assumption of a static universe. It was removed when observations contradicted this assumption, suggesting instead that the universe is expanding. It was then reintroduced when additional observations showed that the universe is accelerating. The cosmological term has been used to model the evolution of the universe near its beginning and near its end. However, all attempts to measure this energy directly have failed. We call it "dark energy" because we can't detect it. We can't detect it because we don't know what it is.

# 6

# A New Kind of Vision

Sometimes a colony that once
yielded a rich harvest is cast aside.
Today's truths may tomorrow be disproved,
and that is why, from time to time,
we must look backwards in order to find
the path that we must take tomorrow.

— Hideki Yukawa

I am often buried in the past. I choose my books from lists with titles like "The Hundred Greatest Novels of all Time." I make my way slowly through these lists. I am not a particularly fast reader, nor am I usually eager for a book to end. I read mostly old books not because I think there is nothing good written now, but because time is a good filter to separate the good from the great.

The same is true of my research, that I continually go back to old works. This is one of my strengths, that I take the time to read original papers and dig into the history instead of just reading summaries of summaries, which become increasingly distorted as they are retold, like in a child's game of telephone. And so I know that in his 1916 paper "The Foundations of the General Theory of Relativity," Einstein aimed at creating a theory independent of reference frame. As he wrote in italics, *The laws of physics must be of such a nature that they apply to systems of reference in any kind of motion.*" This is why he called his new theory of gravity a *general* theory of relativity; it is in contrast to special relativity, in which the laws of physics take their simplest form only in special inertial reference frames. In fact, he coined the term "Special Relativity" in this paper. Before, it was simply called relativity theory.

Although Einstein started with a system of general equations applicable to motion in any reference frame, he then restricted them to free-falling frames. These are the equations of general relativity that we know and use today. They are not general, applicable to motion in any kind of reference frame; they only apply in free-falling frames. I also know that Einstein never explained *why* this simplification works. He defended this choice of reference frame by claiming that it simplified the mathematics and produced a theory consistent with experiment.

Another thing I know through my reading is that Einstein did not believe that space is curved. As he wrote in a letter, "Ultimately, according to my theory, inertia is simply a reaction between masses, not an effect in which 'space' of itself were involved, separate from the observed mass. The essence of my theory is precisely that no independent properties are attributed to space on its own." This was his goal, to create a theory independent of reference frame, in which space has no role. Einstein did not propose that *space* is curved. Rather, the path that light takes when acted on by gravity is curved.

I furthermore know that when Einstein introduced the cosmological term in 1917, his primary purpose was not to create a static universe. It is true that, at the time, he did believe in a static universe, as did most other scientists. But his stated motivation was, as in his original paper, to create a *general* theory of relativity — a theory that is independent of reference frame, in which space has no role and the inertia of a body is entirely determined by other masses. In other words, Einstein wanted to create a theory consistent with Mach's principle, a term he would coin the following year. The cosmological term was Einstein's mathematical attempt to ensure that the inertia of a body infinitely far away from all other masses is zero.

From my readings, I know that people from all different walks of life have made important contributions to science. I also know that there are many ways to contribute, that you don't have to be a Newton or an Einstein to make a difference. The castles that they created were built with bricks that had been formed from the hard work of many that came before them. What all of these people have in common is their passion and persistence. And I know that each person that puts forth a new idea is taking a risk.

My strength is also my weakness; I don't keep up with the latest research. My particular path does not require it as I am trying to pull together threads, some of which have existed for over a hundred years. As far as I can tell, no one is trying to do what I am trying to do. It suits my research style that I am not in competition with anyone. I know that I can take my time.

Although I don't go out of my way to keep up with the latest research, I did hear the news on February 11, 2016 that scientists had, for the first time ever, detected gravitational waves. And they were sure of it. Scientists from LIGO had observed "chirps" in both of their detectors five months earlier. They worked intensely to confirm that these signals were really gravitational waves. They didn't want to make a hasty announcement and have to retract it later. They wrote up their results and submitted them to the *Physical Review Letters* for peer review. Their paper was accepted, and on February 11, they broadcast their news to the world.

One thousand and four scientists were celebrated as authors of this paper. It is nearly impossible to imagine the complexity of the problem of detecting gravitational waves — how many experimental, technological, analytical, theoretical, and computational innovations were necessary for this achievement, not to mention the dedication and patience of all the participants in the project. But this number gives us some idea of the effort: *one thousand and four individual authors.*

The search for gravitational waves began over a half-century earlier in the 1960s. Although Einstein had predicted their existence in 1916, scientists didn't immediately begin to search for them, at least in part because there was no consensus on their physical reality. In 1957, prominent scientists from around the world met in Chapel Hill, North Carolina for six days to attend a conference whose entire focus was gravity. Among the topics debated were various attempts at quantum gravity, questions of cosmology, and the existence of gravitational waves. Although they by no means settled the question of gravitational waves, significant support was given in favor of their existence and detectability. Furthermore, their importance was established, that they might shed some light on problems in cosmology and quantum gravity.

One of the participants at the conference was an engineer from the University of Maryland, Joseph Weber. He had studied general relativity with Archibald Wheeler while on sabbatical at Princeton the previous academic year. During the conference, he listened to talks and participated in discussions. Soon after, he began to design what would become the first gravitational wave detector.

Weber proposed that gravitational waves should be detectable similarly to how a radio antenna detects electromagnetic waves. He created a large antenna that consisted of a twenty-four-hundred-pound aluminum cylinder about one and a half meters long and a little over a half-meter in diameter. He suspended it in a vacuum chamber over an acoustic filter to reduce seismic noise. He used a transducer system to convert the slight mechanical strains that should result from its interaction with gravitational waves into an electrical signal.

Weber's gravitational wave detector was operating with good sensitivity and isolation by the beginning of 1965. Two years later, he reported a possible gravitational wave signal. The next year, with an improved design — two aluminum cylinders tuned to the same frequency spaced about two kilometers apart — he again reported a signal. In 1969, now with the antennas spaced about 1000 km apart, one at the University of Maryland and the other at the Argonne National Laboratory in Illinois, Weber reported positive results with certainty. These results were received with enthusiasm, and several groups began building their own detectors following Weber's design. However, none were ever able to confirm his results. Weber's efforts toward the detection of gravitational waves ultimately ended in failure, but he had started the ball rolling.

Soon after the invention of the laser in 1960, several scientists including Weber conceived of another kind of detector, one that would detect gravitational waves by their effect on light rather than matter. Specifically, they could construct a laser interferometer similar to what Michelson and Morley used in their attempts to detect the influence of ether on the speed of light. The most important difference was that an interferometer used to detect gravitational waves would have to be much, much larger.

The first prototypes of laser interferometers were built in the late 1960s and early 1970s. Afterwards, scientists worked to develop an instrument with suitable sensitivity that might have some chance of getting a positive result. Physicists had been able to estimate the strength of gravitational waves; they should be very, very weak — so weak that only cataclysmic cosmological events like a collision between two massive stars would be detectable. It was also necessary to have multiple detectors in different locations so that they could isolate a gravitational signal from instrumental or environmental noise and determine the location of the source.

Scientists worked together in an international effort to create a network of gravitational wave detectors around the world. Laser interferometers were built in Germany, Italy, Tokyo, and at two sites in the United States. They had arms that ranged from three hundred meters to four kilometers. By the early 2000s, the first network of interferometers was operational. Throughout that decade, experimentalists worked on various techniques and analytical tools to improve sensitivity and eliminate noise in the detectors, while theorists continued their work using general relativity, making approximations and performing numerical calculations to predict the waveforms that they should observe if gravitational waves were ever actually detected.

It soon became clear that these early gravitational wave detectors would not be able to detect even the strongest signals, which led to a proposed upgrade of all the existing detectors. The upgrades included increased laser power, larger mirrors with better coatings, reduced thermal noise, and better seismic isolation. LIGO, the United States system of detectors that consists of one interferometer in Washington State and another in Louisiana, was the first to complete the improvements. Their construction was finished in 2014, and by August of 2015 both sites were operational.

On September 14, 2015 at 9:50.45 UTC, the two interferometers nearly simultaneously detected a brief but clear signal. They estimated the false alarm rate of the signal to be less than one event per two hundred thousand years. After subtracting out the noise, they got a smooth curve that oscillated up and down about eight times while increasing in amplitude and frequency. After reaching its maximum amplitude, the wave quickly decayed. The entire signal lasted only one-fifth of a second.

The shape of the signal was consistent with the waveform of two orbiting masses spiraling into each other. The high frequency of the signal, which reached 150 cycles per second, suggested that the masses were black holes, the only objects compact enough to reach such frequencies before merging. The decay of the waveform after it peaked was consistent with the predicted "ringdown" of a merged black hole as it settles into a stable state. And so, not only was this the first direct detection of gravitational waves, it was also the first-ever observation of a black hole merger — a huge cosmic event that had until then always occurred in complete darkness.

In August of the following year, LIGO detected another event, this time in collaboration with the upgraded Virgo interferometer in Italy. The signal was clearly visible in the two LIGO detectors but not in the Virgo detector. The timing of the signal in the two LIGO detectors along with the absence of a signal in the Virgo detector helped determine an approximate source location. Within this area of the sky, 1.7 seconds later, a short gamma ray burst was observed by the Fermi GBM space telescope. Subsequent analysis of the signals, gravitational and electromagnetic, provided strong support that they came from the same source — the merging of two neutron stars.

And this is just the beginning. Our network of ground-based gravitational wave detectors is slowly expanding throughout the world; two additional detectors, one in Japan and the other in India, should join LIGO and Virgo by the middle of the 2020s. In addition, plans are underway to build a sky-based laser interferometer using three spacecraft arranged in an equilateral triangle with two-and-a-half-billion-meter-long sides. A new age of astronomy is dawning in which gravitational wave detectors work alongside telescopes, giving new eyes through which to view our universe.

As I was reading about the exciting developments in the detection of gravitational waves, I was led back into the past. Most articles on gravitational waves point to an article written by Einstein in 1916, where he predicted their existence. Some also point to a corrected version he published two years later. Interested, I looked up these papers and others discussing them. In one article, the author made an interesting claim: Einstein was unable to derive the existence of gravitational waves using his simplified equations of general relativity, the ones that assume a free-falling frame. In order to derive the existence of gravitational waves, he had to return to the general form of his equations — the ones valid in all reference frames, the ones consistent with Mach's principle.

What an exciting thought! I read Einstein's two articles more carefully and began searching through all his correspondence from this period. The first mention I found of gravitational waves was in a letter written to Karl Schwarzchild, a German astronomer who was at that time fighting at the Russian front. Schwarzschild had recently written to Einstein with an exact solution to general relativity's field equations in the case of a spherical, non-rotating body. They continued to correspond, and in February of 1916, Einstein sent him a mathematical proof that there are no gravitational waves analogous to light waves. He reasoned that this is probably related to the non-existence of the gravitational "dipole," since mass does not come in "positive" and "negative" as does electrical charge.

Meanwhile, Einstein's friends in Leyden, especially Ehrenfest and Lorentz, were challenging him about the specialized, free-falling coordinate system that he had assumed when creating general relativity. Ehrenfest questioned whether Einstein's theory, expressed in the special reference frames, fulfills the general covariance requirement. Lorentz also had some objections to it. In one letter, Einstein wrote, "My specialization of the frame of reference is not based *solely* on laziness." He added that he might present the material also without the specialization, similar to what Lorentz had done in a recent paper.

Another colleague in Leyden, the astronomer Willem de Sitter, also became interested in the generally covariant field equations of relativity. A letter he had written inspired a paper in which Einstein derived the existence of gravitational waves using the general equations, with the simplifying assumption of a weak gravitational field. Einstein justified the general form of the equations for this application by claiming that his specialized reference frame is "not advantageous for the calculation of fields in first approximation." In this paper, Einstein demonstrated that gravitational fields always

309

propagate at the speed of light.

On the same day that he presented his paper, Einstein wrote to de Sitter summarizing his results. He claimed that the gravitational equations to first-order approximation can be solved exactly if the condition for the special coordinate system is abandoned. With this result, he continued, one might think that the specialized coordinate system of the free-falling frame was not at all natural. However, Einstein continued, in the more general formulation three kinds of waves appear, only one of which carries energy. But, when restricting the results to the special coordinate system, only the energy-carrying waves remain. In this way, he justified the specialized, free-falling reference frames.

Einstein continued to apply and extend general relativity, working with many scientists in their efforts to do the same. In all of these efforts, including his cosmological model, he used the specialized coordinate system. Nevertheless, he returned to the general equations once again when writing a second paper on gravitational waves, completed early in 1918, to correct a calculation error he had made in his original paper. His corrected results seemed to imply that the waves in his specialized reference frame carry no energy. He concluded that a static, spherically symmetric mass distribution, which was the source in his calculations, cannot radiate.

Looking at this last paper that Einstein wrote on the topic, it is no wonder that for decades many people doubted the existence, or at least the detectability, of gravitational waves. But there was another paper written in 1922 by the English astronomer Arthur Eddington that added clarity to Einstein's earlier work. Eddington began by reviewing the three types of gravitational plane waves. He demonstrated that the first two types have no fixed velocity, casting doubt on their existence. They cannot be detected by any conceivable experiment and, as Eddington wrote, are merely an "analytical fiction."

However, when unrestricted coordinates are used, a third type of wave appears. These waves travel at the speed of light *in all reference frames*; they are the only physical wave predicted by the theory. And so, while practically every application of Einstein's theory of general relativity, every attempt at unifying gravity with the other forces, and every attempt at creating a quantum theory of gravity makes use of Einstein's specialized free-falling coordinate system, the prediction of gravitational waves only emerges from the general field equations, the ones that take the same form in *every* reference frame, the ones consistent with Mach's principle.

# A NEW KIND OF VISION

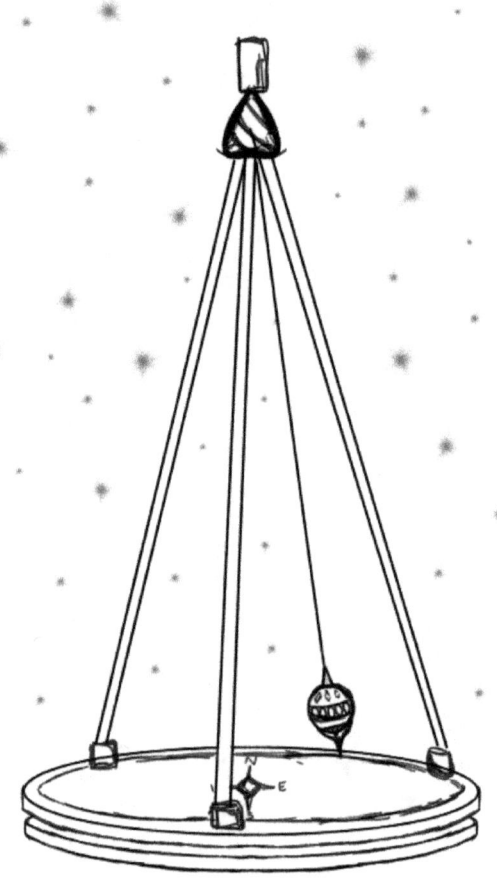

# BOOK IV

# Mach's Principle

# 1

# Ernst Mach

Philosophers who have a greater extent of thought,
and juster notions of the system of things,
discover even the earth itself to be moved.
In order therefore to fix their notions,
they seem to conceive the corporeal world as finite,
and the utmost unmoved walls or shell thereof to be
the place, whereby they estimate true motions.
If we sound our own conceptions, I believe we may find
all the absolute motion we can frame an idea of,
to be at bottom no other than relative motion thus defined.

— George Berkeley

Ernst Mach's father was a tutor in Vienna, almost always away from home. However, when he was home, he took an active role in his children's lives. He entertained Ernst and his sisters with his excellent storytelling and magic shows. He introduced young Mach to science through simple experiments using materials he found around the house. Ernst then began doing his own experiments, but he was not always careful. Once, he tested camphor to see if it would burn and singed his eyebrows.

When Mach was nine, he was sent away to attend a Benedictine gymnasium. He had difficulty, especially with Greek and Latin grammar. In particular, he resisted learning the sentence, *Initium sapientiae est timor domini,* "The beginning of wisdom is the fear of God." The only class he enjoyed was geography. The Benedictine fathers considered him to be without talent and completely unteachable. They passed him but recommended that he learn a trade or be prepared for a career in business.

His father brought him home and began the toughest tutoring job of his life. He taught his son Greek, Latin, history, geometry, and algebra. When progress was slow, he shouted insults. Eventually, Ernst began to enjoy some of the ancient works once he realized that they weren't all about aggressive kings fighting wars. He also listened eagerly to his father's stories about Archimedes, Vitruvius, and other early scientists.

At fifteen, Mach made a second and slightly more successful attempt at a formal education, this time at a Piarist gymnasium. He had difficulties socially, largely because he had no experience in this area; his only playmates had been his younger sisters. Furthermore, despite the fluency he had gained while working with his father, he still disliked ancient languages. Nevertheless, Mach did well enough to graduate, and in the fall of 1855, he entered the University of Vienna to study his favorite subjects — mathematics and physics.

Mach entered the University already behind. His courses assumed knowledge of both differential and integral calculus, neither of which he had studied. Not able to find either introductory classes or an easy textbook, he used his meager allowance to pay for tutoring and studied the subjects on his own. He did passably well in his courses. He also became skilled in laboratory techniques, working with Andreas von Ettinghausen, the head of the Vienna Physical Institute. After five years, he earned his doctoral degree.

Mach remained in Vienna, and the year after graduating he became a *privatdozent*, an unsalaried lecturer. He earned some money by giving public lectures to medical students. He also gave popular lectures on the new field of psychophysics and Helmholtz's latest acoustic discoveries. In addition, he was able to do original research by shifting his interests toward physiology, in which he could conduct experiments with simple or no equipment.

Although he published his lectures, Mach received no professional recognition for them. Despite this, he did not go unnoticed during these years. He often joined a group of scholars, artists, and doctors of medicine and law who met daily at a Vienna coffeehouse known as the *Cafe Elefant*. People would come and go from early afternoon until late into the night — a continuous stream of wit and argument on various topics in science, philosophy, and art. Mach, with his "profound understanding and reflective manner," seemed to preside over the gatherings.

In 1864, Mach was appointed the Chair of Mathematics in Graz, the capital of the Austrian province of Styria. Recalling his own mathematical deficiencies, he offered classes in introductory calculus. He also taught advanced mathematics, physics, physiology, and psychology. His lectures on psychophysics drew large crowds and even other professors. He also continued to write and research; during his years in Graz, he published three books and twenty-seven articles on various topics in experimental physics and physics applied to psychology and physiology. Although he was the mathematics chair, Mach didn't publish a single article on mathematical theory. His interests were in the sciences, and when he was offered the position of Chair of Physics at the University of Prague, he accepted.

Mach moved to Prague in the spring of 1867. It was a beautiful city, rich in architecture, inventors, reformers, and all kinds of characters. A reflection of its history could be seen in the languages of its street signs — Czech, German, French, Russian, Turkish, and Greek, even on the same street. And its complex history was giving way to a complex present as Czech nationalism grew and the German minority fought to defend themselves, especially to retain German as the dominant instructional language. Mach, although born in the Czech region of Moravia, considered himself a German-speaking Austrian. His Czech countrymen tried to pull him to their side, but he refused to be drawn into this political debate.

The summer after moving to Prague, Mach took a trip back

to Graz to get married. Afterwards, he and his new wife settled into a fifth-floor apartment in the laboratory building, which sat in the center of Prague. Before long, they were joined by five children, each of whom acquired at least one nickname. They had cooks, nursemaids, servants, and numerous visitors. They also had an assortment of animals; Mach's favorites were the half-tamed sparrows.

In his experimental physics classes, Mach took a historical approach and supplemented his lectures with many demonstrations, often using instruments that he designed himself. His original laboratory equipment included a pendulum specially designed to demonstrate the relationship between period and acceleration, a tobacco smokebox used with a prism to visually display the separation of light into colors, and a wave machine. He invented, experimented, and wrote. In his first twelve years in Prague, he published four books and over sixty articles.

Mach was well respected at the University, and in 1879, he was elected *Rector Magnificus*. That same year, Count Eduard von Taaffe came to power in Vienna with a plan to reconcile Slavic minorities with the state. The Czechs at the University of Prague immediately demanded full language parity, which Taaffe supported. Mach, after consulting with university deans, openly proposed that the Czechs establish a separate university. Furthermore, in an attempt to appease both sides, he secretly recommended that every historic building be partitioned into two halves, with separate entrances for each. This was eventually implemented, but no one was satisfied.

The following spring, a German nationalistic student group decided to celebrate the twentieth anniversary of its founding. They invited professors, dignitaries, and special student groups from Germany — about three hundred people in all. As rector, Mach gave a speech about the importance of moderation. Others were not so sober. Specifically, the dean of the medical students, Professor Kleb, incited the guests: "The enemies of Germandom do not even want their own university, which has been offered to them, because they know they are incapable of maintaining it."

Two days after the celebration ended, hundreds of Czech students gathered before Professor Kleb's house and threw insults and stones. The group then moved on to Mach's laboratory and home where they did the same. The next morning, Mach posted a notice on the main student bulletin board with the message that student demonstrations such as those held the previous night would not be tolerated. Newspapers in Prague and throughout central Europe got word of the incident, and Mach became known as the "defender of German rights." Two months later, he received regular membership in the Austrian Academy of Science, the most prestigious scientific society in the deteriorating Austro-Hungarian monarchy.

After a year, the rectorate was passed on to the next deserving professor, and Mach had more time for other activities. He continued to teach, experiment, and write. He also supervised the construction of a new science building away from the center of Prague, where students continued to riot.

In 1881, after hearing a lecture at the First International Electrical Exhibition in Paris, Mach became interested in supersonic bullets. The lecture was given by the Belgian artillerist Louis Melsen, who presented a theory accounting for crater-like gunfire wounds experienced during the recent Franco-Prussian War. Because of the nature of these wounds, the French had been accused of using explosive bullets in violation of the International Treaty of St. Petersburg. Melsen disputed the charges, claiming that the cause of the wounds was the impact of compressed air, not prohibited bullets.

Mach wanted to test Melsen's hypothesis, but he was also interested in the problem for another reason. It was commonly reported by artillerymen that *two* bangs were sometimes heard as a result of a gunshot. No one could explain the existence of the second sound. Mach thought that he might be able to explain the two bangs and also the reason for the crater-shaped wounds if he could take a clear photograph of a rapidly moving bullet. He also suspected that both of these occurrences were related to a discovery he had made several years earlier — the existence of supersonic spark waves.

Mach had made this discovery while trying to create an accurate timing device. He designed the timer assuming that sparks are like tiny explosions that produce light and sound. By discharging two sparks at the same time over soot-covered glass, a soot line should be produced where the sound waves meet and reflect, halfway between the sparks. If the sparks are discharged at different times, the position of the soot line should indicate where the waves met and, with the speed of sound, could be used to calculate the precise discrepancy in their discharge times.

To test the device, Mach discharged the sparks using a flying bullet. A soot line appeared, as expected. In addition, he saw two curious V-shaped angles on each end of the soot line. By analyzing the patterns, he determined that the spark wave must have traveled faster than the speed of sound! It could not be a regular acoustic wave, even if it did make a sound. This observation ruined his idea of a timing machine, but it led to an interesting series of experiments on these supersonic spark waves, which became known as "shock waves."

Mach performed various experiments with shock waves to understand the unusual reflection that caused the V-shaped soot marks. He then attempted to photograph the shock wave. However, this proved to be very difficult, and he still had not succeeded in obtaining a good image in 1881 when he heard of Melsen's theory. He remained busy for a few years after that, writing and performing various other experiments. In 1883, he was again called upon to act as University Rector, although the worsening political situation compelled him to resign shortly after. In 1884, he finally succeeded in photographing a spark shock wave, after which he attempted to photograph the shock wave of a speeding bullet.

In his setup, the bullet was shot through a ring, which forced air through a small tube. A candle placed at the end of the tube would then flicker, shorting a circuit that would cause a Leyden jar to discharge and produce a brief spark. By adjusting the length of the tube, the spark could be discharged just as the bullet reached the center of the camera's field. In this way, the spark acted as both the light source and the camera shutter. He photographed the bullet using the same techniques that had been successful in photographing a spark shock wave. Although he obtained clear photographs, he did not see the expected image of a shock wave.

Mach guessed and soon verified that the bullets had not exceeded the speed of sound. He would need faster bullets! He then worked with Peter Salcher, a physics professor at the Austrian Naval Academy, who had access to such bullets. They exchanged hundreds of letters over the course of a year, with Mach giving detailed instructions on how to conduct the experiment. Finally, they were successful, and on June 10, 1886, Mach deposited a note to the Austrian Academy of Science with two clear photographs of the shock wave from a supersonic bullet.

The Austrian Navy soon made a large cannon available to Salcher for further research. Prussia, an ally of Austria-Hungary, responded by offering Mach a place to continue his own experiments. Mach accepted and began working with his eldest son, Ludwig. Their photographs were well publicized in scientific journals, popular magazines, and newspapers.

With these photographs, Mach could confirm Melsen's claim that the crater-shaped bullet wounds experienced in the Franco-Prussian War were not due to exploding bullets, and he also gave a more accurate explanation for their appearance. Furthermore, he could explain why two bangs often result from a gunshot. The first is from the supersonic shock wave that travels at the head of the bullet; the second is from the sound waves coming from the exploding powder. He therefore concluded that when artillerymen hear a double gunshot, it is because the bullet is traveling faster than the speed of sound.

Mach was a teacher and an experimental scientist; as such, he became interested in questions of education and knowledge. Although fragments of these themes had appeared in some of his earlier writings, his first important philosophical work was a fifty-eight-page pamphlet published in 1872 called *The History and Root of the Principle of Conservation of Energy.*

As a youth, Mach began, he had particular difficulty understanding two things. First, he couldn't understand why anyone would want to be ruled by a king, even for a day. Second, he couldn't understand why, borrowing the words from the German writer Gotthold Lessing, "That wealthy folk upon our sphere, Alone possess the riches." Confronted with such difficulties, a person is faced with two choices: just grow used to the ideas or try to understand them through history.

Similar difficulties arise, he continued, when we go to school and ideas are presented to us as self-evident, some which have been developed over thousands of years. Mach's solution to this was the same; these ideas should be understood through history: "One can never lose one's footing, or come into collision with facts, if one always keeps in view the path by which one has come." Therefore, to learn an idea was to learn its origin.

Mach then offered a critique of classical education, which was supposedly a historical education. First, he suggested that one does not gain historical perspective through *eight years of grammatical studies.* Moreover, to the extent that classical education was historical, its scope was too narrowly focused on the Greeks. Instead, it should include all cultured peoples of the past so that students learn to see the world from points of view other than those in which they have been brought up.

Having argued for a historical approach to education, Mach began to examine the principle of energy conservation in a historical context. This principle was usually considered the "flower of the mechanical view of the world, as the highest and most general theorem of natural science." However, it was not as new as was usually assumed. In fact, it existed even before Galileo with the work of the Flemish mathematician and military engineer Simon Stevinus. Its roots were located even deeper within the ancient law of causality, the assumption that an effect is determined by a cause.

More generally, Mach suggested that what we can learn about nature is only the connection of appearances with each other. How we represent these observations depends on our theories, which rank certain facts as "fundamental" according to convenience, custom, and history. This presentation — the organization of facts into fundamental relationships — has the benefit of economy of thought. Its value is in its convenience; it is easier to memorize facts and therefore make predictions if the facts are organized efficiently. Nonetheless, Mach insisted, the way of presenting the facts does not alter the facts themselves.

As a consequence, our physical theories have at most provisional value in that they aid in the organization of many diverse facts. However, if our theories are so limited and inflexible that they no longer allow us to see the many-sidedness of phenomena, they will become a hindrance in our efforts to gain knowledge of the natural world. It is therefore important, he suggested, that we periodically review our theories to give us the opportunity to cast aside what is worthless. In conclusion, theories should be like "dry leaves which fall away when they have ceased to be the lungs of the tree of science."

One of the "dry leaves" that Mach found particularly problematic was the atomic theory of matter. This theory had recently gained support with the rediscovery of a paper written by the Italian scientist Amedeo Avogadro, in which he had proposed a proportional relationship between the volume of a gas and its number of molecules. Josef Loschmidt, an Austrian grade school teacher, used this relationship to estimate the number of molecules in a given volume of gas. He also estimated the diameter of a molecule of air, assumed to be a sphere. For many scientists, this was considered good evidence not only of the *utility* of atomic theory, but also of the physical existence of atoms.

Mach believed that matter is much more complicated and much more abstract. If oxygen and hydrogen are mixed, they explode and are replaced with water. Consequently, scientists claim that water *consists* of hydrogen and oxygen. Mach claimed, rather, that hydrogen and oxygen are merely two thoughts or names kept ready to describe matter that is not currently present but will appear again when the water is decomposed. Furthermore, he did not believe that the hypothesis of the indestructibility and conservation of matter was testable; it was weight, not matter, which was shown to be conserved.

Mach's criticisms of atomic theory were initially ignored. Meanwhile, scientists in Vienna — notably Josef Stefan, Loschmidt, and the young graduate Ludwig Boltzmann — began to make great progress applying atomic theory to the behavior of gases. However, the more they succeeded, the stronger Mach's opposition grew. Mach granted provisional value to the concept of an atom. Nevertheless, he argued that science should describe and relate observations in the simplest possible way, and atomic theory was *not* the simplest possible way. He believed that atomic theory would eventually be abandoned in favor of a completely mathematical description of matter.

In 1883, Mach published another book, *Science of Mechanics*, in which he more clearly and strongly communicated his philosophy of science. He organized the book chronologically as a history, starting with Archimedes and ending with contemporary formulations of Newtonian mechanics. Through history, Mach explained the principles of mechanics and how they were developed, often going into great detail describing the experiments upon which they were based. Although he made significant use of mathematics throughout the book, especially in the later chapters, his aim was to make his treatment as clear and interesting as possible, rather than "buried and concealed beneath a mass of technical considerations."

At the center of Mach's history stood Newton, who had formulated the laws of mechanics so completely that in the two hundred years since, no new physical principle needed to be introduced. The only developments that had occurred were of a formal and mathematical nature. However, the novelty of the subject had forced Newton into a certain disconnectedness. Mach compared him to a commander of an army, who "cannot stop to institute petty inquiries regarding the right by which he holds each post of vantage he has won." Two hundred years later, now that the subject of mechanics has settled into a kind of tranquility, we should pause and examine its philosophical foundations.

Specifically, Mach aimed to rid Newton's theory of "metaphysical obscurities," elements in the theory that were unsupportable by experiment. Furthermore, he wanted to get rid of redundancies to make the theory as economical in its presentation as possible. Toward these ends, he critically analyzed Newton's ideas regarding mass, absolute time, and absolute space.

Newton had defined mass as something distinct from weight; specifically it is the product of an object's density and volume. Mach called this definition "unfortunate" since the only way to define density is mass divided by volume, leading to a circular definition. Newton also identified mass as the "quantity of matter," which Mach claimed lacked clarity.

Mach proposed instead that we define mass in terms of acceleration. According to this, if two bodies acting on each other experience equal and opposite accelerations, they have the same mass. If two masses acting on each other accelerate differently, the ratio of their masses is defined as the inverse ratio of their accelerations. In this way, the relative masses of all bodies can be determined without appealing to any theory. In addition, this definition eliminates the need for a separate law stating that when masses interact, the forces between them are equal and opposite; rather, this law is embodied in the new definition of mass. Finally, Mach demonstrated that a consequence of this definition is that mass can be determined through weight, eliminating yet another redundancy.

He next criticized Newton's views on space and time. Newton had pronounced these as absolute, without regard to anything external. Mach called absolute time an "idle metaphysical conception" without either practical or scientific value. We can only measure time using the motion of another body such as the ticking of a clock or the rotation of the Earth. Similarly, he thought that absolute space and motion are "pure things of thought, pure mental constructs, that cannot be produced in experience."

Newton did not ignore these difficulties, at least with regards to absolute space. He acknowledged that "space" cannot be directly observed with our senses. However, he claimed that we do have evidence of it *indirectly*. As an example, the centrifugal effects due to the Earth's absolute rotation cause it to bulge around the equator. If we assume that the Earth is at rest while the rest of the universe rotates, we have no explanation for this observation.

Newton had suggested an experiment to indirectly measure acceleration with respect to space by observing water placed in a spinning bucket. As the bucket rotates, the centrifugal force on the water causes it to move outwards toward the sides of the bucket — the greater the absolute acceleration, the higher the water rises. In this way, Newton claimed, we can determine the acceleration of a body with respect to space.

Mach challenged these conclusions. He suggested that centrifugal forces are produced by the relative motion of the water with respect to the Earth, stars, and other bodies. Although from these observations it appears that the bucket produces no noticeable centrifugal forces, who can say how the experiment would turn out if the sides of the bucket were several leagues thick?

In summary, all of the principles of mechanics had been built on experiments that only dealt with relative positions and motions of bodies. Mach showed this in detail with the development of its history. In terms of relative quantities, the motion of the universe is the same whether we allow the Earth to rotate and the stars remain fixed in the background, or whether the Earth is stationary with the stars rotating. Both are equally correct in *describing* our universe. If our theories don't reflect this relativity of motion, then we should reexamine our theories.

Mach then began to reexamine mechanical theory starting with the law of inertia. According to this law, a body tends to retain its motion with respect to absolute space. Mach suggested that absolute space is simply an abbreviated reference to the entire universe. Even more generally, the law of inertia should apply to all pairs of masses so that every mass resists a change in motion with respect to every other mass. Since the bulk of the mass in the universe is in the distant stars, bodies *seem* to preserve their motion with respect to the stars. Mach then suggested a mathematical equation to represent the inertia between masses, written only in terms of relative quantities. Applying this equation to a universe where enough of the masses appear to be fixed in space, the usual law of inertia is recovered. However, he admitted, it is impossible to say whether this expression would be a correct description if the stars moved rapidly among one another.

With the publication of *Science of Mechanics*, Mach's philosophical views steadily gained acceptance. Several editions of the book were published during his lifetime, with additional material and commentary in numerous appendices. Many scientists accepted

Mach's redefinition of mass in terms of acceleration, which seemed to be valid when considered from many different points of view. Even Boltzmann, who continued to disagree with Mach about the existence of atoms, adopted his economy of thought; he defended atomic theory as the most "economical" way to describe and relate physical observables. More and more scientists became interested in the foundations of mechanics, and many productive controversies arose as a consequence of Mach's work.

Meanwhile, Mach became involved in educational reform, a movement that had emerged in Europe and America as a result of democratic ideals and a rise in industry. By that time, many schools had already made significant changes in their curricula, including less Greek and Latin and more mathematics and science. However, most secondary school graduates in Prussia came from gymnasiums, which had retained their Greek and Latin-oriented programs.

In April of 1886, Mach delivered an address before the Union of Realschule Men in Germany. He began by reviewing the history that had led to the current educational system. His purpose was not to say much that was new, but he did put forth some proposals.

One of his most important suggestions was to greatly decrease the amount of material, number of school hours, and amount of work done outside of school. Mach claimed that overwork was not only harmful to the body, but also extremely injurious to the mind. He continued, "I know nothing more terrible than the poor creatures who have learned too much. Instead of that powerful judgement which would probably have grown up if they had learned nothing, their thoughts creep timidly after words, principles, and formulae, constantly by the same paths. What they have acquired is a spider's web of thoughts too weak to furnish supports, but complicated enough to produce confusion."

How then can we better educate students in mathematics and science if we decrease the subject matter? Mach's solution was to abandon systematic instruction, at least as it is required by all students. Students should be able to choose different programs of study depending on their intended professions. "Uniforms are excellent for soldiers, but they will not fit heads," he declared. Both students and teachers need to have freedom if they are to give good results.

Mach's speech received attention in newspapers and journals throughout Germany, Austria, and even France. The following year, Mach became a coeditor of a new educational magazine in which he wrote several articles sharing his own educational philosophy as well as articles about teaching physics, physical experiments, laboratory equipment, and thought experiments. He also wrote several textbooks for use in secondary schools, which were reprinted and revised for use throughout central Europe for many decades.

By 1890, the gymnasium reform movement in Germany had grown so strong that a conference was called in Berlin to settle the matter. Wilhelm II, the young and recently crowned Kaiser, attended the meeting and proclaimed that reform was the only answer: "We should bring up young Germans and not young Greeks and Romans." Two years later, the gymnasiums in Germany were reformed with more German, less Greek and Latin, and fewer hours overall.

After a decade of theoretical and experimental research in the field of electromagnetism, Heinrich Hertz began writing a book on mechanics. This would be his last scientific work. He had been working on it for three years when, in poor health and fearing that his end was near, he sent the greater part of the manuscript to his publisher. He then asked a colleague, Philip Lenard, to edit the book in case he would not be able to. Lenard accepted and, studying Hertz's notes and earlier manuscripts, completed the book. It was published in 1894, shortly after Hertz's death.

Mach referenced this book in one of the appendices to the 1902 edition of *Science of Mechanics.* He praised the work, which he felt nearly expressed his own position. Although Mach summarized the main parts of the book, he recommended that anyone interested in the foundations of mechanics read Hertz's book in its entirety.

Following a general suggestion made by Mach, Hertz had rejected the concept of forces, which he called "empty-running wheels," not observable to the senses. Similarly, and with equal vigor, he criticized energy concepts in terms of their physical reality. Instead, he attempted to describe all mechanical laws using only three quantities — position, time, and mass — with the aim of only giving expression to what was directly observable.

As a starting point for his mechanical theory, Hertz assumed a generalized law of inertia according to which every mass resists a change in motion with respect to every other mass. In addition, he assumed the principle of least action in which a mathematical expression called the "action" is minimized or maximized to obtain equations of motion for the objects in the system. In his theory, every deviation of a mass from straight-line, uniform motion is due, not to forces, but to direct interaction with other masses.

In order to create a realistic theory, Hertz had to introduce hidden masses with hidden motions, similar to the cells that Maxwell had assumed in one of his models leading up to what eventually became a completely mathematical theory of electromagnetism. In this way, Hertz eliminated action at a distance from gravitational theory as Maxwell had done for electromagnetism, which was undoubtedly one of the motivations for his research in mechanics. Mach called Hertz's theory "simpler and more beautiful," though not more practical, than the present theory of mechanics.

Ultimately, Hertz was also not satisfied with his own theory, but he had made a start. In his book, he anticipated not only future theoretical advances in the study of mechanics, but also new experimental discoveries. A few years after its publication, the French mathematician and physicist Henri Poincaré discussed Hertz's book, using the words "classical" and "modern" to contrast current theory with Hertz's vision of a future theory. Boltzmann used the same terms when referring to Hertz's book in 1899. As a result, when Planck and Einstein came out with their revolutionary theories at the beginning of the 1900s, scientists already had words to describe them — "modern physics," a term that originally was largely based on Mach's philosophy of science.

# 2

# Einstein and Mach

I believe even those who consider themselves
opponents of Mach are hardly aware of how much
of Mach's way of thinking they imbibed
— so to speak, with their mother's milk.

— Albert Einstein

Mach made a great impression on physicists during his time, especially the younger ones. They enjoyed his skepticism, independence, and intelligent critique of Newton. Not least among his admirers was Einstein, who read Mach's *Science of Mechanics* as a teenager after it was recommended to him by his friend Michele Besso. Einstein wrote about its influence years later, saying that it shook his confidence in Newtonian mechanics as the final basis of physics. And so, as Einstein entered the field of physics, despite claims by many that all the great theories had already been formed, he believed that there was still interesting work to be done.

Einstein was in good company in his admiration of Mach. After graduating from the Zürich Polytechnic, he joined other youth in a reading group where they discussed science and philosophy. It was a band of individuals; they all had their own ideas. "But outside of that," recalled one of its members, "one person thought for all of us: Ernst Mach. The great Vienna physicist and natural philosopher was the central Sun for us. In his name we had collectively founded a quasi organized society. We had made it our task to spread the teachings of the master inside and outside of the academic professions and in so far as possible to employ their fruitfulness in our own investigations."

By 1909, Einstein had gained a reputation in the scientific world for his theory of relativity, and in that same year he received a book from Mach — a new German edition of *Conservation of Energy*. In the book, Mach referred to Einstein's theory: "Space and time are not here conceived as independent entities, but as forms of the dependence of the phenomena on one another. I subscribe, then, to the principle of relativity, which is also firmly upheld in my *Mechanics* and *Thermodynamics*." Einstein replied that he had read these books with care, adding that Mach had had a great influence on the younger generation of physicists. Mach's response was lost, but it must have been favorable, as Einstein wrote a second note only eight days after his first: "Your friendly letter has pleased me enormously as has your treatise."

At this time, Einstein also sent Mach one of his own papers on Brownian motion. In this paper, he suggested an experimental method to measure the size of an atom, offering scientists a way to verify the existence of atoms. Soon after, he visited Mach, attempting to convince him of the physical reality of atoms. In this regard, the visit was a failure; Mach adamantly held onto his skepticism. Whatever the conversation was regarding relativity, Einstein left confident that Mach remained an ally on this front, if not on the matter of atoms.

Einstein made no mention of Mach in his earlier work on relativity. However, as he struggled to create a relativistic theory of gravity, Mach's ideas grew to play a central role. In a 1912 paper, Einstein demonstrated that the inertia of a mass in a spherical shell is higher than if the shell were absent. Therefore, he suggested, "the *entire* inertia of a mass point is an effect of the presence of all other masses, which is based on a kind of interaction with the latter." Along with this statement he included a footnote giving credit to Mach, writing that this was exactly the point of view he had advanced in his *Science of Mechanics*. Einstein also began to view his equivalence principle – that the effects of acceleration are indistinguishable from a gravitational field – in terms of Mach's desire to replace an approach based on forces with one that only included directly measurable quantities.

In June of 1913, Einstein wrote a letter to Mach about a recent attempt at a relativistic theory of gravity that he had finished after "unending labor and painful doubt." He told Mach excitedly about the plan to test whether light rays are bent by the Sun, in other words, whether the equivalence between acceleration and gravity really holds. If so, wrote Einstein, "your brilliant investigations on the foundations of mechanics will have received a splendid confirmation."

Around the same time, Einstein shared a new idea with his friend Paul Ehrenfest. "The conviction to which I slowly forced my way through," he wrote, "is that *privileged coordinate systems do not exist at all.*" At the time, he had only partly succeeded in working out this position formally. Two and a half years later, he succeeded, developing general field equations that take the same form in all reference frames. But then, in the same paper that introduced these equations, he placed a restriction on the theory so that the equations only apply in free-falling frames.

Initially, Einstein believed that general relativity was consistent with Mach's rejection of absolute space and his principle of the relativity of inertia. In a letter written to Karl Schwarzschild responding to a series of questions that he had asked, Einstein suggested that inertia is simply an interaction between masses, and no independent properties are attributed to "space" on its own. "It can be put jokingly this way," he continued, "If I allow all things to vanish from the world, then following Newton, the Galilean inertial space remains; following my interpretation, however, *nothing* remains."

Unfortunately, this correspondence was cut short; while in Russia fighting with the German army in the world war, Schwarzchild became ill and died. However, Einstein returned to the idea of the relativity of inertia later that year while visiting friends in Leyden. By that time he had realized that according to his theory, the inertia of a body is *not* determined exclusively by other masses. Rather, it is determined by both nearby masses and the requirement that space become geometrically flat at infinity, like the spacetime of special relativity. He justified this boundary condition by suggesting that far beyond the observable universe, unseen masses exist that nonetheless produce observable effects.

After returning home from Leyden, Einstein continued to debate these ideas through letters, especially with Willem de Sitter. De Sitter expressed his own doubt regarding Einstein's views. Where are these distant masses and of what are they composed? How does the inertia reach over here from all the way out there? Although he admitted that the principle of relativity would still hold with these boundary conditions, it would be at the cost of returning to the old concept of absolute space with the ether. De Sitter suggested that Einstein's explanation of inertia was not really an explanation from known or verifiable facts, but rather that it stemmed from an invention of masses that he believed would go the same way as the ether.

Einstein clarified his position with the claim that he did not assume the existence of any special new inertia-generating matter and simply attributed inertia to the influence of stars and other massive bodies. He admitted that his explanation was only compatible with observations if the portion of the universe visible to us were extremely small compared with the entire universe. He explained that his ideas on inertia had played an important psychological role in developing the general theory of relativity, giving him the courage to work on the problem when he was unable to obtain the covariant equations. Now that he had found these equations, perhaps there was no reason to place such importance on the total relativity of inertia.

However, Einstein was not yet ready to give up on the idea that the inertia of a body should be completely determined by its interaction with other masses. Although he *must* and *can* make do with the theory as it is, he wrote to de Sitter, "you must not scold me for being curious enough still to ask: Can I imagine a universe or the universe in such a way that the inertia stems entirely from the masses and not at all from the boundary conditions? As long as I am clearly aware that this whim does not touch the core of the theory, it is innocent; by no means do I expect you to share this curiosity!"

Einstein continued to pursue this idea of the relativity of inertia. Specifically, he suggested that the metric tensor of general relativity, which determines all spacetime properties and therefore the motion of objects, should "*be fully determined by matter and not be able to exist without the matter.*" As long as this requirement remained unfulfilled, he claimed, the goal of general relativity had not been completely achieved.

Einstein eventually succeeded in satisfying this requirement, at least he believed that he had, with the addition of a cosmological term to his original equations. Assuming a spatially finite universe, he did not need to place any boundary conditions at infinity. However, de Sitter quickly found a finite matter-free solution to the new theory. A test mass placed in this otherwise massless universe has inertia, contradicting Einstein's claim that the entire inertia of a body is determined by its interaction with other masses.

While Einstein struggled with de Sitter's universe, another challenge came from a former schoolmate at the Zürich Polytechnic, Friedrich Adler. Adler was a passionate idealist who had assassinated the Austrian prime minister in an attempt to reopen the parliament and end Austria's involvement in the war. His efforts were unsuccessful, but now, awaiting his trial in prison, he had ample time to study Einstein's theory and share his thoughts.

Adler was a loyal disciple and friend of Mach, and as such, he was enthusiastic about the "relativistic" elements of special relativity. However, he was disturbed by the absolutist principle of the constancy of light in a vacuum. He termed this *Naturwunder* and considered it suspicious enough to reject the entire theory, at least in its present form. He wrote several letters to Einstein and sent him a long paper, going on and on about Mach.

Einstein complained about Adler's ramblings to Besso, writing that he "rides Mach's poor horse to exhaustion." Besso, loyal to Mach, responded, "As to Mach's little horse, we should not insult it, did it not make possible the infernal journey through the relativities?" Einstein replied, "I am not complaining about Mach's little mount; you know what I think of it. However, it cannot bring forth any living thing but only exterminate harmful vermin."

In the spring of 1918, Einstein wrote a short paper clarifying the foundations of general relativity in response to publications that had questioned or challenged certain aspects of his theory. In the article, he coined a new term: "Mach's Principle." According to this principle, all inertial and gravitational effects are completely determined by mass, or more specifically, by the energy tensor of matter. He explained why he chose the term Mach's principle — it is a generalization of Mach's claim that inertia is an effect of the interactions among bodies. Though the necessity to uphold the principle was not shared by all of his colleagues, Einstein felt that it was "absolutely necessary to satisfy it." Finally, he claimed that although his original theory did not satisfy Mach's principle, it *seemed* to be satisfied with the addition of the cosmological term.

However, there was still de Sitter's empty universe. Einstein thought he had found a fault in this model: a discontinuity on a surface. De Sitter's world, therefore, did not act like an empty universe but rather one in which all the mass was concentrated on the surface. De Sitter disagreed, claiming that this did not rule out his solution *physically* since, as he had shown earlier, the problematic surface is physically inaccessible. Einstein came back again, insisting on the existence of a surfacelike singularity.

Then a colleague, the German mathematician Felix Klein, demonstrated that the singularity that Einstein had noticed could be transformed away. Einstein gave in. Yes, de Sitter had found a finite solution in an empty universe. After this, Einstein never again claimed that general relativity was consistent with Mach's principle.

Throughout the late 1880s and early 1890s, Mach's laboratory experienced a great decline. The number of physics students at the German University in Prague had fallen to only a handful, and financial support for his laboratory had decreased accordingly. His instrument maker had become a drunkard; Mach kept him on anyway because of his past work and large family. In addition, his principal assistant found his duties onerous and refused to do the work. Mach's eldest son, Ludwig, took over most of what should have been done by the other two, but unfortunately this caused more discord. At one point, Mach even had to intercede to prevent a sword duel.

During the winter of 1893-1894, Mach became seriously ill with the flu; he therefore declined to employ any laboratory staff that semester. His assistant took the opportunity to leave the University. In parting, he wrote Mach a friendly letter, thanking him for his patience and apologizing for his own behavior. Ludwig, no longer working in the laboratory, resumed his medical studies, which had been cut short so that he could work for his father. Mach's second oldest child, Caroline, was getting married. His third child, Heinrich, a brilliant and hardworking chemist studying in Göttingen, passed his doctoral examination that summer. In his dissertation, he wrote, "To his dear parents, dedicated in gratitude by the author." The date on his diploma was his twentieth birthday. Seven days later, he took an overdose of sleeping medicine and died.

Mach insisted that they leave Prague and move to Vienna. He contacted the University of Vienna, even offering to teach as an unsalaried "honorary" professor. No physics chair positions were available, but there was at least one open philosophy chair. He paid a brief visit to Vienna to give a lecture, long remembered afterwards, attacking the notion of cause and effect. He broke off in the middle of the speech, unable to finish. He continued in his efforts toward employment in Vienna, and in May of 1895, the Emperor of Austria and King of Hungary officially gave Mach a position at the University of Vienna as a philosophy professor.

Mach's return was triumphant. The people of Vienna were captivated by his lectures and numerous publications, both old and new. His words, *Das Ich ist unrettbar* — "the I is unreal," were adopted by the youth as a credo. He had overthrown the ego, the last of the idols! With this, they had won the highest freedom. Mach was a philosopher, but, as he emphasized in his inaugural lecture, he remained a scientist. He would build no speculative systems for them.

For two decades, Mach had been teaching the same five-hour course in experimental physics with only minor adjustments. Now in Vienna, every course he taught was different. He taught courses on the history and development of mechanics, acoustics and optics, thermodynamics, and electricity. He taught a course on the psychology and logic of scientific investigations and another on scientific instruction. His lectures were enthusiastically received, as were his lecture notes, which he compiled in the book *Popular Scientific Lectures*.

During the summer of 1898, Mach departed from Vienna to visit his son Ludwig, who had finally finished his medical degree and was employed at Zeiss Optical Works in Germany. While on the train, he suffered a stroke that paralyzed the right side of his body. Ludwig met his father and accompanied him back to Vienna. Following the stroke, Mach's memory suffered, and he had difficulty speaking. He stopped lecturing, as it became torture for him and his audience. He composed his will and prepared to die.

But Mach didn't die, at least not for many years. He learned to walk with the help of a cane. He began to write again, using a typewriter with his left hand. He kept up with current developments in physics, including quantum theory and Einstein's theory of relativity, and began to experiment again. He also continued his correspondences and wrote forwards to over a dozen books. He revised and added content to his old works and wrote many new articles on philosophy, popular science, and his experimental work.

Meanwhile, Ludwig had become wealthy through several inventions. He used some of his money to build a laboratory-house for his father in an isolated forest by the town of Haar in Bavaria. Mach was reluctant to go at first, and then he was eager. Then, he fell and injured his hip and was unable to go. Finally, in May of 1913, he got ready to depart Vienna for his last home. Before leaving, he wrote a farewell to the Austrian Academy of Science: "Should this letter be my last, please merely assume that Charon the Rogue has carried me off to a station which has not yet joined the postal union."

Upon settling in his new home, Mach's health suddenly improved. By July, he was welcoming visitors and getting back to serious writing. His most important work was a book on optics. At this time, Ludwig encouraged his father to publish the first half of the book, which was already finished. Mach agreed, on condition that they perform a number of additional experiments. And so, father and son busied themselves with experimental work. Sometimes, if they had an especially difficult problem to solve, they stayed in the laboratory for many hours at a time, living on chocolate, until they solved it. Apparently, they once stayed in the laboratory for two whole days. Finally, in 1916, the optics book was ready for printing, but then upon Mach's death it was put off, not to be published for many years.

Ernst Mach died on February 19, 1916. On this occasion, Einstein wrote a scientific obituary, praising him as a "man of rare independence of judgment." He continued, "In him, the direct joy of seeing and understanding — Spinoza's *amor dei intellectualis* — was so prevalent that he could look into the world with the curious eyes of a child, even in his old age, and thus be perfectly happy to enjoy himself by understanding interwoven connections."

Einstein more specifically praised Mach's historical-critical writings, in which he "follows the evolution of individual sciences with so much love." As the most important example, he described Mach's critique of Newton's bucket experiment, with multiple quotations from both scientists. This example, claimed Einstein, showed that Mach had come close to a general theory of relativity almost a half-century ago! Einstein complimented him, suggesting that if physicists had been questioning the implications of the constancy of the speed of light when he was still "in his fresh and youthful spirit," perhaps Mach might have come up with the relativity theory himself.

In 1921, Mach's book on optics was finally published, and undoubtedly, Einstein was eager to obtain it. In the preface, he read the following, dated July 1913:

I am compelled, in what may be my last opportunity, to cancel my views of the relativity theory. I gather from the publications which have reached me, and especially from my correspondence, that I am gradually becoming regarded as the forerunner of relativity. I am able even now to picture approximately what new expositions and interpretations many of the ideas expressed in my book on Mechanics will receive in the future from this point of view. It was to be expected that philosophers and physicists should carry on a crusade against me, for, as I have repeatedly observed, I was merely an unprejudiced rambler endowed with original ideas, in varied fields of knowledge. I must, however, as assuredly disclaim to be a forerunner of the relativists as I personally reject the atomistic doctrine of the present-day school, or church. The reason why and the extent to which, I reject the present-day relativity theory, which I find to be growing more and more dogmatical, together with the particular reasons which have led me to such a view – considerations based on the physiology of the senses, epistemological doubts, and above all the insight resulting from my experiments – must remain to be treated in the sequel.

By this time, Einstein had already somewhat dissociated himself from Mach's ideas. Now, he called Mach a "deplorable philosopher." He ridiculed Mach's idea that science is the study of the relationships among phenomena, calling this view of science a "catalog," not a system. He embraced metaphysics, declaring: "You will be astonished about the 'metaphysicist' Einstein. But every four- and two-legged animal is *de facto* in this sense metaphysicist." He also stubbornly rejected the dominant interpretation of quantum theory, which was essentially Mach's view, that all we truly *know* about nature is what we can measure.

In Einstein's *Autobiographical Notes*, written at the age of six-ty-seven, he again spoke of Mach's idea that the inertia of an object depends only on its interaction with other bodies. He admitted that he had held this view for many years until he realized that it does not fit into a consistent field theory. He again praised Mach for his "incorruptible skepticism and independence" that had shaken the dogmatic faith of the scientific community in classical mechanics and that had made a profound influence on himself.

# 3

# An Early Machian Theory

After the secular precession of the perihelion of Mercury
was deduced, in amazing agreement with experiment,
from it, every naive person had to ask: With respect to *what*,
according to the *theory*, does the orbital ellipse perform
this precession, which according to *experience* takes
place with respect to the average system of the fixed stars?

— Erwin Schrödinger

Eight years before Mach's triumphant return to Vienna, in that same city, Erwin Schrödinger was born. His father, who came from a cultured Viennese family, was his friend and teacher, never tiring of the boy's lively curiosity. His mother's family was from England; with her influence, he learned almost perfect written and spoken English. He loved studying languages, the ancient languages as well as classical German. He also enjoyed poetry and the theater. His main interests, however, were mathematics and physics.

Schrödinger entered the University of Vienna in 1906, where he studied physics under Friedrich Hasenöhrl, whose research at the time focused on blackbody radiation. In 1910, Schrödinger began publishing his own papers on various topics including statistical mechanics, electromagnetism, and x-ray diffraction. His research was interrupted for a time because of the world war, during which he served as an artillery officer in Italy. Afterwards, he became interested in Einstein's general theory of relativity and wrote several papers in this area.

In 1925, Schrödinger published a paper on Machian relativity and its application to classical mechanics. According to Mach's principle of relativity, he began, the laws of physics should hold in every coordinate system, independent of its motion. This idea follows from the recognition that we can only measure relative quantities. In classical physics, the principle is violated since the laws only hold for special inertial frames. Empirically, these inertial reference frames are all moving at constant velocity with respect to the fixed stars, but classical physics offers no reason for this.

General relativity, Schrödinger continued, also could not fulfill this relativity requirement in its original form. The theory could be used to calculate the perihelion precession of Mercury, but this invites the question: With respect to what does Mercury's orbit precess? From our perspective on Earth, it is with respect to the fixed stars. However, the fixed stars do not appear at all in the calculations. Instead, according to the theory, the precession takes place with respect to a coordinate system that satisfies certain boundary conditions at infinity, and there is no clear relationship between the boundary conditions and the stars.

Einstein attempted to overcome this difficulty by requiring a spatially closed universe, avoiding boundary conditions altogether. However, Schrödinger suggested, the theory still presents conceptual difficulties, and its mathematics makes it difficult to understand. In this way, he claimed, the concept of the fixed stars being responsible for the inertia of matter has been replaced by a theory that few can follow, within which it is "truly difficult to distinguish truth and fantasy."

Therefore, Schrödinger believed that it was worthwhile to create a classical model in an attempt to understand the relationship between the distant stars and the inertial properties of matter. Toward this end, he created a simple model of the universe: a spherical shell with mass evenly distributed on its surface. Near the center of the shell, he placed a single mass.

In order that the theory be consistent with Mach's relativity principle, all the quantities in the theory must depend only on relative quantities. The classical expression for the gravitational potential energy between two bodies already meets this requirement since it depends only on their masses and relative separation. However, the classical expression for kinetic energy does not; it is the energy of a single mass in a specific reference frame. The object's speed is an absolute speed, not relative to that of any other mass. Schrödinger modified the expression for kinetic energy to have the same dependence on mass and separation as the potential energy between two masses. Furthermore, like its classical counterpart, he let it be proportional to speed squared, where speed refers to the relative speed.

Using energy principles, Schrödinger derived an expression for the total kinetic energy of a mass placed near the center of the universe due to its interaction with the shell, which represented the mass of the universe. It equals the classical expression if the constants are related in a certain way. Using the same universe, he derived the perihelion precession of a planet orbiting a star. He showed that his result is consistent with the expression derived from general relativity and used the known precession rate to further fix constants in his theory.

Schrödinger pointed out several deficiencies in his model, the most important of which was that he assumed gravity acts instantaneously at a distance. However, the object of his paper was not to present a complete theory, but to demonstrate the existence of one that fulfills the Machian requirement of relativity and explicitly accounts for the influence of the fixed stars on inertial forces. In doing so, he was able to show that the perihelion precession of Mercury, which before had been derived only in the context of general relativity, is actually a Machian effect.

Schrödinger was also able to do something with his model that could *not* be done with Einstein's. He was able to draw conclusions pertaining to the distant mass in the universe, matter too far from us to see. Specifically, he deduced that of all the mass in the universe, our Milky Way contains only the tiniest fraction necessary to create the observed inertial effects on Earth and in our solar system. Furthermore, he predicted that the asymmetric disk-like matter distribution of the Milky Way galaxy should cause bodies located in the galactic plane to have greater inertia than those outside it. His calculations demonstrated that this effect should be just below the limits of observability.

Schrödinger was not attempting in this paper to overthrow general relativity. He only saw one flaw in Einstein's theory — that it did not, or did not seem to, include the effects of the distant stars on the inertia of a body. In conclusion, he suggested that through further development and modifications of the ideas presented in his paper, it was probable that one would arrive at Einstein's theory.

Within a year of writing this paper, Schrödinger developed a quantum wave model of the atom, after which his research was almost exclusively in this area. His paper on Machian relativity, which never received a great deal of attention, was lost even to those who researched in the field. Almost seven decades later, it was rediscovered and translated into English for a conference on Mach's principle.

# 4

# Tübingen Conference on Mach's Principle

*Hypothesen sind Netze, nur der wird fangen der auswirft.*
Hypotheses are Nets, only those who cast them will catch.

— Novalis

In July of 1993, a group of scientists gathered together at the Max Planck House in Tübingen, Germany for a five-day conference devoted to Mach's principle. Subjects of the talks included the history and development of Mach's principle, whether and how general relativity reflects Machian principles, the experimental status of Mach's principle, and quantum gravity. The papers that were the basis of these talks were published in a book, *Mach's Principle: From Newton's Bucket to Quantum Gravity*.

The book begins with a quotation by the astronomer Fred Hoyle, who attended the conference: "I think it was Hermann Bondi who once said that physics is such a consistent and connected logical structure that if one starts to investigate it at any point and if one pursues correctly every issue that branches away from one's starting point, in the outcome one will be led to understand the whole of physics. With Mach's Principle it seems something like that." This established the importance of Mach's principle, which did not seem to be challenged throughout the conference.

Although important, Mach's principle had remained "elusive." It is a guiding physical principle like energy conservation, the relativity principle, the equivalence principle, and the cosmological principle. Like other such principles, it can be believed but never proven. Its significance is that it can help shape the theories we create. But first, we must ask: What is Mach's principle?

Throughout the conference, the participants defined and redefined Mach's principle. They demonstrated that Mach's principle had appeared implicitly before Mach formulated it, even as far back as Aristotle, who defined all motion with respect to either the center of the Earth or the celestial sphere of stars. They combed through Mach's and Einstein's writings, comparing what they said, debating what they meant, and discussing how Einstein had misinterpreted Mach and how Mach had misinterpreted Newton.

They discussed the evolution of Einstein's ideas that eventually led to the coining of the term "Mach's Principle." They discussed alternate formulations of the principle after Einstein: a selection principle within general relativity, the metric determined entirely by matter *and* gravitational degrees of freedom, the need for covariant boundary conditions or spatial closure, and so on. In fact, according to the index at the back of the book, *twenty-one* different formulations of Mach's principle were offered, most of which were referenced multiple times throughout the conference.

Largely due to the ambiguity surrounding its definition, the question of whether general relativity is consistent with Mach's principle was intensely debated. As evidence of Machian effects in general relativity, presenters pointed to "frame-dragging," wherein the presence of matter "drags" an inertial frame. This indicates that in general relativity mass *influences* inertial frames even if it doesn't create them.

Nonetheless, as the conference co-organizer Julian Barbour pointed out, there remained the "bogey man" — the idea that an object placed in an empty universe still has inertia, meaning that it will resist a change to its motion with respect to space, contradicting *any* version of Mach's principle. While acknowledging this problem, Barbour still insisted that general relativity is "wondrously Machian, as perfectly so as any mortal could construct." However, according to the straw polls taken at the beginning and end of the conference, very few other participants agreed.

They discussed Machian alternatives to general relativity, including several theories developed in the early decades of the century that were virtually unknown until the conference. They also discussed more recent models. Ultimately, all alternate theories or attempts to modify general relativity to make it consistently "Machian" failed; time and time again, observational evidence favored Einstein's theory. In this way, although Mach's principle had been used to claim that general relativity is wrong, or at least incomplete, general relativity had likewise been used to claim that Mach's principle is incorrect.

One of the strongest points in favor of Mach's principle was made toward the end of the conference regarding the flattening of the Milky Way. This flattening is an inertial effect due to the galaxy's rotation, interpreted according to Mach's principle as a result of the acceleration of our galaxy with respect to distant matter. In fact, the flattening is with respect to distant stars far beyond our galaxy.

The rotation of the Milky Way is very slow; it completes one full rotation in about a hundred million years. However, if the Milky Way rotated much faster, its rotation would have been detectable in Mach's day, and Mach could have used this observation to predict the presence of enormous amounts of extragalactic matter *fifty years before it was actually discovered.* Herein lies the importance of Mach's principle, which is why discussions are still alive today — it gives us the hope that a theory that embodies this principle can help us gain knowledge about the universe.

# Is the Universe Machian?

The universe is not *twice* given, with an earth at rest
and an earth in motion; but only *once*, with its *relative* motions,
alone determinable. It is, accordingly, not permitted us to say
how things would be if the earth did not rotate. We may interpret
the one case that is given us, in different ways. If, however,
we so interpret it that we come into conflict with
experience, our interpretation is simply wrong.

— Ernst Mach

Deep within the human experience is an attraction to symmetry. This can be seen in architecture, musical compositions, poetry, and dance movements. We can see it in the rules of chess and the construction of a deck of cards. It even enters our daily lives in basic actions like the way we dress or arrange furniture in a room.

It is no real mystery why we are attracted to symmetry. If objects or ideas are symmetrically arranged, they are easier to process and remember. It is less stressful for our brains to have things organized symmetrically. And so it is no surprise that when trying to understand nature, we both consciously and subconsciously are attracted to symmetry when forming our physical theories.

The symmetry of Aristotle's universe depended on the Earth being at its center. The Copernican revolution replaced this ancient model with a universe with no fixed center, in which moons rotate about their axes while orbiting around planets that are themselves also rotating about their axes as they orbit around the Sun. Arguments were made at this time for the relativity of motion — that *all* reference frames were equally valid. Accordingly, we should be able to consider motion from the perspective of a stationary Earth or a stationary Sun. Yes, these thinkers conceded, the description of motion is *simpler* from the perspective of the Sun, but there is no physical way to determine absolute motion.

Newton disagreed. He suggested that we can *feel* absolute motion. When a bucket filled with water is rapidly spun about its axis of symmetry, the water in the bucket moves toward the walls of the bucket in its attempt to maintain a constant velocity *with respect to space*. Therefore, we can observe the water in a bucket to determine its absolute motion without reference to anything external. In the same way, the flattening of the Earth along its rotational axis is evidence that it is spinning. A rotating Earth in a stationary universe is not equivalent to a stationary Earth in a rotating universe, for in the latter case, the Earth would be an unflattened sphere.

This view was met with opposition by many contemporaries who were ready to embrace a universe with no special reference frames. But nature seemed to contradict this view; the laws of physics take their simplest form only in a set of reference frames in which, in the absence of a net force, an object does not accelerate and the water in a stationary bucket remains still. Newton's view of absolute space won out. Simplicity triumphed over symmetry.

Mach revived the idea of the relativity of all motion and added strength to this position. The laws of physics should only depend on *measurable* quantities. We can only measure relative quantities, such as the separation between masses and their relative velocity and acceleration. Therefore, a stationary Earth in a rotating universe should be equivalent to a rotating Earth in a stationary universe since these two models agree on all relative quantities. He used this to criticize Newton's mechanical theory since it predicts a flattening of the Earth in the second case but not the first. Mach set forth as a goal a theory in which the laws of physics are the same in *all* reference frames.

However, as Einstein developed his relativity theory, nature again seemed to challenge this ideal of symmetry among reference frames. According to special relativity, the laws of physics take their simplest form only in inertial reference frames. Although Newton described these reference frames with respect to absolute space and Einstein described them with respect to the fixed stars, they are the same special frames in which an object subject to no net force experiences no acceleration.

But gravity didn't fit. The laws of electromagnetism take their simplest form in classical inertial frames, but the laws of gravity do not. Inspired by Mach, Einstein worked toward developing a theory of gravity valid in *all* reference frames. However, in order to obtain agreement with observations, he had to make use of special reference frames; the laws of gravity take their simplest form in *free-falling* frames. Again, simplicity triumphed over symmetry.

Almost all agree that general relativity is not consistent with Mach's principle. But there is another question that must be asked: Is the universe Machian? *Philosophically,* it seems that it must be. If the only quantities we can measure are relative quantities — the distance and relative motion between objects — it seems that it must be possible to write the laws of physics only in terms of these quantities, which would necessarily yield a theory independent of reference frame. The question remains, then, does *observation* support this idea?

This is a very difficult question, and in fact may be impossible to answer. Mach's principle is a *principle,* a conceptual idea that can help guide our reasoning and the development of our theories, but it cannot be tested directly. However, we can examine a specific assumption that is included either implicitly or explicitly in many formulations of the principle — the idea that all inertial effects are determined by matter.

We see the influence of mass on inertia in the coincidence between inertial reference frames and distant stars. Foucault's pendulum keeps a constant plane with respect to the stars. The flattening of the Sun and planets are all with respect to distant stars. The flattening of the Milky Way is with respect to stars way beyond the galaxy. In general relativity, too, the boundary conditions that help define motion are always taken to be the distant stars.

Our understanding of the universe has changed greatly since Mach. The universe we talk about today is many orders of magnitude larger. We now know that all of the stars, which were previously assumed to be "fixed," are actually moving relative to each other. Furthermore, we have observed the entire universe to be expanding at an ever-increasing rate. Still, the observation stands that the inertia of a body — its resistance to a change in motion — is with respect to distant stars.

Even more compelling is what we don't observe: the flattening of the universe. If inertial effects are caused by matter and there is nothing outside the universe, the universe should not experience any flattening. And it doesn't. As far as astronomers can tell, at the largest scales the universe is completely isotropic.

However, although it is reasonable to expect that distant matter — because there is more of it — dominates over nearby masses when considering inertial effects, nearby masses should at least have some influence. One of the earliest predictions based on this view was that a Foucault pendulum should be *ever so slightly* influenced by the mass of the Earth. This should cause the rotation of the pendulum to be impeded because of the inertial resistance between it and the Earth. General relativity also predicts such an effect; unfortunately, all estimates suggest that it is so slight as to be completely undetectable.

A larger effect should come from our disk-shaped galaxy, the Milky Way. If inertial effects are due exclusively to the resistance between masses, an object's resistance to motion should be slightly greater in our galactic plane than in the areas perpendicular to it, which should affect solar system dynamics. Schrödinger came to this conclusion almost a hundred years ago, and with his model, he estimated that the magnitude of this effect should be just below observational limits. Now, with updated estimates of the mass and radius of our galaxy, it has been shown that the effect *should* be observable, comparable in size to Mercury's perihelion precession. And yet it has not been observed.

Similarly, variations in inertial mass that might be observable in atomic energy levels have not been found. Rather, the inertia of a body, to an extraordinary degree of precision, is completely independent of its orientation in space. Although on the largest scales, matter seems to determine inertial effects, variations in inertia due to local mass distributions such as our disk-shaped galaxy appear to be completely absent. This result, incidentally, is in perfect agreement with general relativity.

So where does this leave us? Every attempt so far to modify or replace general relativity with a theory consistent with Mach's principle has failed. Meanwhile, general relativity has survived every observational and experimental test.

However, as Mach said, the Earth isn't twice given, with an Earth at rest and an Earth in motion. *All of the evidence* that has come together to build our theories is based on *relative motions*, alone determinable. General relativity, which makes different predictions for an Earth at rest with the stars rotating around it and an Earth rotating with the stars at rest must be wrong. Or it must at least be incomplete.

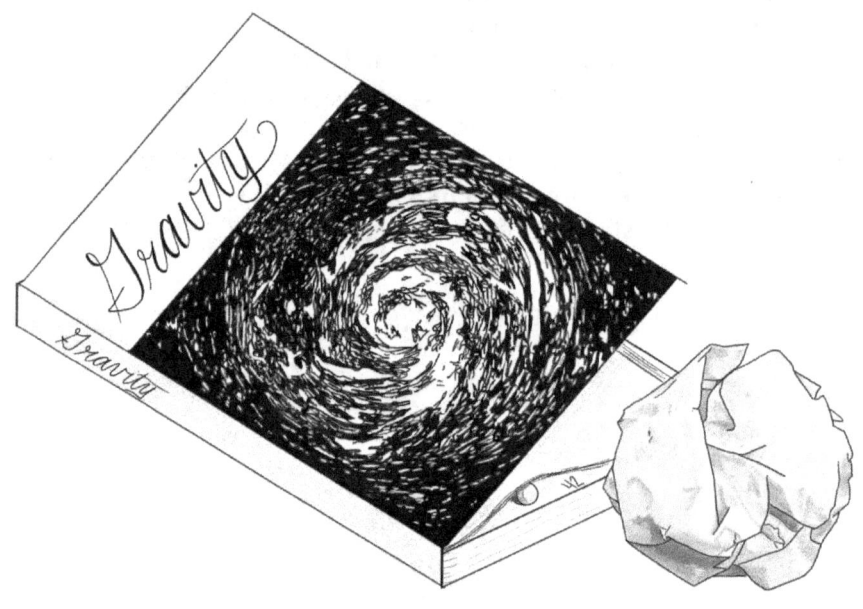

# BOOK V

---

# Dead Ends
# &
# Loose Ends

---

# 1

# A Maxwell-like Theory of Gravity

Here end my trials for the present.
The results are negative.
They do not shake my strong feeling of the existence
of a relation between gravity and electricity,
though they give no proof that such a relation exists.

— Michael Faraday

Although it was nearly twenty years ago, I clearly remember digging through research articles related to Mach's principle as the basis for a theory of gravity, looking for anything that would indicate that this approach was a dead end and hoping that it wasn't. I can even reconstruct my path somewhat from the articles I had collected in a spiral binder, each with a timestamp of when I printed it.

I also remember searching through my old electromagnetism textbooks for a derivation of the equation that was the basis of Sciama's law of inertial inductance, the law allowing the calculation of the density of the universe that had made such a great impression on me. I didn't find the derivation then, and in fact I did not find it for many years. However, I found something even better. I found a chapter on Thomas precession.

Thomas precession was discovered in the mid-1920s as a result of the experimental analysis of atomic spectra. Pauli had introduced a two-valued quantum number to explain the splitting of some spectral lines, which the Dutch physicists George Uhlenbeck and Samuel Goudsmit identified as the intrinsic spin of the electron. Using this assumption, Uhlenbeck and Goudsmit calculated the splitting of the relevant spectral lines. They wrote up their results in a paper and sent it to Bohr.

Bohr doubted their hypothesis because it predicted splitting that disagreed with experiment by a factor of two. However, a young British graduate student working with him, Llewellyn Thomas, suggested that the discrepancy might be resolved by taking into account special relativity. Specifically, he demonstrated that the spin axis of the electron should precess, an effect now called Thomas precession. Mathematically, this precession provided the missing factor of two.

Although Thomas precession was derived in the context of quantum theory as a way of explaining the observed spectral lines of hydrogen, I came across it in the context of relativistic electromagnetism. Here, the phenomenon was presented simply as a relativistic consequence of a body being acted on by an inverse-square central force. In Thomas precession, the central force is the electric force. I wondered whether this precession had anything to do with Mercury's perihelion precession. The assumptions seemed to apply to this problem as well: a body acted on by a central, inverse-square force according to the principles of relativity.

I began studying, and I began calculating. I worked through several different derivations of Thomson precession, carefully examining all the assumptions. I then made appropriate substitutions in order to estimate Mercury's perihelion precession. I obtained an expression for the precession whose dependence on the mass of the Sun and the radius of Mercury's orbit agreed with that derived using general relativity. However, the size of the precession was a factor of six too small.

I continued to read, think, and calculate. I knew that in special relativity, electricity and magnetism were one force and that it was actually possible to predict the equations of magnetism by assuming an inverse-square law of electricity together with the postulates of special relativity. I worked through this derivation and examined the assumptions. If an inverse-square law of gravity is assumed together with special relativity, there should be a magnetic-like counterpart to gravity. I began to read about "gravitomagnetism," something in general relativity analogous to the magnetic field that results from the motion of masses.

Meanwhile, I began studying general relativity. I did little "play" calculations, trying to use a gravitational theory modeled after Maxwell's electrodynamic theory to calculate effects that I was learning about in general relativity — not only the perihelion precession, but also precession when the orbiting body is spinning, precession when the central body is spinning, precession when the orbiting *and* central body are spinning. I could always predict the correct direction of the precession and also the dependence of the precession on such variables as mass and radius, which was encouraging.

These calculations helped me gain intuition for general relativity. Gravity is like electromagnetism, more or less. However, all of my values obtained using a Maxwell-like gravitational theory were too small. I wondered if the problem was that I was ignoring the universe. I went back to my old classical mechanics books and restudied the derivations of inertial forces on a rotating body. At some point, I calculated an effective gravitomagnetic field of the universe that would make my numbers work out.

In the end, all of this came to nothing. However, I did learn something. I came to understand why gravity doesn't "fit" into special relativity. If we assume Newton's inverse-square law of gravity and the postulates of special relativity, we get a perfectly good, consistent theory. In fact, we get a theory that looks very much like Maxwell's electromagnetic theory. But the numbers don't work out. This approach just doesn't yield a theory consistent with our universe.

# Lorentz's Transformation

Einstein's theory has the very highest degree of aesthetic merit: every lover of the beautiful must wish it to be true.

-Hendrik Lorentz

Like many physicists during the late 1800s, Hendrik Lorentz became deeply interested in electromagnetism. Maxwell's theory appeared in book form for the first time in 1873, about the same time that Lorentz began his graduate work at the University of Leyden. His doctoral research focused on Maxwell's theory, comparing and contrasting it with other theories regarding the reflection and refraction of light. He concluded that Maxwell's theory was preferable and outlined a program of study to similarly reconsider other optical phenomena, the majority of which he later carried out himself.

In 1878, Lorentz was appointed as Chair of Mathematical Physics at the University of Leyden, a position that would give him leisure to conduct independent theoretical research. This position was not only the first of its kind in Holland, but also one of the first in the entire world. And he used it well. He extended his doctoral research on Maxwell's electrodynamic theory and began studying atomic theory. Most, if not all, of Lorentz's early work was in these two areas: electromagnetism and the atomic theory of gases.

After Hertz's discovery of electromagnetic waves in 1888, Lorentz began to focus exclusively on electromagnetism. What was the nature of the charge that produced this radiation? Maxwell's theory did not have an answer, and most scientists at the time assumed it was a kind of fluid. But Lorentz wondered: If matter is made of particles, might not charge also be made of particles? He then began to develop a particle theory of electricity.

Throughout the early years of his career, Lorentz was isolated from international physics except for the papers he read and wrote. A discovery made in 1896 by a young colleague, Pieter Zeeman, quickly changed this. Zeeman worked with Lorentz as a graduate student and shared many of the same research interests. With a newly obtained apparatus to measure spectral lines, he observed a small but distinct widening of one of sodium's spectral lines when he applied a magnetic field. He shared this observation with Lorentz.

Lorentz began to calculate. He had Maxwell's theory and his own extensions of that theory, including an equation for the magnetic force on a charge. He understood the interaction between electromagnetic fields and light. He had his own growing theory of electricity. He had over two decades of theoretical research to prepare him for this opportunity.

Lorentz hypothesized that the emitted light was due to the vibration of small charged particles. In this way, he was able to explain the observed widening of sodium's spectral line. The application of a magnetic field should alter the frequency of the light, which would widen the line. He also predicted that the widening was actually the result of three distinct lines and that light on opposite sides of the widened line should be polarized differently.

Zeeman confirmed these predictions with additional measurements, and using Lorentz's theory, he was able to demonstrate that the vibrating charges responsible for the spectral lines are negatively charged. He was also able to calculate the charge-to-mass ratio of these particles. The following year, J.J. Thomson found these same particles in cathode rays and obtained a similar charge-to-mass ratio, thereby confirming the existence of negatively charged subatomic particles, which became known as electrons.

With his part in the discovery of the electron, Lorentz began his international career. In 1897, he traveled to Düsselforf, Germany to speak about his work. In 1900, he gave a lecture at the first international physics conference held in Paris. In 1902, he and Zeeman shared the Physics Nobel Prize for their discovery and interpretation of what became known as the Zeeman effect.

In 1904, Lorentz published a paper that was the culmination of many years of work, presenting a theory to account for the negative results obtained by the American physicists Michelson and Morley in their attempts to detect a change in the motion of light with respect to the ether. Lorentz believed in the ether, but he also believed in Michelson and Morley's results. In his paper, Lorentz showed that it was possible to repair classical theory by assuming that the ether contracts in the direction of motion. He also made use of a "local time," which was affected by the motion of the ether. He developed equations implementing these ideas that were consistent with Michelson and Morley's result that the speed of light is the same in all inertial reference frames.

Soon after, Einstein reinterpreted Lorentz's theory without the ether, and it became known as the theory of relativity. This was Einstein's theory, but Lorentz's contributions were celebrated in the terminology. The set of equations that relate position and time in one inertial frame to another was called the "Lorentz transformation."

Quantities that have the same value in all reference frames, like the speed of light, became known as "Lorentz invariants." Equations that take the same form in all inertial reference frames were called "Lorentz covariant."

In the years that followed, Lorentz kept up with developments in relativity theory as well as in practically every other area of theoretical physics, including quantum theory. He learned and extended the new theories and then taught them to others in a Monday morning series of lectures in Leyden that were given to colleagues and advanced students. He conversed with Planck about his energy quanta. In 1909, Einstein reached out to him regarding his own challenges in this area.

Einstein sent Lorentz a paper on radiation theory in which he proposed that light sometimes acts like a particle and other times a wave. He called his paper "the trifling result of years of reflection." He still had not been able to achieve a real understanding of the matter and wanted Lorentz to take a quick look at it. Two weeks later, Einstein followed up with a short note expressing how much he admired a derivation that Lorentz had presented in a recent paper.

Lorentz eventually responded, apologizing that he had taken so long to do so. He had been rather busy and had to wait for a quiet day. He wrote a long reply that included some of his own work, wanting Einstein's opinion before he published. In conclusion, Lorentz expressed how glad he was that this problem of radiation theory had given him the opportunity to enter into a personal relationship with Einstein, whose papers he had admired for such a long time.

They continued to correspond, and in 1911, Einstein accepted an invitation to come to Leyden. He would give a lecture, but what he looked forward to most of all was making the personal acquaintance of Lorentz, hoping for a "lively give and take dialogue." After the visit, Einstein enthusiastically thanked him: "You radiate so much goodness and benevolence that the troubling conviction that I did not deserve the great kindness and honors bestowed upon me could not even enter my mind during my stay in your house."

They met again that year in Brussels at the first Solvay conference. Lorentz was the chair of the conference, praised by many for his clarity, intelligence, and ability to, even amid differences of opinion, project a cheerful yet serious tone. Soon after the conference, faced with too many responsibilities, he decided to reduce his university position to part-time, keeping only his Monday morning lectures. For his replacement, Lorentz wanted Einstein.

Lorentz wrote on behalf of his colleagues asking Einstein to accept this position, but he also wrote from himself personally. It had been a long cherished wish — a "beautiful dream" — that Einstein would become his successor and colleague. How much he would enjoy maintaining constant contact and hearing about his work and his thoughts, and it would be even better since the scientific relationship would also be a friendly one. Lorentz was getting older, now nearly sixty; he would enjoy the cheerful and creative energy of a younger person. He understood that there may be circumstances that would prevent Einstein from accepting the position, but for now he would give himself over to the hope that his wishes, and those of his colleagues in Leyden, would be fulfilled.

Einstein wrote back immediately. He had refused an offer for a position in Leyden only months before, and he would refuse again. Einstein had given excuses — he worried that he would be seen as a foreigner, that he would have to struggle with a foreign language, that another candidate was a better fit. Now, he had to refuse in such a way that he would not be asked again.

Einstein began with a strong and sincere expression of the honor he felt by this offer, coming from "the most admired and the dearest man of our times." Not only was Lorentz offering him a place close to him, but also the prospect of a friendly personal relationship! Einstein could imagine nothing more beautiful than this, experiencing the mysteries of the universe in conversation with this great man. And the fatherly kindness that Lorentz bestowed on all people would more than compensate for his own feelings of intellectual inferiority.

However, to occupy Lorentz's chair would be "inexpressibly oppressive." He could not analyze this in greater detail, but he added that he had always felt sorry for his colleague who had occupied Boltzmann's chair. In any case, Einstein concluded, this would not be a great loss for Leyden — his knowledge was only modest and his mathematics was not good enough to make him a worthy representative at Holland's most important university. He would brood over his few "scientific eggs" in a "less exposed and illuminated niche." He devoted the rest of the letter to the radiation problem, sharing a couple of eggs that had hatched since they had last met.

And that was that. Lorentz graciously accepted Einstein's refusal and found a suitable successor, Paul Ehrenfest — an inspiring, passionate teacher who could clarify even the most difficult concepts. Einstein returned to teach at his old school in Zürich for a year and then took a position in Berlin with Max Planck, where he would remain for the next two decades. Even so, Einstein's most important intellectual home, especially during the formative years of general relativity, was in Leyden.

In Berlin, much of the theoretical work revolved around the development of quantum theory. Although Einstein's original relativity theory was embraced, his efforts at extending this theory to gravity were not. Planck, in particular, took a negative stance: "As an older friend, I must advise you against it... In the first place, you won't succeed, and even if you do, no one will believe you." Most of their scientific colleagues in Berlin agreed.

Meanwhile, those in Leyden urged him on. Lorentz and Ehrenfest, especially, pushed him forward with their interest and critique. After Einstein succeeded, they not only believed in the theory but were among the first to understand it. Lorentz, in his Monday morning lectures, taught it to students and colleagues, many of whom went on to help develop the theory.

During his frequent visits, Einstein enjoyed the warmth and support of the big Leyden family of scholars, headed by "Papa Lorentz." He would return home reinvigorated, refreshed, and cheered. The feeling was mutual — as Ehrenfest wrote to him after one visit, "You have left great waves of sympathy and friendship behind in this land, which will lap back and forth between us all, a long time from now. You really made everyone *completely* happy and 'boundlessly optimistic.'"

Lorentz remained an important figure in Einstein's life and in the scientific community. In the aftermath of WWI, he became the president of the International Committee on Intellectual Cooperation in the League of Nations. He continued to act as the chair of the Solvay conferences held in Brussels every three years, including the one held in 1927 devoted to quantum theory. He died the following year, but he was not forgotten; twenty-five years later, on what would have been his 100th birthday, Einstein wrote in celebration, "He meant more to me personally than anyone else I have met on my life's journey."

With this, we end the present story, a small tribute to a great scientist. Now, regarding *our* story, we return to Lorentz's transformation equations. These equations grew out of experiment, out of the observation that the speed of light, the mediator of the electromagnetic force, is independent of inertial reference frame. The transformation went hand in hand with the relativity principle: the laws of physics — specifically the laws of *electromagnetism* — take the same form in these frames.

After Einstein determined that the laws of gravity were incompatible with special relativity since they could not be made invariant under a Lorentz transformation in a classical inertial frame, he abandoned these preferred reference frames, though he did keep the transformation equations. In his search for the equations of gravity, he required that they be Lorentz covariant, that they take the same form before and after a *Lorentz transformation*. And, as far as I know, he never asked *why* this works.

# 3

# Inertia and Induction

By the way, I have proved that if there be nine coefficients of magnetic induction, perpetual motion will set in, and a small crystalline sphere will inevitably destroy the universe by increasing all velocities till the friction brings all nature into a state of incandescence...

— James Clerk Maxwell

When we use the word "inertia" in physics, it is usually in reference to mass. In Newton's theory, inertia is the tendency of an object to maintain constant motion. The greater the mass of an object, the greater its resistance to a change in motion. The greater its resistance to a change in motion, the larger the force required to impart a given acceleration. Here, motion is defined in reference to an inertial frame that is moving at constant velocity with respect to absolute space.

Mach, objecting to the concept of absolute space, suggested that "space" in Newton's theory is actually an abbreviated reference to all of the mass in the universe. And he questioned, "What would become of the law of inertia if the whole of the heavens began to move and stars swarmed in confusion?" Under Mach's influence, inertia became a resistance to acceleration *between* masses; every mass resists motion with respect to every other mass in the universe.

In Einstein's special theory of relativity, the concept of inertia evolved in a different direction. As in Newton's theory, the inertia of a body is defined with respect to an inertial reference frame. However, following Mach, inertial frames are in constant motion with respect to fixed stars rather than absolute space. Additionally, according to special relativity, the inertia of an object depends not only on its mass but also on its speed in an inertial reference frame. As its speed approaches the speed of light, its inertia approaches infinity, placing a speed limit on its motion.

In general relativity, the concept of inertia evolved once more. Here, it is a manifestation of the geometry of spacetime, and the geometry of spacetime is affected by mass. Conceptually, this is evident in frame-dragging – the dragging of an inertial frame in the presence of a large, accelerating mass. An example of this is the Lense-Thirring effect, whereby a large, rotating mass causes a precession of the spin axis of a small, orbiting mass in the same direction as the rotation of the large mass. We can interpret this in classical terms as resistance to an acceleration between the large rotating mass and the small orbiting mass.

In all of these examples, inertia is related to mass. Although different versions of this concept appear in our various theories, it is always in some sense the resistance of a *mass* to a change in motion. But the concept of inertia is broader than this, applying in any situation where an object resists a change in motion or circumstance. In physics, this same concept appears very clearly in electromagnetism. We call this kind of inertia "inductance."

Electromagnetic inductance was discovered in 1831 by Michael Faraday while attempting to create electricity from magnetism. Inspired by musical resonance in which a note played on one instrument induces a sound in another instrument, he passed a current through a coil of wire, hypothesizing that it would produce a current in another concentric coil of wire. This is in fact what he observed. However, he was surprised to find that the induced current only lasted for a moment. It would appear briefly when he connected the battery, and it would appear again in the opposite direction when he disconnected the battery.

These observations and other similar ones formed the basis of Faraday's law of induction. According to this law, the rate of change in the magnetic flux through a closed loop is proportional to the induced current. This language, with magnetic flux and closed loops, obscures the similarity between induction and inertia. However, if we abandon our theories for a moment and just look at the phenomena, we find that they function in exactly the same way.

In his experiment, Faraday found that the induced current always opposes the *acceleration* of the charges that make up the current. When he connected the battery to initiate a current in the primary circuit, a current was induced in the opposite direction in the secondary circuit. When he disconnected the battery so that the current in the primary circuit slowed down and came to a stop, a current was again induced in the secondary circuit. Now the induced current was in the same direction as the current in the primary circuit, again in the opposite direction to the acceleration.

With a little creativity, we can set up a similar experiment with mass. We can use a train to act as the primary circuit and place wooden blocks on the floor of the train to act as the secondary circuit, allowing for an induced mass-current. In place of the many loops of wire we have in the electrical example, we have the distant stars. When the train begins to move, the blocks move in the opposite direction to the train. Once the train has reached a constant velocity, friction causes the blocks to quickly come to rest with respect to the floor. As the train slows down to a stop, the blocks again move, this time in the same direction as the train. Just like in the electrical example, the blocks always oppose the change in motion of the train, acting against its acceleration.

364

Faraday's work on induction soon led to a new field of physics, electronics, and here we find an even clearer connection to inertia with a device called an "inductor." An inductor is designed to exaggerate the effects of magnetic induction that arise from Faraday's law — like Faraday's device, an inductor is made of many coils of insulated wire. An inductor inserted into a circuit functions like a classical source of inertia, opposing any acceleration of charge.

With an inductor, we can find connections to classical inertia. Even more importantly, with the phenomenon of inductance, we can find connections to Mach's relativity of inertia, which is manifested in general relativity as frame-dragging. However, whereas gravitational frame-dragging is such a slight effect that any experimental observation of it must be done on astronomical scales, this same phenomenon in electromagnetism is so large that it can be demonstrated with simple equipment in a laboratory or even in one's home.

Such experiments were first performed by the French scientist François Arago; Faraday then repeated the experiments and offered an explanation. In one experiment, a copper plate is rotated close to a magnetic needle suspended in such a way that the copper plate and needle can both rotate parallel to each other about a common axis. If the copper plate is motionless, the needle points toward magnetic north. However, if the plate is rotated, the magnetic needle follows the motion of the plate.

As a variation of this experiment, the roles of the magnet and copper disk can be reversed. If the magnet is spun, the copper disc rotates in the same direction as the magnet. In yet another setup, a copper disc is spun over a magnet that is kept stationary. The presence of the magnet impedes the movement of the copper disc, making it more difficult to spin.

These experiments can be interpreted in terms of frame-dragging; the large, rotating object drags, or attempts to drag, the inertial frame of the nearby object. Even more simply, all of these are examples of the relativity of inertia applied to electromagnetic phenomena. Charges resist acceleration with respect to other charges just as masses resist acceleration with respect to other masses. The first we attribute to induction; the second we attribute to inertia.

And so, despite how different gravity and electromagnetism are according to modern theories, we see that they have something else in common. One big difference, however, is that we have a theory to explain inductance, but we don't have a theory to explain inertia.

# 4

# The Origin of Inertia

Mr. Einstein's Theory of Relativity does not supersede the Newtonian law of Gravitation or of Inertia. It only says, Beware! The law of Inertia is not the simple ideal proposition you would like to make of it. It is a vast com-plexity. Gravitation is not one elemental uncouth force. It is a strange, infinitely complex, subtle aggregate of forces. And yet, however much it may waggle, a stone does fall to earth if you drop it.

— D. H. Lawrence

Although Dennis Sciama entered Trinity College in 1944 with the intention of studying mathematics, he also studied physics to qualify for an academic military deferment. After graduating, he was drafted into the British army. During his two years of military service, he worked on the quantum mechanics of photoconductive materials, research believed to be useful for the detection of enemy airplanes. Afterwards, because of the strength of his work during this time, Sciama was accepted back at Trinity as a graduate student. Initially, he worked on problems in statistical mechanics, but his interest was soon drawn away from the tiny world of atoms and molecules; he wanted to understand the universe.

Sciama began to study general relativity. He worked through Eddington's writings and Schrödinger's recently published book, *Space-time Structure*. He also took a class in cosmology by Hermann Bondi, one of the originators of the steady-state model of the universe. In this model, matter is continually created, keeping the density of the universe constant as it expands.

Sciama also became interested in Mach's principle, and this research led him to write an article titled "On the Origin of Inertia," published in 1953. He began the paper with a brief overview of the development of Mach's principle: the idea that inertial effects are not due to a body's acceleration with respect to fixed space but rather its interaction with the entire universe. General relativity was, in fact, inspired by and created to incorporate this principle. However, as Einstein himself emphasized, it ultimately failed to do so.

In his paper, Sciama presented a theory of gravity to account for the origin of inertia. For simplicity, he modeled it after Maxwell's theory of electromagnetism. Furthermore, he assumed that a body in its own rest frame experiences no force. Using Newtonian language this implies that in an object's rest frame, the gravitational force is balanced by an inertial force. However, unlike the inertial force in classical mechanics, this inertial force is not "fictitious" but a real force due to the object's acceleration with respect to the universe. The inertial force arises out of Maxwell-type equations as an inductive effect — a mass resists acceleration with respect to other masses just as in electromagnetic theory a charge resists acceleration with respect to other charges. Sciama called this effect "inertia-induction."

In applying his theory, Sciama assumed that at every point there exists a preferred reference frame given by nature in which the observed redshift of distant galaxies is isotropic. According to an observer in this reference frame, the matter in the universe is isotropic, homogeneous, and expanding at a constant rate according to Hubble's law. Since the inertia-induction effect is reciprocal, a rotating Earth in a stationary universe is equivalent to a stationary Earth in a rotating universe, consistent with Mach's principle.

This was a "tentative" theory, not intended to replace general relativity. Sciama had used the simplest set of equations that could be used to model inertia — a full theory would have to be more complicated. He also ignored any relativistic effects that might result from speeds approaching the speed of light. However, his model had two important advantages over general relativity. First, the equivalence between inertial and gravitational mass is a consequence of the theory rather than an assumption. Second, with his theory it was possible to estimate the mass density of the universe.

When Sciama estimated the density of the universe, he obtained a number much larger than the observed density. He suggested that there may be great amounts of interstellar matter, as evidenced by the large average densities within some star-forming regions. If we accept Mach's principle, he concluded, then on the basis of this theory we should *predict* that the universe contains enormously more matter than that which had been observed. In contrast, general relativity makes no such prediction and is therefore, in this respect, unfalsifiable.

The year after he published this paper, Sciama completed his doctoral thesis on Mach's principle and the origin of inertia and then continued at Trinity as a fellow. He spent the academic year from 1954 to 1955 abroad at the Institute for Advanced Study in Princeton, where Einstein was working. Sciama had already reached out to Einstein a few years before during his first attempt to do something in gravitational theory and had received a lengthy response. Now, he arranged to meet Einstein in person.

Sciama started out a bit nervous. He had read that Einstein had a hearty laugh and a simple sense of humor and set out to put both of them at ease with a little humor. He also knew that Einstein had embraced Mach's ideas on the relativity of motion and even coined the phrase "Mach's Principle," though he had later disowned it. And so he began: "Professor Einstein, I've come to talk about Mach's Principle and I've come to defend your former self against your later self." It worked — Einstein said, "Ho, ho, ho, that is gut,

Ja!" He really laughed! Then Sciama talked about his way of using Mach's principle, and Einstein talked about his work and doubts about quantum theory. It was a wonderful experience, and Sciama had arranged this meeting just in time; Einstein died only a week later.

After his fellowship at Princeton was over, Sciama moved on to Harvard and afterwards returned to England to finish up his fellowship at Trinity. All the while he continued to think about Mach's principle. In 1958, he completed a book on cosmology, *The Unity of the Universe*. It was a broad overview of the subject written in simple language to be enjoyed by the general public. A theme that ran throughout the book was that the universe is a single unit; the behavior of the here and now is influenced by the faraway in space and time. In the introduction to the book, he celebrated this idea with a quotation from the English poet William Blake:

> *To see a world in a grain of sand,*
> *And a heaven in a wild flower,*
> *Hold infinity in the palm of your hand,*
> *And eternity in an hour.*

Throughout the book, Sciama emphasized the relationship between the near and the far in our universe, which are connected, first of all, optically. We can *see* distant planets, stars, galaxies, and other astronomical objects with our eyes and the help of telescopes. In this way, we receive an enormous amount of information about our universe. In addition, he suggested, distant regions of our universe are connected to us *mechanically* by the influence of distant matter on motion according to Mach's principle.

With this introduction, Sciama began a history of Mach's principle. He began before the principle was named — even before Mach — with the writings of the Irish philosopher Bishop George Berkeley. A contemporary of Newton, Berkeley had challenged the idea of absolute space on the grounds that it is not observable. Furthermore, if every place is relative, every motion should also be relative. He suggested that centrifugal forces, which in Newton's theory result from an object's acceleration with respect to absolute space, were actually due to the action of the fixed stars.

Sciama continued his history with Mach's writings and Einstein's attempt to create a theory incorporating Mach's principle. He wrote about the skepticism toward the principle; specifically, scientists objected to the idea that distant matter creates inertial effects because no one had been able to explain how this interaction takes place. An even stronger argument was made by such scientists as Arthur Eddington and more recently the British mathematical physicist Edmond Whittacker, who both suggested that there is not enough matter in the universe to account for observed inertial effects. Whittacker argued that if we adopted Mach's principle we would have to postulate an enormous quantity of matter that has never been detected, which he felt was unjustifiable.

However, it was just this that Sciama felt was the strength of Mach's principle — with Mach's principle we can make *predictions* about the universe. As an example, he considered the rotation of the Milky Way. It was discovered in 1926 that the stars in our galaxy rotate very slowly, completing one orbit around its center every 100 million years. This rotation, defined with respect to the much more distant stars, causes the galaxy to bulge. If it rotated faster, this rotation might have been discovered during Mach's time, in which case Mach could have predicted, on the basis of the rotation, large amounts of matter beyond the Milky Way fifty years before it was actually detected.

Sciama believed in Mach's principle, but he still had not succeeded in creating a complete theory of gravity incorporating it. He shared what he had formulated with his readers: a theory of gravity based on electromagnetism that could account for the observed effects of inertia, and he gave a short history of this theory.

In 1872, the French astronomer Félix Tisserand had proposed a theory of gravity based on Maxwell's theory. He had no interest in inertia when developing the theory; rather, he was inspired by the success of Maxwell's electromagnetic theory in accounting for the influence of electromagnetic forces without appealing to action at a distance. With his theory, Tisserand predicted a deviation from Newton's planetary orbit — a small precession of the perihelion. Such a deviation had in fact been detected in Mercury's orbit — an extra 43 arcseconds per century. However, the value that Tisserand calculated for this precession was much smaller than the one observed.

It was not until Einstein's theory of general relativity that the precession in Mercury's orbit was explained. Interestingly, Einstein's original aim in developing the theory was not to explain this deviation but rather to account for inertia according to Mach's principle. It was still a matter of much debate whether and to what extent general relativity embodies this principle. The reason for such ambivalence, according to Sciama, is that general relativity is non-linear, meaning that the combined gravitational force of two bodies is not just the sum of the individual forces that would be present if the other object were ignored. Therefore, to determine the influence of the stars on the motion of objects, one cannot simply add up the influences of all the stars. This was not a defect in the theory; in fact, it was an essential part of its conception. Sciama did not go into any more detail here, concluding with the words, "All we can say is that an intriguing problem awaits solution."

When Sciama published *The Unity of the Universe*, he was a research associate at King's College in London. He next took a position at Cornell University and then returned to Cambridge first as a lecturer in mathematics and then as a research fellow at Peterhouse. He was an excellent teacher but reluctant to take on a traditional academic post, preferring to devote his best energies to research. And for those interested in the universe, what an exciting time!

It was in the mid-1960s, while Sciama was at Cambridge, that astronomers made the first measurements of the cosmic microwave background radiation — low-temperature radiation coming uniformly from all parts of the universe that had been predicted by the big bang model. This was closely followed by the first observation of high-redshift quasars, which gave additional support to the big bang theory. Next came the discovery of neutron stars and the first results from space astronomy.

Sciama was in the middle of it all. He drew talented students into the field of cosmology, directing them toward the most recent research. He did this through Saturday morning seminars in which he encouraged the open discussion of ideas and placed a strong emphasis on observational testing. He also instituted daily coffee time, creating a friendly space where students and colleagues could interact and learn from each other. Here, he would hand out and comment on the latest preprints he had received from associates around the world. When an interesting discovery was published, he would assign a student to read about and share it with the others. When an interesting lecture was given in London, he would take a group of students or send some to go and report back.

In this way, Sciama built up a powerful research group at Cambridge. While encouraging and facilitating the research of those around him, he continued to pursue his own projects. He was initially a passionate supporter of the steady-state model of the universe, following the lead of his former professor Hermann Bondi. Sciama struggled to incorporate new observations of the cosmic background radiation and strongly redshifted quasars into the theory, but he eventually concluded that too much evidence contradicted it. With great honesty and courage, he abandoned the steady-state model of the universe and became a passionate supporter of the big bang theory.

Though he abandoned the steady-state model, Sciama held onto Mach's principle. He had used Mach's principle to develop a theory of gravity modeled after electromagnetism that could reproduce the main properties of inertia. In other respects, however, the theory remained incomplete. Specifically, although it could conceptually predict the observed bending of light in a gravitational field, his numerical values did not agree with observation. In his earlier paper on the origin of inertia, he had promised another paper, a full treatment that would use the more elaborate equations of a tensor potential that were used in general relativity.

However, when Sciama attempted to create such a theory, he discovered what had already been discovered independently by a number of people: such a theory had to be *very* close to general relativity. He finally despaired of discovering field equations that would do better than Einstein's, even though these did not embody Mach's principle. He embraced general relativity as a correct theory, but still he held onto Mach's principle. And he gloried in the strength of the inertial forces produced while he sped around corners in his Jaguar — what a remarkable effect the "fixed stars" have upon the passengers!

Sciama would not abandon Mach's principle on philosophical grounds, but he could not abandon general relativity on observational grounds. He hoped that it might be possible to find Mach's principle in the boundary conditions of general relativity. In 1969, over fifteen years after his first paper on Mach's principle, he published a second paper attempting to do this — to use Mach's principle as a selection principle for the boundary conditions. Several years later, one of his students took this idea further, determining that if the universe is homogeneous and Machian, then according to the boundary conditions it cannot spin, a result that was supported observationally. The mathematics had gotten very complicated by this point, and Sciama was never fully satisfied with the approach. By this time, he had turned his attention to other problems.

In 1970, Sciama left Cambridge for Oxford, where he built up another research group. He encouraged research in quantum gravity as a link between big bang cosmology and particle physics. When his former student Stephen Hawking announced the remarkable result that the temperature of a black hole should be proportional to its surface gravity, Sciama immediately recognized its importance. He hailed it as a revolution in our understanding of the connection between general relativity and quantum theory. He quickly arranged a conference devoted to quantum gravity at Oxford during which Hawking publicly presented his result.

In 1983, Sciama left Oxford to become the head of the Astrophysics Sector at SISSA, an international school for advanced studies in Trieste, Italy. During his time there, he developed a theory to explain dark matter based on massive decaying neutrinos, linking together cosmology, astrophysics, and high energy, though it was later contradicted by observations. He also wrote papers on an assortment of other topics: the emission of bursts of light from imploding bubbles when excited by sound, the nature of the quantum vacuum, and the deceivingly simple question: Does a uniformly accelerated quantum oscillator radiate?

A conference was held at SISSA in 1992 to celebrate Sciama's sixty-fifth birthday. The conference was titled "The Renaissance of General Relativity and Cosmology," paying tribute to the enormous role he had played in the resurgence of these fields since the early 1950s. Sciama's influence was apparent through the long list of publications highlighting four decades of steady, productive work. Even more important than his personal achievements, his insights into and enthusiasm for gravitational theory and cosmology generated *interest* in these subjects. And for me personally, Sciama had an enormous influence; it is because of his work that I wrote this book.

# 5

# Einstein's Castle in the Air

The search was not so much a search as a
groping in the gloom of a mathematical jungle
inadequately lit by physical intuition.

— Banesh Hoffmann

"**B**ut the best thing," wrote Einstein to his friend Michele Besso, "which I have been pondering and figuring out for days on end and half the night, is now complete before me, condensed into seven pages and titled 'A Unified Field Theory.'" The Prussian Academy printed one thousand copies of Einstein's little pamphlet, an unusually large number, which sold out immediately. A London department store displayed all of its pages in a window — crowds of people pushed forward to read it. About a hundred reporters camped outside Einstein's home in Berlin, hoping for a word from the great scientist. "Now, but only now, do we know," he announced to one interviewer, "that the force that moves electrons in their ellipses about the nuclei of atoms is the same force that moves our earth in its annual course about the sun."

The public shared Einstein's excitement, but Einstein wasn't sure his colleagues would feel the same. As he admitted to Besso, the theory looked a bit "old-fashioned." Specifically, it did not anywhere include Planck's constant. Without Planck's constant, there was no wave-particle duality, no quantum uncertainty, no quantum theory.

This was fine as far as Einstein was concerned. He believed that *nature* was deterministic, even if our theory to describe it was not. He saw the uncertainty in quantum mechanics as a flaw in the theory, believing that quantum mechanics must be a special case of some more general theory that did not have this flaw. When scientists eventually gave up what he called the "statistical mania" of quantum theory, the equations he presented in his little pamphlet would form a starting point for a new theory.

It was a starting point, but it was not yet a theory. Einstein had come up with a set of equations that he was convinced would correctly describe both gravitational and electromagnetic fields. However, in January of 1929 when he published his pamphlet, he had not yet been able to demonstrate this connection. By the fall, he believed he had solved all the difficulties, but he was alone in his confidence. "I have now completed the marvelous theory," he wrote to his sister, "to the lively mistrust and passionate rejection of my colleagues in the field."

Among Einstein's colleagues, Wolfgang Pauli stepped forward to express this skepticism. Not only did the new theory not include the many successes of quantum theory, Pauli protested, it could not even replicate the success of Einstein's own theory of general relativity.

"But I stick to that fine theory," he continued, "even if it is being betrayed by you. With your remark that you are still far from being able to assert the physical validity of the derived equations you have in effect silenced your critics among the physicists! All that is left to them now is to congratulate you (or had I better say: express their condolences?) on your having gone over to the pure mathematicians."

Einstein wrote back to Pauli that he found his letter "quite amusing." He admitted that the road he was going down was not necessarily the correct one. *Intellectually*, though, it seemed the most natural road to follow given all that he had considered. Moreover, until the physical consequences had been thoroughly worked out, it was not yet justifiable to dismiss these efforts. He encouraged Pauli to immerse himself in the problem without prejudice, as if he had just come down from the Moon and still had to form a fresh opinion.

Einstein had been working on this problem for a long time, and it was only with great reluctance that he had, as Pauli had written, "gone over to the pure mathematicians." He had been working on it even before 1918, when he received a paper from his former colleague Hermann Weyl attempting to do the same — to create a theory unifying gravity and electromagnetism.

Weyl had witnessed firsthand the early development of general relativity, and now he attempted to extend it. In his paper, he proposed that not only was space curved, but that the sizes of space elements also changed as a result of motion. With this new geometric structure of spacetime, he could link Einstein's theory of gravity to Maxwell's electromagnetism; according to Weyl's theory, electromagnetism is an aspect of geometry.

Einstein was enthusiastic about Weyl's idea, but he quickly found a flaw in the theory. If the sizes of space elements change as a result of motion, the length of an object depends on its past. Since in spacetime "length" refers to time as well as space, the frequency of light emitted by atoms would also depend on the past. If this were the case, the spectral lines of atoms would appear as smears rather than the observed sharply defined lines.

Einstein continued his own search for a unified theory, and the following year he published a paper suggesting that it might be possible to account for the structure of the elementary particles of matter with the help of gravitational forces. His starting point was a slightly modified version of the field equations for general relativity, where the energy-tensor of matter was replaced with the energy-tensor of the electromagnetic field. He demonstrated the plausibility of its application in the realm of atoms. He then applied it in the cosmic realm, demonstrating that with this new equation, the cosmological term he had added earlier to allow for a static universe no longer required a special cosmological constant. It was a speculative paper; in fact, the title of the paper was a question: "Do Gravitational Fields Play an Essential Part in the Structure of the Elementary Particles of Matter?" At the end, he expressed his own doubts about the idea, and soon after publication he rejected it altogether.

In 1921, Einstein reached out to Theodor Kaluza, a German colleague who had sent him a paper two years earlier proposing to unify gravity and electromagnetism by adding an extra spatial dimension. Einstein had been supportive but asked him to clarify one point, and Kaluza didn't end up publishing it. Now, Einstein urged him to publish with no reservations and offered to present the paper before the Academy.

Soon after Kaluza published his paper, Einstein wrote a follow-up article. He began by emphasizing the importance of the subject matter: "Surely the most important current issue of the general theory of relativity today is the essential unity of the gravitational field and the electromagnetic field." Until a short while ago, he continued, the only progress that had been made in this direction was with the work of Hermann Weyl, but Weyl's solution was plagued with rods, clocks, and atoms having prehistories, in contradiction to experience. Kaluza's solution had no such problems, and Einstein suggested that no special physical assumptions needed to be made for the extra spatial dimension. However, the theory did have weaknesses. First, the requirement of general covariance with the extra dimension appeared to lack justification. Furthermore, Einstein could find nothing in the theory that could be reasonably interpreted as an electron.

Although Einstein had rejected Weyl's theory, he and Weyl continued to correspond, united in their shared goal of unifying gravity and electromagnetism. In the summer of 1922, Einstein sent him a letter commenting on a paper that Eddington had recently published extending Weyl's proposal for unification. Eddington's theory, Einstein wrote, has a "pretty frame but one absolutely cannot see how it has to be filled in." He added that Kaluza's theory felt closest to reality, but in order to include the electron, one would have to permit the possibility of infinity at a point; this did not seem like the correct approach. In conclusion, he wrote, to really make progress on this subject one would have to find another principle "eavesdropped from nature."

Shortly after, Einstein had another idea. He found a link between Eddington's formalism and the principle of least action that he believed "leads to a theory almost free from arbitrary steps, one that conforms with what we know at present about gravity and electricity, and which unites both types of field in a truly perfect manner." He finished writing up a paper on the subject on a boat while traveling home from Asia. He was so excited about the idea that he sent the paper to Max Planck as soon as they docked so that Planck could submit it to the Academy on his behalf. After returning to Berlin, he lectured on the idea and wrote two more papers, but he was slowly realizing that his hopes for the theory would come to nothing. In fact, he had gone down a path similar to that taken by Weyl earlier and reached the same dead end. In the wake of this failure, Einstein wrote to Weyl questioning their formal approach to the problem, which lacked any kind of physical anchor. "The mathematics is fine and good," he wrote, "but Nature is leading us on a dance."

Einstein had struggled before. It had taken him almost a decade of trying and failing before he came up with his final version of general relativity. But with gravity, he had the perihelion precession of Mercury to test his theory and the equivalence principle to guide him. Now he had nothing except an unshakable belief that the two most important fields in nature must be unifiable. However, lacking any experimental result or physical principle, his only guide was mathematical simplicity, which he admitted was "not free from arbitrary aspects."

In the summer of 1925, Einstein had a breakthrough. He wrote up his ideas in a paper titled "Unified Field Theory of Gravity and Electricity." In the introductory paragraph, he admitted that his previous attempt at a unified theory, which was based entirely on Eddington's ideas, was not the "true solution." But now, after an unending search over the past two years, he had finally arrived at what he believed to be the "true solution." With his new formalism it might be possible to not only unify gravity and electromagnetism but also derive the existence of elementary particles.

A few weeks after submitting his paper, Einstein wrote to Besso, excitedly telling him about the "new egg" he had laid. After expressing his regrets that he could not share this egg in person, he launched right into the details. The starting point of his theory was to independently introduce an affine connection ($\Gamma^{\alpha}_{\ \mu\nu}$) and a metric tensor ($g^{\mu\nu}$). Assuming that these are both symmetrical, applying the variational principle yields the old gravitation equations for empty space. However, if the assumption of symmetry is abandoned, one obtains to first-order the laws of gravity and Maxwell's field equations with the antisymmetric part of the metric corresponding to the electromagnetic field. "This really is a magnificent possibility that surely could correspond to reality," he wrote to Besso. The theory was certainly correct in the macroscopic range. The question remained as to whether it agrees for atoms and quanta. If only the calculations were easier!

Besso quickly responded. He regretted that he lacked the knowledge of the $g^{\mu\nu}$s to understand Einstein's message. Where could he find this knowledge? He then continued based on what he thought he understood. First, he wondered if in the $g$s the gravitational and electromagnetic fields are of the "same kind," as opposed to the $\Gamma$s that represents the geometry. He asked Einstein to write down a few equations so that he could see what was being varied and how the results emerged. The mystery of the matter, Besso suggested, is at its foundation. What should be regarded as the container, and what should be regarded as the content — the *things* in space? Tell me more, he urged, and also what I may share with whom.

Einstein replied a couple days later with a short note, writing that Besso could talk about this scientific matter with anyone. He didn't send him any equations or clarifications, but he did continue, with the help of a mathematical collaborator, to calculate. A month later, he still had not obtained anything decisive, but the balance was tilting toward the "contra side." A month after that, Einstein received a long letter from Lorentz, who had been doing his own calculations based on Einstein's theory. Lorentz concluded that the functions that Einstein had connected with the electric and magnetic fields conflict with Maxwell's equations. Within a few days, Einstein submitted a paper retracting the theory altogether.

In this paper, Einstein proved that if the electromagnetic field is represented by an antisymmetric tensor, as he had attempted in his most recent unified theory, then there cannot be general covariant equations describing the electron or even a positive particle with the same mass. He claimed that this result is of such a simple nature that he did not think it could be new. However, it was new to him and therefore, he believed, would be welcome to some. In conclusion, he suggested that electromagnetism and gravity differ fundamentally, and it therefore no longer seemed possible to merge the two.

Even so, Einstein did not give up. In the spring of 1928, during the tranquility of an illness, he laid another "wonderful egg." The old mathematics had not worked, so he developed new mathematics. And since general relativity seemed to be incompatible with the unification of gravity and electromagnetism, he gave up general relativity! With his new mathematics, he created new kinds of tensors and invariants that led to new physical constants.

The idea was to merge the Euclidean geometry of flat space with Riemannian curved space by extending the concept of "parallel" to large distances, which in curved space is only defined for infinitesimally small distances. Pauli had chided him for this idea — for abandoning general relativity and going over to the pure mathematicians. And he teased him, predicting that Einstein would give up on this idea of distant parallelism within a year.

Einstein, however, did not give it up within a year. Writing about their happy summer of 1930, his cousin and second wife Elsa wrote, "Albert is working as he has hardly ever worked before... Has thought up the most wonderful theory. It's getting more beautiful every day. If only it proves to be *true!!!*" Einstein continued to think, calculate, and try to connect his beautiful mathematics with the physical world. Finally, after more than three years of vain effort, he abandoned the theory. With good humor he acknowledged his defeat to Pauli: "So you were right after all, you rascal."

Einstein did not mourn the death of this idea for long. He was over fifty years old but still filled with seemingly boundless youthful optimism. He became fascinated with a five-dimensional formalism that "psychologically" linked up with Kaluza's theory through its extra spatial dimension while avoiding an extension of physical space. This was achieved by linking five-dimensional vectors with each point in four-dimensional spacetime. The gravitational and electromagnetic field equations could then be derived through a process called "rejuvenation." He was excited about this new idea, but his colleagues were not. The older ones kept silent, and the younger ones started to make jokes about him.

Einstein *had* gone over to the side of the pure mathematicians, and he did so without apology. He shared his views during a lecture given at Oxford, "On the Method of Theoretical Physics," delivered in June of 1933. He suggested that since Galileo, natural philosophers had fallen into the error of believing that the foundations of physics are grounded in experience — that the concepts and postulates of physics can be "abstracted" from experience. Einstein claimed that this method was doomed to failure, as demonstrated by his invention of general relativity; he had been able to account for a wider range of empirical facts more satisfactorily and completely than Newton, who had deduced his laws from experience.

Einstein then posed the question: If the laws of physics cannot be deduced from experience, but rather must be "freely invented," how can we have any hope of finding them? He answered without hesitation. Experience teaches us that the laws of nature are the simplest possible realizable laws. Therefore, we can discover these laws by means of "purely mathematical constructions." Of course, experience still plays a *role* in scientific discovery. Experience might suggest the appropriate mathematical concepts and remains the sole criterion for evaluating the usefulness of a mathematical construction. However, the *creative* part of scientific discovery lies in mathematics. "In a certain sense," he concluded, "I hold it true that pure thought can grasp reality, as the ancients dreamed."

Einstein continued to give birth to new mathematical constructions and write hopeful papers, but none of his theories passed the test of experience. He also kept raising the standards for such a theory. A correct theory should be able to predict the existence of the electron and matter in general as a logical necessity. It should also be able to predict the sizes of universal constants like the speed of light so that they would not have to be obtained from experiment. He began to wonder whether there was any *choice* in creation, whether the demand for logical simplicity left any freedom at all.

All the while, Einstein continued to reject quantum theory. He believed that there must exist some more general theory for which quantum theory was a special case. With such a theory, one would find that the uncertainty and wave-particle duality embedded in the present theory were merely "temporary obstacles during a transitional stage of theoretical confusion." In 1939, John Wheeler invited him to discuss this point of view at one of the Princeton colloquia. During the talk, Einstein maintained that the laws of physics are simple. A question came from the audience, "But if they are not simple, what then?" to which Einstein replied, "Then I would not be interested in them."

Ten years later, in March of 1949, Princeton held a symposium to honor Einstein's seventieth birthday. Most of the three hundred participants had already taken their seats by the time he arrived. His entrance occasioned a moment of respectful silence. Even while others were celebrating his achievements, he felt the weight of his own failures: "You imagine that I look back on my life's work with calm satisfaction," he wrote two weeks later. "But from nearby it looks quite different. There is not a single concept of which I am convinced that it will stand firm, and I feel uncertain whether I am in general on the right track."

Nevertheless, Einstein never ceased to "sing his lonely old song." Year after year, he submitted new papers with new attempts at a theory to unify gravity and electromagnetism — a theory that would predict universal constants, a theory of everything that would make even quantum theory obsolete. However, he did not reject the idea that perhaps everything was different. In the summer of 1954, he wrote a letter to Besso, the last letter to his dear friend of over fifty years. He admitted that it was entirely possible that the laws of nature were not built on continuous fields, as he had been assuming in every attempt. "In that case *nothing* remains of my entire castle in the air, including the gravitation theory."

# 6

# A Book Conceived

People say to write about what you know.
I'm here to tell you, no one wants to read that,
cos you don't know anything.
So write about something you don't know.
And don't be scared, ever.

— Toni Morrison

Many years ago while outdoors at a high school retreat, I heard a beautiful melody playing in the distance. I hummed it over and over inside my head, determined not to forget it. About a year later I heard the same melody in a music theory class — it was the theme from the second movement of Beethoven's piano sonata, *Pathetique*. I had taken some piano lessons, and so I decided that I would try to play it.

Over the next twenty years, I kept coming back to that same piece, working through most of the second and third movements, never getting it exactly as I would have liked but always enjoying it. I didn't hold it as a goal to master it, and I certainly had no intention of performing the piece for anyone. I wanted to experience the music, and the best way I knew to do that was through the slow process of learning to play it, note by note.

When I began my research, it was the same way. It was a beautiful problem — to try to discover the fundamental laws that guide our universe. I focused on one part, then another and another, always trying to understand how the different parts were connected. I struggled with it for years, never expecting to "figure it all out," but all the while enjoying the work and holding a small hope that, through my efforts, I might be able to discover some new insight.

I first set out to learn general relativity. I read many reviews of textbooks on the subject and finally picked out *Gravitation* by Charles Misner, Kip Thorne, and John Wheeler. It was highly regarded; it seemed to be the standard textbook in the field. Besides, it is enormous. A book so large it carries its own gravitational field, one reviewer mused. Through this book, I began my studies.

General relativity is by no means an easy subject, but Misner, Thorne, and Wheeler did an excellent job making the subject accessible. The book seems intentionally designed so that it can be used without a teacher. It has clear explanations along with worked examples and practice problems, many of which include answers. It is really an introductory textbook, making no assumptions about the student's previous knowledge of the subject. While it is an introductory textbook, it is also an intermediate and advanced textbook, intended to bring motivated graduate students, in twelve hundred pages, up to a level where they can begin research in the field.

I slowly developed my skills, gaining comfort with the mathematics of four-dimensional curved spacetime. I carefully worked through the chapters on electromagnetism in special relativity. I read and reread the chapter discussing the incompatibility of gravity and special relativity. I noted that the authors claimed that trying to incorporate gravity into special relativity presented *difficulties;* they did not say it was impossible.

I worked through several hundred pages of the book and still had not gotten to a general relativistic treatment of the perihelion precession of a planet. Looking through the table of contents now, I can see that this problem is not considered until almost halfway through the book. I did not get that far. Although I was developing intuition for the mathematical formalism of general relativity, I was having a difficult time making connections that would allow me to transition to another approach. I didn't want to just learn general relativity; I wanted to *change* it.

After a little more searching I found another highly recommended book, *Gravity: An Introduction to Einstein's Theory* by James Hartle. This book takes a more conceptual approach with many connections back to Newtonian gravitational theory. It focuses on the physical applications, teaching the mathematics efficiently in order for the student to be able to apply the theory as quickly as possible. This was exactly what I needed, so I purchased the book and renewed my studies.

With the background I had already gained, I was able to rapidly move past the introductory chapters into the applications, starting with the geometry outside a spherically symmetric body described by the Schwarzschild metric. Here, I learned how to do many of the same problems I had solved using Newtonian mechanics, like calculating the escape velocity from the surface of a planet and the radius of a stable circular orbit. I was intrigued by the fact that, although the methods of solving the problems were completely different, these two particular values agreed exactly with the results from Newton's theory.

I also studied precession in general relativity and then tried to derive the same effect using an electromagnetic-like theory of gravity. I studied gravitational lensing, Schwarzchild black holes, the metric outside a slowly rotating body, and the dragging of inertial frames. I solved problems using the methods of general relativity with symmetries, Killing vectors, and Christoffel symbols. Then, whenever possible, I solved the same problems using a Maxwell-like theory of gravity. I was able to qualitatively derive the relativistic effects, but my numbers were always too small.

At this point, I still believed that special relativity was the correct framework for the gravitational problem, and it occurred to me that I didn't understand that theory well enough. Specifically, I wanted to *deeply* understand relativistic electromagnetism. So I bought and began to work through another book, Wolfgang Rindler's *Introduction to Special Relativity*.

I jumped around a bit in this book. I already knew special relativity fairly well, but I wanted a stronger foundation. I spent a lot of time studying the beginning of the book, where the concepts and mathematics of special relativity are developed. I also spent a lot of time near the end of the book, where the theory was applied to electromagnetism. By that point I felt like I *really* understood special relativity and its connection to electromagnetism — what a beautiful theory! But still, the connection between gravity and electromagnetism eluded me. I'm not sure what I expected — for some equation to just drop down from the sky revealing this mysterious connection?

Throughout my studies I recorded my work in a series of notebooks. My typical work is hardly legible with arrows connecting one equation to another and many mathematical errors that I would then scribble out and correct. I would use scratch paper and then, after I had arrived at a satisfactory conclusion to any problem, I would rewrite my work neatly in a notebook. At some point I started to date my notebook entries. In my daydreams, these notebooks would be treasured by posterity — the notebooks that preserved the work of the woman who figured out how to unify gravity and electromagnetism into a quantum theory of everything! Realistically, these notes were for my future self to refer back to after I had forgotten what I had worked so hard to learn.

After I finished working through the book on special relativity, I returned to Hartle's book on general relativity, starting an entire new notebook and working through the same chapters a second time. Then I returned to Misner, Thorne, and Wheeler's book with more confidence and understanding. I was making progress. My notebooks include scattered questions — looking back, I see in one place a question that I asked in 2006. By 2010, the question seemed trivial, and I had written an answer to my past self in the margin. I was no closer to figuring out anything new, but both my understanding and admiration of Einstein's theory of gravity had grown immensely.

Around that time my studies took a new direction. I had reached out to a professor with whom I had worked in graduate

386

school to tell him about my research interests, and he recommended the book *Mach's Principle: From Newton's Bucket to Quantum Gravity*, a book collecting articles from a conference held in 1993 dedicated to Mach's principle. This was by far the best resource I had come across on Mach's principle, and I began to read it from the beginning.

I enjoyed the early articles, which were mainly a conceptual and historical introduction to Mach's principle. Then I began working through early Machian models. I spent a lot of time on a paper written by Hans Reissner in 1914, in which he presented a theory to account for inertia. To do this, he borrowed a Lagrangian from classical physics, modifying the kinetic energy term to reflect the relativity of inertia. I worked through his calculations, filling in omitted steps and making sure I understood all of his assumptions. I then wondered: What if instead of borrowing a Lagrangian from classical physics, I borrowed one from electrodynamic theory?

I started with the relativistic Lagrangian for a charged particle. It has two main parts, the first of which is the Lagrangian for a free particle and depends on its inertial mass. I set this to zero, being only interested in the electromagnetic part. I then changed all the constants to their gravitational counterparts, and I also changed one of the signs to reflect the attractive nature of gravity. It was similar to Reissner's Lagrangian — the most important difference was that my kinetic term was inversely related to distance, whereas his kinetic term had no distance dependence at all. With this Lagrangian I determined the equations of motion in a two-mass universe.

The result was similar to Reissner's in many ways. The equation could be rearranged so that one side was the classical expression for the gravitational force between two masses and the other side was the inertial force, proportional to the relative acceleration between the two masses, consistent with the relativity of inertia proposed by Mach. The inertial term was inversely proportional to the separation between the two masses, as in Sciama's law of inertial induction. This was a satisfying though not surprising result since he too had based his gravitational theory on electromagnetism.

My inertial term, however, had something that neither Reissner's nor Sciama's terms had — a factor that predicted the increase in inertial mass with speed, consistent with special relativity. This was something! I continued, solving the equation assuming that the two masses are released from rest at a given distance apart. I found that the masses accelerate toward each other at a constant rate determined by their initial separation. After some time, also determined by their initial separation, the two masses meet and presumably crash into each other. At this point, *independent of their initial separation*, they would be traveling at the speed of light.

I knew that my little theory didn't represent reality. It didn't include light or even a way for the two masses to communicate their influence on each other — no curvature of space or quantum exchange of particles. And besides, how would it be possible to make any measurements with only two masses? But still, what an interesting result — two masses released from rest at any initial separation always approach the speed of light just before meeting! What was the speed of light in this theory? Was it a fundamental constant or a speed limit for a given universe?

I never reached a dead end with this approach. I saw so many possibilities. I did a quick estimation of the density of a spherical universe using this model and got infinity. This didn't concern me. I had ignored the expansion of the universe and the finite propagation of influence, presumably through gravitational particle-waves as in electromagnetism. I was also pretty sure I didn't correctly include the speed-dependent inertial term I had found; what was simple in a one-dimensional universe with two masses became much more complicated in a three-dimensional universe. I wrote up the results that I had obtained in my notebook — my two-mass universe and my failed attempt at calculating the density of the universe. I made a mental note to return to these calculations and pushed forward.

The next article in the book was written in 1925 by Erwin Schrödinger, who presented another Machian theory to account for inertia. Like the kinetic part of my electromagnetic-inspired Lagrangian, his kinetic term was inversely related to distance so that the inertia between two objects infinitely far apart would be zero. However, his term was a factor of three greater than mine. I knew that factor of three well. When I compared results derived from general relativity to those derived from a Maxwell-like theory of gravity, I was often a factor of three or six too small. In fact, Schrödinger included the factor of three so that the perihelion precession of Mercury would come out correctly.

However, if his equation for kinetic energy is applied to a two-mass system, the maximum relative speed between the masses is the square root of one-third the speed of light. That just didn't seem right. I began imagining myself in a friendly debate with the great Schrödinger!

Of course, Schrödinger would have agreed with me, at least on this point. His kinetic equation was classical, only valid for small relative speeds between the masses. When he was writing, empirical evidence seemed to justify the assumption of small speeds when accounting for the interaction with distant stars. Nevertheless, later in the paper he suggested another Lagrangian that could be applied to near-light speeds.

When applied to a single mass in a universe filled with matter, this new Lagrangian reduced to the Lagrangian for a single mass in special relativity. It only included relative quantities, consistent with Mach's principle. It also reduced to his classical expression, with the factor of three in the kinetic term, for speeds much smaller than the speed of light. In this way, it was made to be consistent with special relativity and observational results of the perihelion precession of a planet. Maybe he was onto something here!

And yet, I didn't understand it. I understood it *mathematically*, and I satisfactorily worked out all the steps to verify everything Schrödinger claimed about the Lagrangian. Despite this, I still didn't understand it *physically*, and Schrödinger had made no attempt at a physical or intuitive explanation. It seemed clumsy or clunky, like a patchwork equation that checked all the boxes but didn't mean anything. I wondered, though, could I break it apart? What if I could dissect the expression into its parts and then compare them to the relativistic, covariant equations of electrodynamic theory? Maybe then I could figure out why, when I used a Maxwell-like theory to derive gravitational effects, my numbers were often exactly a factor of three too small.

I couldn't quite wrap my head around how to do this, so I left it as something to come back to and moved forward. Schrödinger had used his relativistic Lagrangian to model a single mass near the center of the universe. He modeled the universe as a spherical shell with a given radius and surface mass density, eliminating the arbitrary quantities describing the universe with results he had derived using his classical model. What if I modeled the universe more realistically?

When Schrödinger wrote this paper, the "radius" of the universe would have been something arbitrary in need of eliminating. Astronomical observations of the redshift of distant galaxies were just starting to come in, and we were still a few years away from Hubble's discovery of the expanding universe. Today, the Hubble radius — the limit of the observable universe due to its expansion — would be a reasonable assumption for the radius of the universe.

I reworked Schrödinger's calculations using his relativistic Lagrangian, modeling the universe as a sphere with uniform density and radius equal to Hubble's radius. All of the difficult mathematics I borrowed from Schrödinger; without much effort I arrived at an expression for the density of the universe in terms of the Hubble constant and the gravitational constant. And then, without taking the final step of plugging in the numbers to obtain a value, I put my books away.

I did eventually plug in the numbers, but even now it surprises me how long I took to do it. I had finished the calculations near the end of the summer, just before school was about to begin again. I'm a high school teacher, and my research is my hobby. It was time to stop playing and get back to work! Yet it would have taken hardly any time to finish the calculation. I had to figure out where to insert the speed of light into my expression, having worked in units in which it was taken to be one. I also had to look up Hubble's constant. Finally, I had to pull out my calculator or decide to do the calculation by hand.

A psychologist might have fun analyzing my behavior, but it is really quite simple. I don't like finishing things. No, that's not quite true. I am a quick grader. I always finish my reports on time. Anything I have to finish, I finish quickly and efficiently. I am good at finishing *work*. But anything creative, I seem unable to finish. It's as if I have a pile of Legos that I'm using to build a castle, but as soon as I finish building the castle, I knock it down so I can build a better one. And when that one's finished, I knock it down to build an even better one. I would never dare to cement the pieces together because it could never be...perfect.

That isn't quite right either. I begin to sense that the castle isn't perfect even before completing it. Or I realize that I can never make it perfect and abandon the project altogether. Or better yet, I just build a castle in my mind, where it can remain perfect in my imagination, full of beautiful possibilities. It was probably at least in part this last reason that kept me, for almost a year, from taking that final step of plugging in the numbers needed to estimate the density of the universe using Schrödinger's Lagrangian.

When I finally finished the calculation, the result was as close to the current estimate as I could ever hope for. I was ecstatic. I quickly wrote up a paper, titling it "On Machian Theories of Gravity." Without even an introduction to Mach's principle, I launched into a discussion of various Machian theories. I discussed the work of

Reissner and Schrödinger. I discussed the results I obtained with the Lagrangian derived from electromagnetism and my estimate for the density of the universe calculated using Schrödinger's Lagrangian. I included a concise summary of all my calculations, beautifully written up with the help of an equation editor.

The paper was neither perfect nor complete. As I typed up my work, I noticed an omission in one of my calculations. I wrote "OOPS!!!" and made a note to return to it later. I made another note to check something in Reissner's paper. After reviewing all the Machian theories, I scribbled out a conclusion — a few sentences on their possible importance to dark matter, quantum gravity, and a unified theory. I made a photocopy of my handwritten calculations to keep at school so that if a fire destroyed all of my notebooks, I would still have these important papers.

A year passed and then another and another. During this time I didn't open up my books on general relativity, Mach's principle, or electromagnetism. I didn't write anything in my notebooks. I suppose on some level I realized that I hadn't gotten anywhere, that I wasn't getting anywhere, and that I would probably never get anywhere.

My life was so full that I hardly even noticed that I had stopped my research. I had a new baby, and certainly this contributed to my contentment. My two oldest children were practically grown up, and I was determined now not to let time pass too quickly. I had my job, my friends, and my family. And I had found a new way to satisfy the part of my brain that needs to solve puzzles. I had started to program again.

The science department at my school was restructuring its curriculum, and I took the opportunity to redesign my eleventh grade physics course to incorporate programming. I created programs to simulate one-dimensional motion, projectile motion, and motion under the action of forces. I turned the projectile motion program into a kind of video game by putting in a target. The students solve the problems and then insert the required numbers into the program. If they solve it correctly, when they run the program the projectile hits the bullseye, and the target explodes!

I put vectors in my force program so that students could visualize the size and direction of the forces. I created several collision programs: one-dimensional and two-dimensional, where I used an online resource discussing the mathematics behind billiard ball collisions to simulate them realistically. I added multiple balls colliding elastically to simulate an ideal gas. I created a loop-the-loop program and another to simulate wave motion, which could be used to demonstrate constructive and destructive interference and beats. I also created a program using Newton's law of gravity to simulate the motion of artificial satellites, planets, and moons.

This was only the beginning. I wasn't creating finished programs but rather starter programs for my students to use and develop. I assigned programming homeworks and programming projects. I was happily busy troubleshooting and helping my students implement their designs. When they chose an extra-challenging project idea, I would program the project myself. I programmed two masses connected by a string over a pulley that sat on an inclined plane. I programmed a projectile subject to air resistance. I programmed Pluto and its largest moon, Charon, orbiting around their center of gravity. With Newton's law of gravity, it was easy to model any number of interacting masses. And so, for fun, I added an asteroid to my program that hit Pluto's surface and exploded, with the light from the explosion fading into the background...

I was having so much fun that I could almost keep the philosopher inside me quiet — the part of me that always needs to know why, that questions everything, that is disturbed by broken symmetry or parts that don't fit together. I taught my students about inertia, only rarely mentioning that *observationally* a body resists motion with respect to the distant stars and that we *have no theory* that satisfactorily explains this influence of the stars on our motion. I introduced my students to Maxwell's theory, praising it for its unification of electricity, magnetism, and light and for its role in the development of special relativity, without telling them that I believed a theory of gravity modeled after electromagnetism might act as a model for quantum gravity.

However, every winter when I began to teach Faraday's law of induction to my seniors, I couldn't resist making a comparison between magnetic induction and inertia. I would then inevitably go off on some tangent and start talking about my research, often continuing after class with an interested student. And when I talked, I could hear the excitement in my voice. I could *feel* the excitement in my body, and I'm sure my students could see the excitement in my face. I loved this project, this *possibility* of pulling together theoretical physics into a beautiful and consistent whole. I didn't want my ideas to die with me.

I wanted to give this project to my students, but how? I often gave a talk on my work, "Mach's Principle and the Origin of Inertia" or alternately "Quantum Gravity," depending on the focus. I had my students stand up and turn in a circle while looking at the flecks in the ceiling, pretending that they were stars. What do you see? The stars moving in a circle. How do you feel? A little dizzy. We have *no theory* that adequately explains this connection.

I talked about the graviton, a hypothetical particle that, according to quantum theory, must exist to mediate the gravitational force. I talked and answered questions. But it was too much, too many connections to make in a fifty-minute period. I wanted to give this problem to my students; perhaps they could complete what I was unable to. I recommended books, but there was no book that contained all the ideas I wanted to share. And so I decided to write this one.

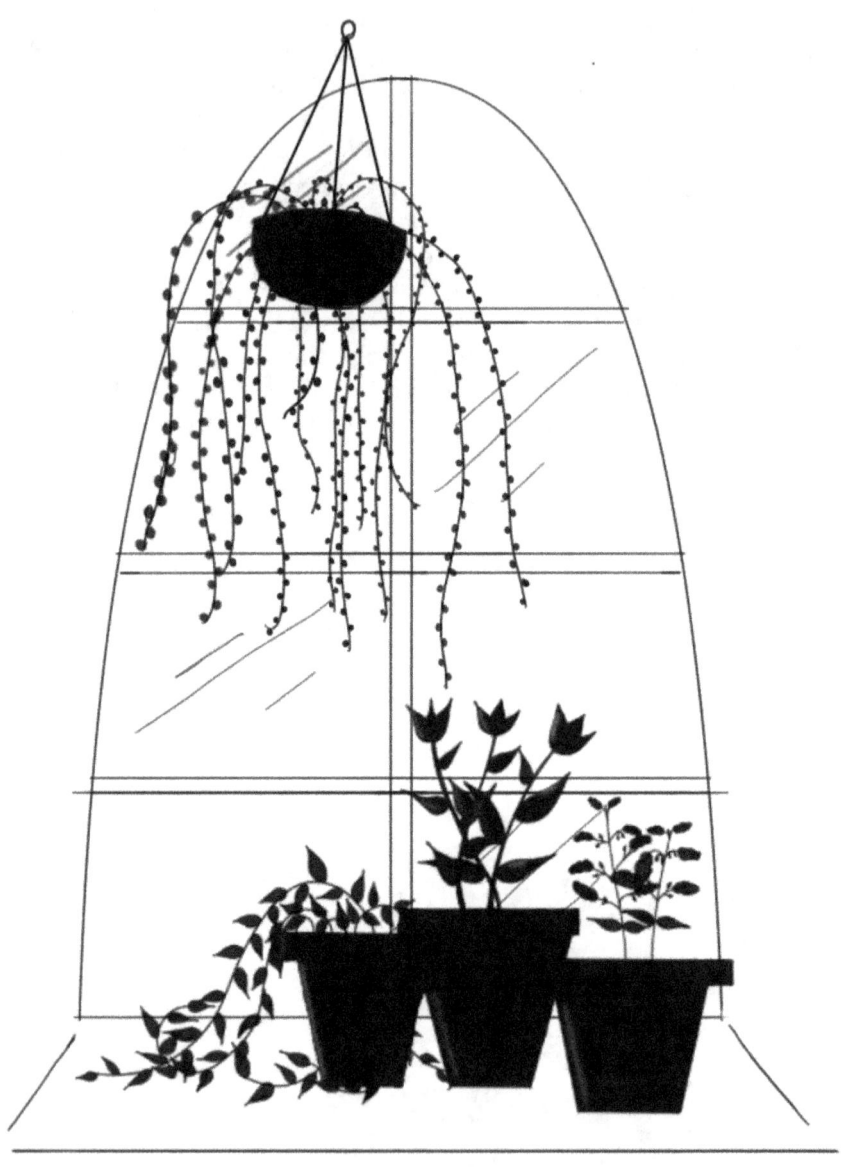

# BOOK VI

---

# The
# Search
# for a
# Conclusion

---

# 1

# An Outline

For, after all, every one who wishes
to gain true knowledge must climb the Hill Difficulty alone,
and since there is no royal road to the summit,
I must zigzag it in my own way. I slip back many times, I fall,
I stand still, I run against the edge of hidden obstacles,
I lose my temper and find it again and keep it better,
I trudge on, I gain a little, I feel encouraged,
I get more eager and climb higher and begin to
see the widening horizon. Every struggle is a victory.
One more effort and I reach the luminous cloud,
the blue depths of the sky, the uplands of my desire.

— Helen Keller

I am not a writer. I am a reader, a teacher, and a thinker, and I hoped that this, along with great determination, was enough to write a book. After two years and two months, I had finished five pages. In addition, I had pages and pages of research notes and two chapters that I had decided didn't work but hopefully had salvageable parts. I also had an ever-growing document that I named "Stuff that didn't make it." I finally came to the conclusion that if I was going to finish this project, I needed to figure out a better method of writing.

I spent the next six weeks outlining the entire book. The first five parts were fairly easy. I had given talks about my research to interested students; I used these presentation notes as the basic structure for my book. With a thirty-page outline, my confidence was growing. But how would I end the book?

I had prepared myself and the reader for an inconclusive ending. After all, the title is *An Incomplete Theory*. I had written in the opening of the book that I wanted to convince the reader that we need to abandon general relativity as a starting point for a theory of quantum gravity; I did not claim that I had figured out how to move forward from that point. In my head, I had sketched out many inconclusive endings. I kept coming back to the image of a detective carefully laying out all the pieces of evidence in order to solve a mystery. Then I would just end the book.

As I began to outline the final section of my book, nothing I had come up with was satisfying. I wanted a better ending. I may not be a writer, but I am a reader. I know that a book without a good ending is not a good book. If I couldn't come up with a good ending, there was no point in writing the book. I tried to find an important scientist whose main contribution was knocking down another theory, but I couldn't find any. I eventually realized that if I was to achieve anything meaningful, I would have to figure out a way to move forward. So, like a climber faced with a tall and steep mountain, I began to climb.

But the process of scientific discovery is not like climbing a mountain. It is more like grasping about in the dark, not being able to see a top, not even knowing whether any kind of top exists. When I look back on the outline that was to begin the last part of my book, I find only three things. First, quantum theory has been successful. Second, general relativity has been successful. Finally, I had written that we don't actually *need* to reconcile them, but that at the end of a unification, there always seems to be a surprise. I had no direction, no foothold, no special anchor. All I had was a tentative hope.

# 2

# Another Look at $E = mc^2$

But if every gram of material contains this
tremendous energy, why did it go so long unnoticed?
The answer is simple enough: so long as none of the energy
is given off externally, it cannot be observed.
It is as though a man who is fabulously rich should never spend
or give away a cent; no one could tell how rich he was.

— Albert Einstein

As I was going through the overheads for my research talk, turning it into the outline of my book, I came across one that didn't seem to go into any of the first five parts. It crossed all the boundaries of the earlier parts, and so I saved it for the book's ending. There weren't many words on it — most of the overhead was covered with Einstein's famous equation written in large hand-written letters: $E = mc^2$.

The equation comes from special relativity, where it represents the energy of a free mass. It was obtained by requiring that if energy is conserved in one inertial reference frame, it be conserved in every other inertial reference frame. In this equation, "m" is its inertial mass, which approaches infinity as the object's speed approaches the speed of light, setting a speed limit for massive objects in an inertial frame. The "c" is the speed of light.

Often, this equation is considered in a reference frame in which the object is at rest; here, the object's inertial mass takes its lowest value, called its rest mass. The corresponding energy is called its rest energy. Einstein interpreted this rest energy as the potential energy an object has because of its mass. It is an enormous amount of energy because of the largeness of the speed of light — a single gram of matter contains as much energy as the explosion of an atomic bomb.

Energy from mass is enormous, and it is everywhere. Tiny bits of mass are converted into energy during digestion, respiration, and combustion. Mass is converted into energy during the radioactive decay of heavy nuclei, which heats the core of the Earth, and through nuclear fusion, which heats the core of our Sun. The energy that is generated can be converted back into mass — in the growth of plants, the charging of batteries, and the slow formation of fossil fuels. In fact, all the matter in the universe is believed to have come into being in this way, created from energy according to Einstein's equation.

It is fascinating how this one equation, derived through the transformation between inertial frames, pervades almost all of physics and, by extension, all of science. What makes it *intriguing*, and why I saved this for the final section of my book, is that within this equation we see the unification of gravity with the electromagnetic and nuclear forces. On one side of the equation, we have mass. Although this is technically the inertial mass, it is proportional to the gravitational mass, the source of gravity. On the other side, we have what Einstein identified as potential energy generally, but it is more specifically the electromagnetic and nuclear potential energies.

According to this equation, the *source* of gravity is potential energy for the other forces. But what is potential energy? We define potential energy as *stored* energy, but it is really nothing. Or rather, it is simply a theoretical construction to represent *another* physical reality, the creation and annihilation of light. When an electron jumps to a higher energy state in an atom, the atom gains mass by absorbing light. When an electron falls into a lower state, the atom loses mass by emitting light. When a nuclear fusion or nuclear decay reaction occurs that decreases the total nuclear potential energy, some of the mass of the nucleons is transformed into high-energy light called gamma rays.

We see in these examples that the *physical* manifestation of a change in electric or nuclear potential energy is the conversion of mass into light or light into mass. In this way, mass, the source of gravity, is intimately and inextricably linked to the other forces. But what about gravity? If stored electric or nuclear energy is mass, what is stored gravitational energy? According to our theories, nothing. A rock can be lifted, increasing its potential energy, and nothing is absorbed or gained. A star can collapse into a black hole, decreasing its gravitational potential energy at a tremendous rate, and nothing *physical* is emitted, no quantity is lost – there are just ripples through the incorporeal spacetime.

# 3

# A Visualization of Quantum Reality

The layman always means, when he says "reality"
that he is speaking of something self-evidently known;
whereas to me it seems the most important and
exceedingly difficult task of our time is to work on
the construction of a new idea of reality.

— Wolfgang Pauli

It seems obvious to me that a quantum theory of gravity would have to include gravitons, hypothetical particle-waves that mediate the gravitational force like light mediates the electromagnetic force. It seems so obvious that it surprises me that this is not a requirement and that many attempts at quantum gravity have not included this particle. I would rather miss something complicated than something that is obvious, and so I will assume that gravitons exist and see where, conceptually, this leads.

In quantum theory, all massive particles are described with the help of probabilistic wavefunctions that embody the wave-particle duality of matter. These particles interact with other particles through mediating quanta that embody this same duality. The quanta have different ranges and coupling constants. The nuclear forces are both extremely short-ranged, which means that their mediating quanta are only passed between particles that are very close to each other. The strong nuclear force responsible for the binding of nuclei has a large coupling constant, which means quanta are passed back and forth very frequently. The weak nuclear force, responsible for nuclear decay, has a small coupling constant; these quanta are thrown and caught much less frequently.

Particles of light, or photons, are the mediators of the electromagnetic force. These quanta have a long range; they can travel indefinitely without being reabsorbed. They have a large coupling constant, although not as large as that of the strong nuclear force. Furthermore, charges emit photons in such a way that only opposite charges can catch them, presumably through the direction of spin. If two like charges are near each other, the emitted photons crowd each other since they can't be caught, causing repulsion. If two unlike charges are near each other, an intimate game of catch draws them together. Of course, this is a greatly simplified picture, and some of it is of my own invention. I will, however, continue in the hopes that being able to *imagine* the quantum world in terms of pictures may lead us forward to a new way of thinking about quantum gravity.

For a spherically symmetric charge distribution in an inertial frame, the light quanta will be thrown out equally in all directions. This leads us naturally to the inverse-square law of electricity; the density of the light quanta will decrease as they spread out over a larger and larger spherical surface. With this same picture we can understand the magnetic field; relative motion among charges would create asymmetries that produce magnetic-like circular forces. Relative acceleration could be detected through an increase or decrease in the rate of interaction. The energy of the particles could be detected through the frequency of the quanta.

This is at least enough to set the scene before considering quantum theory applied to gravity. The universe is the same universe. The particles are the same particles; all of the particles interacting through the nuclear forces and the electromagnetic force also have mass and therefore interact through the force of gravity. All we need to add is a new particle mediator for gravity, the graviton. But what would it look like?

First, the graviton would have to have an infinite range, like light, to account for the infinite range of gravity. It would have a *very* small coupling constant to account for the weakness of the force. And to account for the attractive nature of gravity, a particle of matter should be able to catch a graviton thrown from another particle of matter. Whether an antimatter particle could catch a graviton thrown from a matter particle, we do not know. If antimatter and matter attract, the answer is yes. If they repel, the answer is no. Gravitons would likely travel with the same speed as light since this constant appears in the equations of general relativity.

Although gravity and electromagnetism are both infinite-range forces, their games of catch look very different. For electromagnetism, since charges can only catch photons tossed from an opposite charge, charges tend to form little stable games of catch — between the positive nucleus of an atom and its electrons or among atoms that come together to form neutral chemical compounds. For gravity, the game of catch extends throughout the universe.

The scope of the games is very different, but their nature should be very similar. Gravitons being thrown equally in all directions would lead to an inverse-square law of gravity. Gravitomagnetism and inertia could be accounted for similarly to magnetism and induction. This visualization therefore leads us to expect similar equations for gravity as for electromagnetism, adjusting for the different coupling constants and the attractive-only nature of matter.

However, guided by the successful theories of electromagnetic theory and general relativity, we would expect one more important difference: their "special" reference frames. The special reference frames of electromagnetism are classical inertial frames, while those of gravity are free-falling frames. What this means is that while a particle at rest in a classical inertial frame throws out photons equally in all directions, a particle at rest in a *free-falling* frame will throw out *gravitons* equally in all directions. Therefore, although we expect a similar set of equations for electromagnetism and gravity, they should each take their simplest form in their own special reference frame.

With this assumption we can also understand something in general relativity that Einstein assumed but was never able to explain physically — why the free-falling frames of general relativity are connected by Lorentz transformations. In special relativity, these equations were chosen to reflect the experimental result that the speed of light in a vacuum is the same in all inertial reference frames. In general relativity, the mathematics of the Lorentz transformation was adopted without physical justification. However, if we assume that gravitons travel in straight lines at a constant speed in all free-falling frames, we have physical justification.

I *believe* in this visualization of a quantum reality applied to gravity. I cannot imagine that the same particles that play catch to mediate the electromagnetic and nuclear forces influence each other gravitationally in a completely different way. I *don't* believe that undetectable "space" can influence the motion of a particle. Furthermore, I believe that the gravitational waves that have been detected are actually these gravitons — particle-waves mediating the gravitational force like light mediates the electromagnetic force. But how can this idea be tested?

For light, its particle nature has been confirmed in a multitude of experiments — by analyzing blackbody radiation, the photoelectric effect, atomic spectra, the Compton effect, and many others. However, the weakness of gravity would exclude all such experiments with the possible exception of the gravitational analog to blackbody radiation, the gravitational wave spectrum from a star in thermal equilibrium. If this were possible, the shape of that curve could be used to test the quantum nature of the waves. And if the quantum nature of gravitational waves were revealed in these curves, it should also be possible to determine a constant analogous to the Planck constant, which would presumably be a new constant of nature.

Could our gravitational detectors gain sufficient sensitivity so that we could observe the ordinary gravitational radiation from a massive astronomical object? Perhaps, but almost certainly not within my lifetime.

# 4

# Measurement

Here and elsewhere in science that view is out of date
which used to say, "Define your terms before you proceed."
All the laws and theories of physics have this deep and
subtle character, that they both define the concepts they use
and make statements about these concepts.
Any forward step in human knowledge is truly creative in the sense:
that theory, concept, law, and method of measurement
– forever inseparable – are born into the world in union.

— Misner, Thorne, and Wheeler

When Robert Kerr set about to translate Lavoisier's *Traité Élémentaire de Chimi* into English, he intended to convert all the French measurements into their English equivalents. However, this task proved to be too difficult and time-consuming considering his time constraints; he was given the manuscript in the middle of September and was supposed to have it ready for use in universities by the end of October. Instead, he included tables allowing readers to perform the conversions themselves.

With Lavoisier's work the quantitative study of matter took a huge leap forward. From a scientific point of view, it became increasingly apparent that further study would benefit from a universal system of measurement. In France, the timing for an overhaul of the current system was perfect. The revolutionists had gotten rid of their king, Louis XVI. Now it was time to get rid of the king's foot, which had been used as a standard unit of length in France since Charlemagne had introduced it a thousand years earlier. Besides, within France there was really no true standard; weights and other measurements that went by the same name varied greatly throughout the country, posing challenges in commerce as well as in science.

The revolutionary Charles Talleyrand took the lead, proposing that the French Academy of Sciences come up with a standard system of weights and measurement. They accepted the challenge and began assembling a committee devoted to the task, which included mathematicians Joseph-Louis Lagrange and Pierre-Simon de Laplace, with Lavoisier taking an advisory role. Talleyrand invited an English representative to sit on the committee, but the invitation was rejected on the grounds that such a reform would be "almost impracticable." A similar invitation was sent across the ocean to the young United States and was met with a similar response. And so, the French Academy set about this reform of measurement alone.

The committee first tackled the question of a standard length. At this time, two competing ideas for such a standard existed, one based on a pendulum and another based on an arc of the Earth's surface. The Academy chose this second way, since defining length in terms of the period of a pendulum would introduce error due to differences in gravity, even if the measurements were limited to those at sea level.

Hence, in 1791, the French Academy of Sciences defined the standard distance as one ten-millionth of an arc representing the distance from the North Pole to the equator passing through Paris. This was named the "meter" from the Greek *metron*, meaning "to measure." They also agreed on a standard unit of mass. The "gram" was defined as the weight of pure water held in a cube with sides equal to one hundredth of a meter. Since volume depends on temperature, they also needed a standard temperature. They initially chose the freezing point of water; afterwards, they decided to use the temperature at which the density of water is the greatest as their reference temperature.

Additionally, they agreed that areas and volumes would be based on the meter. Fractions and multiples of these basic measurements would use base ten to allow for easy conversions. Now, all they had to do was figure out the length of one ten-millionth of the distance between the North Pole and the equator, and all other measurements would be set.

They had a preliminary value but wanted greater accuracy, so in 1792, an expedition was sent to perform the measurement. Rather than measure the complete quadrant from the North Pole to the equator, they chose a sector of about one-ninth of a quadrant from Dunkirk on the English Channel to a site near Barcelona on the Mediterranean coast of Spain, both locations having the advantage of being at sea level. During their journey, they were accused of being foreign spies or royalist agents and consequently suffered detentions and even formal arrests. In the end, they amended the previous estimate of the meter by a mere fraction of a millimeter.

By the time the expedition returned in 1798, the royally founded Academy of Sciences had been disbanded. In its place, the post-revolutionary Academy of Sciences and Arts had been established, which carried on the former Academy's mission to reform the standards of weights and measures. They invited European scientists to help perfect the system. Representatives from nine countries came, validating computations and helping to set up the prototype meter and kilogram. In 1799, these standards, ratified by the national legislature of France, were deposited in the archives at the Academy.

There had been a brief attempt to reform the standard for time as well, having each day divided into three "decidays." However, the proposal was met with great resistance, especially by clockmakers and watchmakers. Therefore, the second — defined as a fraction of the mean solar day and already in use throughout the world — joined the meter and kilogram as the standard units in what became known as the "metric system."

The new units were quickly adopted in France by scientists and bureaucrats, and the youth were educated in the new system in national schools. However, businesses resisted the change until 1837, when fines were imposed for using non-metric units. By this time, Switzerland, Italy, and parts of Prussia had adopted the system. The Netherlands had converted to metric units when it was occupied by Napoleon, and Belgium, when it was united with the Netherlands, adopted the system as well.

Having obtained some degree of success, France again reached out, proposing international collaboration regarding weights and measures. In 1875, they hosted a convention in Paris attended by representatives from numerous European nations, the United States, Brazil, Peru, Venezuela, and the Ottoman Empire. They established an International Bureau of Weights and Measures outside Paris, where the prototypes of the meter and kilogram would be kept for use by all member countries. Two years after the convention, the German government established the Physikalisch-Technischen Reichsanstalt in Berlin for scientific research related to standards of weights and measures. In 1900, the National Physical Laboratory was established in London for the same purpose, quickly followed by the establishment of the National Bureau of Standards in Washington, D.C.

The next decades saw tremendous growth in theoretical physics with the birth and development of relativity and quantum theory. Alongside theoretical developments came the need for high-precision measurements of fundamental constants like the elementary charge, the masses of elementary particles, the speed of light, and the Planck constant. By the 1950s, it had become clear that the current system of weights and measures was inadequate for the needs of the scientific community.

Specifically, the standards chosen to represent time, distance, and mass were changing. The mean solar day on which a second was based slowly increased due to tidal friction. The prototype meter and kilogram were very slowly deteriorating. If scientists were to continue to improve the precision of their measurements, they would need more stable units.

The second was the first unit to change its standard. By international agreement, it would no longer be defined in terms of the average solar day but rather in terms of the tropical year. However, since the tropical year was also unstable, decreasing in value by about three milliseconds per year, they chose a specific year, 1900, and declared the tropical year to be exactly 31,556,925.9747 seconds. The change went into effect in 1960, but it proved to be impractical, based on an event that was entirely unreproducible. It was abandoned seven years later.

Meanwhile, the Eleventh General Conference on Weights and Measures had redefined the meter. Instead of using the prototype kept in Paris, the meter would be based on the wavelength of light. Specifically, they defined one meter as 1,650,763.73 wavelengths of the orange-red line produced by krypton-86, a length reproducible in scientific laboratories throughout the world. In 1967, the second was redefined in terms of the frequency of the light produced during the transition between the two lowest energy states of cesium-133. With this, the units of time and space were no longer derived from the Earth's size and motion but from the tiny microcosm of the atom and the light it emits.

With these new standards scientists determined the speed of light in a vacuum with greater and greater precision. Then in 1983, they fixed this value, redefining the meter in terms of the speed of light. Now, the second was defined in terms of an atomic transition, the meter in terms of the speed of light — both in terms of supposed constants of nature. Still, the kilogram was based on a physical object, "Le Grand K," a cylinder constructed out of platinum-iridium in 1889 and stored under three glass bell jars in France. And this cylinder was losing mass, atom by atom.

Scientists knew that the kilogram had to be redefined, but how? It took many years, but the solution finally came in the form of the Kibble balance, a high-precision device that could be used to measure the Planck constant. By weighing an exact copy of Le Grand K with this balance, scientists calculated the Planck constant to eight decimal places. Then, by fixing the Planck constant to this value, the Kibble balance could be used to determine mass, creating a new standard for the kilogram. Similarly, it was decided that three other units — amperes, Kelvin, and moles — would be redefined in terms of fundamental constants. These new definitions went into effect on World Metrology Day in May of 2019.

As a result of this international system of measurement, there are seven fundamental constants that cannot change, including the speed of light. So now, if the matter distribution affects the speed of light or if the speed of light changes with time, it doesn't matter. We can't even ask that question. The speed of light is *defined* as a constant — independent of everything — a constant that helps us determine the length of a meter.

It is all very well and good to challenge the current theory of gravity, insisting that it is incomplete because it does not take into consideration the effect of the distant stars. We can *imagine* a theory that not only does this but is also compatible with quantum theory, with matter interacting through games of catch that mediate *all* the forces, gravity included.

But if we have learned anything from our long journey through the history of physics, we have learned that theory and measurement are inextricably linked together. We cannot consider a theory without also considering systems to measure and test it. This is really the essence of Mach's principle as *he* conceived it: theory should be based on the measurement of observables, efficiently expressing relationships among phenomena.

And so, although we can *imagine* a theory that brings some kind of symmetry between gravity and the other forces, where particles are exchanged to mediate the gravitational force just as in our quantum theories of electromagnetic and nuclear forces, this symmetry is broken in our systems of measurement. Time is determined through the frequency of light emitted in an atomic transition between two electromagnetic energy states. Distance is defined through light that follows a curved path determined by gravity. The asymmetry between gravity and the other forces is woven into the very fabric of our measuring system. How can we move forward?

# 5

# Between Dreams and Reality

There will hover before him as an ideal
an insight into the principles of the whole matter,
from which accelerated and inertial motions
result in the *same* way.

— Ernst Mach

As I began my outline, I promised myself not to do any research, only to write out what was already in my head. I would get lost in books, I reasoned — I had already spent over two years writing and researching, not knowing where I was going. I needed a structure first, and then I would fill it in. And I already had a lot in my head.

I knew the history of physics well. For years, I began one of my courses with a unit in which students read and discussed excerpts from the works of Aristotle, Copernicus, Kepler, and Galileo as an introduction to scientific inquiry. I had taught courses in astronomy and modern physics, including special relativity and quantum theory. I knew at least enough about dark matter, dark energy, and gravitational waves to include them in my outline. I had *a lot* in my head. Now I needed to collect it into an organized whole.

I started the outline at the beginning of April, just after returning from spring break. I didn't want to wait until my next vacation, so I committed to working on it a little each day. It was easy work because I wasn't researching, and I had a rough structure from talks I had given. And it was just an outline. After five weeks, I had finished outlining five out of the six parts.

But how would I end the book? I knew I wanted to include a chapter on $E = mc^2$. I also wanted to write about gravitons, and initially that chapter also included something about black holes. Finally, I knew that I needed to consider measurement. All of these ideas seemed *open* to me, although I didn't know where they would lead. If I ever felt stuck, I don't remember. All I remember is that while finishing this last section, something finally clicked.

I excitedly shared my thoughts with one of my colleagues, another physics teacher. He said the idea had the "ring of truth" and that it reminded him of de Broglie's matter-wave hypothesis. Well, that was encouraging! I told another colleague and some of my students; they were all supportive. I woke up in the middle of the night, and the idea began to grow. When I examined these new thoughts in the morning light, they held firm. The idea was becoming solid, and I gradually stopped fearing that it would evaporate if I touched it.

My idea was simple. If mass creates the inertial reference frames of electromagnetism, then couldn't *charge* create the special, free-falling reference frames of gravity? A free-falling frame is defined as one in which gravity completely cancels out, where the attractive gravitational force cancels out the inertial force. General relativity is based on the existence of these frames — tiny, infinitesimally small regions covering the entirety of space. But if gravity cancels out in these frames, what's left to influence the motion of an object? The only possible answer given our physical theories is charge.

And charge is *everywhere*, scattered throughout our universe. Although the attractive and repulsive parts of the electromagnetic force would on average cancel out, the *inductive* part — the part analogous to inertia that would oppose acceleration with respect to *any* charge — would not. With this, not only do we have a physical explanation for the free-falling frames of general relativity, but we also have the long sought-after connection between gravity and electromagnetism: they create each other's special reference frames.

But this is not all. Toward the end of the 1993 conference devoted to Mach's principle, Kenneth Nordtvedt identified as a "key mystery or clue" that local inertial frames are almost perfectly isotropic, despite the irregularities in cosmological structure. If mass were *entirely* responsible for the inertial properties of matter, an object should have greater inertia in the plane of our galaxy, which we don't observe. However, if we assume that gravity and electromagnetism work *together* to give matter its inertial properties, the experimental isotropicity of local frames makes perfect sense; locally, electromagnetic forces dominate, erasing the effect of any large-scale mass irregularities in the universe.

As I played around with this idea over and over in my mind, it became simpler and clearer. Matter acts on light, creating special frames where light moves in a straight path with constant speed. Charge acts on gravitons, creating special frames where gravitons move in a straight path with constant speed. The "special frames" are all equivalent, since the mediating particles move identically, connected to each other through Lorentz transformations.

Furthermore, if we can understand how gravity and electromagnetism create each other's reference frames, we should also be able to apply this *quantitatively* — create a mathematical theory that in the limiting case of a free-falling frame yields general relativity and in an inertial frame reduces to Maxwell's theory of electromagnetism. This theory should be independent of reference frame altogether, consistent with Mach's principle. This would be a theoretical success — achieving Einstein's dream of unification using Mach's principle!

Would this idea lead to any new physical predictions? Yes, it would! If charges create the free-falling frames of general relativity, and if the free-falling frames of general relativity are physically based on gravitational waves being emitted uniformly in all directions, then *an electric field should bend gravitational waves.* Furthermore, with a quantitative theory uniting gravity and electromagnetism, *we should be able to predict the size of this bending.*

And while we need the mass of an enormous astronomical object like the Sun to bend light sufficiently for it to be detectable, we should be able to bend gravitational waves easily with controllable electric fields created in laboratories. I know that this is not easy — there are probably hundreds or even thousands of hurdles in the way of actually accomplishing this. However, with the experimental, calculational, and analytical expertise that has already allowed us to observe gravitational waves along with the growing system of gravitational wave detectors around the world and soon in space, this should be possible.

I still struggled with several things, especially understanding *how* charge, which is both positive and negative, should bend gravitational waves. I struggled to connect the idea that charge creates the free-falling reference frames of general relativity with antimatter — does this lead to a definite prediction about whether antimatter rises or falls in a matter field? I could never quite wrap my head around this, but I knew that I already had enough. I had enough to write a book. I had enough to be able to end a book.

I briefly considered taking time off work to write, but I was afraid that if I had too much time I wouldn't be able to write at all. Also, I enjoyed my job and the rhythm of my life. However, to make sure I would actually finish the book, I needed to give myself deadlines. I had sketched out about forty-five chapters. If I wrote a chapter a week, I would be finished with a draft in less than a year. This seemed doable, and so I began.

I already had many pages of notes and writing from my previous attempts. I had books, both physical and digital ones, and I continued to add to my collection, usually with the "one-click" purchase of a Kindle book. I also found many free resources; most of the classic scientific papers and books are available online. The volume of resources available at my fingertips was overwhelming! Fortunately, I had my self-imposed schedule. I spent at most three days reading and taking notes. It would always be a struggle to figure out a good way to begin a chapter, but if I couldn't figure out a good way, I would begin anyway just to get something written. At the end of the week, I would sketch out what I had not gotten to while the ideas were still fresh in my mind. On Monday, I would eagerly move on to the next chapter and begin again.

In this way, my summer break passed happily. My youngest, Max — now five years old — was at camp three mornings a week. This time, along with afternoon "quiet times," gave me just enough uninterrupted time to be extremely productive. By the end of the summer, I had finished a draft of the first two parts. But how would I keep up this pace during the school year? Fortunately, a solution to this problem quickly presented itself.

I began to experience insomnia. My days would end with storytime, where Max and I would take turns with chapter books that I had chosen for his enjoyment as well as my own. After storytime, I would be so comfortably sleepy that I would go right to bed. But then sometime in the middle of the night, I would wake up and be unable to fall back asleep, often for several hours. Fighting my middle-of-the-night wakings seemed pointless; instead, I decided to use these quiet times to work.

Fall turned into winter while I finished chapter after chapter. I bought myself a mini iPad that I kept at hand in my pocketbook so I could continue my nighttime writing on my morning commute, often jotting down thoughts even while trying to balance myself against the irregular movements of the subway. I wrote on weekends, holidays, during the workday when I could, and after school. By spring I had finished a draft of the entire book.

Well, this was an accomplishment! I didn't know if it was any good; I didn't dare to actually read it. But at least when students and colleagues asked how my book was going, I was no longer met with pitying looks when I answered that I had finished five pages; I had a two-hundred-page draft! Now I would have to make it good.

I had read and edited the first five pages of my book so many times that I had memorized much of it. Those pages might not have been perfect, but at this point I couldn't imagine them any other way. I therefore began my editing with the second chapter, the chapter on Aristotle.

Aristotle! I was fascinated with Aristotle. I had spent a summer reading and taking notes on his works, attempting to write a chapter devoted to this ancient philosopher. I started with his *Physics* and *On the Heavens;* these led me to others, which led me to others. His thoughts were like an intricate web, where everything was connected in some way to everything else. For example, in one work he connected his theory of light to dreams, making use of analogies with eddy currents and projectile motion to explain why we see while our eyes are shut. What kind of mind thinks up things like that?

I could go on and on, but the point is that Aristotle *impressed* me. I didn't want to present him merely as the one in the story who got everything wrong about the universe. I tried to include the kind of subtlety and detail that would reveal the profoundness of his thought. However, when I would afterwards go back to read my chapter, I could see that it didn't fit in with *my* story. And nothing sparkled. Nothing I wrote made me, putting myself in the position of my reader, want to continue reading.

I gave up on Aristotle and decided that my second chapter should be on Einstein. This was also a failure. Then, when I began my outline, I returned Aristotle to the second chapter based on chronology. When I wrote my draft, I did a quick revision of what I had written earlier, knowing it wasn't any good. Now that I had my two-hundred-page draft, I was again faced with this unsatisfactory chapter. Not having any idea how to proceed, I read a biography of Aristotle.

Nothing I read surprised me. Aristotle loved books. He enjoyed spending time alone thinking. He was a teacher. No letters or personal writings remain; very few, if any, of the details of his life are certain. Still, the biography gave me a setting for his life and work, allowing me to imagine the journey he took to become who he became.

Now I knew how I had to write my book. I wasn't just acting as a teacher to my students. I needed to give the ideas context and present a historical and personal setting for the thoughts and accomplishments that I was sharing. Even more, in order for the characters in my book to be *inspiring*, they needed to be portrayed as ordinary human beings, like you and me, whose curiosity, passion, and commitment allowed for extraordinary achievements. These realizations destroyed practically my entire two-hundred-page draft.

I love the way the main room of our apartment looks at night. The only windows in the apartment are in this room, and we've packed them with plants — at night, the light from the street lamps shining through the plants makes the room look like a jungle. Everything is quiet except for the water filter in our fish tank. There's nothing to distract me, and my thoughts can roam freely.

I especially love the very early morning, hours before the Sun rises. At the end of the day, my brain is too tired, and I just want to sleep. After I have rested for a while, it becomes fresh again. It is in these early hours when I've done my best research and my best writing. Often, when looking over my work in the light of the morning, I am surprised at how much I have accomplished.

On many mornings, Max finds me working and snuggles next to me, falling back asleep. He is eight now and obsessed with astronomy, especially black holes. He sometimes reads over my shoulder while I'm writing. He says that he likes my book — not so much the stories about the people, but the science. He can't wait for me to be finished so he can read it. I've been working on this book since he was three; he can't even remember a time when I wasn't writing.

I have dreamed for so long of sharing this project, maybe even finding a colleague to work on it with me. I don't like endings or having to say goodbye. There's sadness in finishing something that I've enjoyed so much, but there is also satisfaction and excitement in not knowing what will come next. And that's that. I will now close my computer and rest my eyes a bit. When I wake up, my book will be done.

# Acknowledgments

A big "thank you" to a multitude of students and colleagues at Trevor Day School who have given me encouragement and support throughout the years of researching and writing. I especially want to thank Robert Stoll, my former physics teaching colleague, who was my most important sounding board as I wrote my book. I would also like to thank my former student, Lucia Gordon, whose interest in my research helped spur me on to complete the book and who helped me finalize the manuscript through her thoughtful editing.

Thank you to all of my teachers and other educational influences. In particular, I want to thank John Delos, who introduced me to modern physics and gave me the idea to pursue graduate studies in physics. I also want to thank the physics department at Penn State, whose kind leadership made it possible to succeed despite entering the program with several missing prerequisites. I want to thank my thesis advisor, Roger Herman, whose mentorship and respectful guidance — which sometimes even allowed me to win an argument! — gave me the experience and confidence to pursue the work that is the subject of this book.

Finally, I would like to express my deep gratitude to my family for their support. I want to thank my mother, who was my most important listening ear as I wrote my book. I also want to thank my youngest child, Max, for his enthusiasm about the subject of my book. Finally, I want to thank my two oldest children, Jessi and Emily, who were with me from the beginning, when I was first introduced to Mach's principle. They have encouraged me throughout the years, and now, as my project manager/typesetter and artist, they have been an essential part in giving my book a physical form.

# BIBLIOGRAPHY

## BOOK I

1. Berry, Mike. *Principles of Cosmology and Gravitation.* Cambridge, Cambridge University Press, 1976.
2. Natali, Carlo. *Aristotle: His Life and School.* Princeton University Press, 2013.
3. Aristotle. *Aristotle: Complete Works, Historical Background, and Modern Interpretation of Aristotle's Ideas.* Annotated Classics. Kindle Edition.
4. Chroust, Anton-Hermann. "Estate Planning in Hellenic Antiquity: Aristotle's Last Will and Testament." Notre Dame L. Rev., vol. 45, iss. 4, 1970, pp. 629-662.
5. Heller-Roazen, Daniel. "Tradition's Destruction: On the Library of Alexandria." October, vol. 100, 2002, pp. 133–153.
6. Gutas, Dimitri. *Greek Thought, Arabic Culture: the Graeco-Arabic Translation Movement in Baghdad and Early Abbasid Society (2nd-4th/8th-10th c.).* Routledge, 1998.
7. Goodwin, Jason. "The Glory That Was Baghdad." The Wilson Quarterly (1976-), vol. 27, no. 2, 2003, pp. 24–28.
8. Cohen, H. Floris. *How Modern Science Came into the World: Four Civilizations, One 17th-Century Breakthrough.* Amsterdam, Amsterdam University Press, 2012.
9. Hawking, Stephen. *On the Shoulders of Giants: The Great Works of Physics and Astronomy.* Philadelphia, Pa.; London, Running, 2004.
10. Sobel, Dava. *A More Perfect Heaven.* Bloomsbury Publishing, 2014.
11. Ptolemaios, Klaudios, et al. *Ptolemy's Almagest.* Princeton University Press, 1998.
12. Ferguson, Kitty. *Tycho & Kepler: The Unlikely Partnership That Forever Changed Our Understanding of the Heavens.* New York, Walker & Co, 2004.
13. Galilei, Galileo. *Sidereus Nuncius* translated by Edward Stafford Carlos edited/corrected by Peter Barker.
14. Galilei, Galileo, Stillman Drake, and Thomas Salusbury. *Dis-*

*course on Bodies in Water*. Univ. of Ill. Press, 1960.

15. Reston, James. *Galileo: A Life*. Beard Books, 2000.
16. Galilei, Galileo. *Two New Sciences*. Kindle Edition.
17. Wootton, David. *Galileo : Watcher of the Skies*. New Haven, Yale University Press, 2013.
18. Gleick, James. *Isaac Newton*. Knopf Doubleday Publishing Group. Kindle Edition.
19. Newton, Isaac. *Newton's Principia: The Mathematical Principles of Natural Philosophy*. New-York: Published by Daniel Adee. Kindle Edition.
20. Descartes, René. *The World and Other Writings*. Edited by Stephen Gaukroger (Cambridge University Press, 2004.
21. Einstein, Albert, and John J Stachel. *The Collected Papers of Albert Einstein*. Princeton, N.J., Princeton University Press, 1987.
22. Neffe, Jürgen. *Einstein: A Biography*. Cambridge: Polity Press, 2016.
23. Isaacson, Walter. *Einstein : His Life and Universe*. New York, Simon & Schuster, 2017.
24. Frank, Philipp. *Einstein - His Life and Times*. Read Books, 2007.

# BOOK II

1. Masson, Flora. *Robert Boyle*. Hardpress Publishing, 2012.
2. Webster, Charles. "New Light on the Invisible College the Social Relations of English Science in the Mid-Seventeenth Century." Transactions of the Royal Historical Society, vol. 24, 1974, pp. 19–42.
3. Boyle, Robert, and Robert Sharrock. *New Experiments Physico-Mechanical, Touching the Air*. London: Printed by Miles Flesher for Richard Davis, bookseller in Oxford, 1977.
4. Boyle, Robert. *The Sceptical Chymist*. London: J.M. Dent & Sons, Ltd, 1900.
5. Kohler, Robert E. "The Origin of Lavoisier's First Experiments on Combustion." Isis, vol. 63, no. 3, 1972, pp. 349–355.
6. McKie, Douglas. *Antoine Lavoisier: Scientist, Economist, Social Reformer*. New York, N.Y: Da Capo Press, 1990.
7. Davis, Kenneth S. *The Cautionary Scientists: Priestley, Lavoisier,*

*and the Founding of Modern Chemistry.* New York: Putnam, 1966.

8. Bell, Madison S. *Lavoisier in the Year One: The Birth of a New Science in an Age of Revolution.* New York: W.W. Norton, 2006.

9. Schubert, Hermann. *The Squaring of the Circle: An Historical Sketch of the Problem from the Earliest Times to the Present Day.* Chicago, etc., 1891.

10. Huygens, Christiaan, and Silvanus Phillips Thompson. *Treatise on Light, in Which Are Explained the Causes of That Which Occurs in Reflection, & in Refraction and Particularly in the Strange Refraction of Iceland Crystal, by Christiaan Huygens. Rendered into English by Silvanus P. Thompson.* Chicago, Ill., University Of Chicago Press, 1955.

11. Bell, Arthur. *Christian Huygens and the Development of Science in the Seventeenth Century.* London: Bell Press, 2008.

12. Young, Thomas. *Miscellaneous Works of the Late Thomas Young: 3.* New York: Johnson, 1972.

13. Oldham, Frank. *Thomas Young, Philosopher and Physician.* Edward Arnold, 1933.

14. Young, Thomas, and Philip Kelland. *A Course of Lectures on Natural Philosophy and the Mechanical Arts.* London: Printed for Taylor and Walton, 1845.

15. Watson, Bruce. *Light : A Radiant History from Creation to the Quantum Age.* Amherst, Massachusetts, Levellers Press, 2019.

16. Suter, Rufus. "A Biographical Sketch of Dr. William Gilbert of Colchester." *Osiris*, vol. 10, 1952, pp. 368–384.

17. Brundtland, Terje."Francis Hauksbee and his air pump." 66. *Notes and Records: the Royal Society Journal of the History of Science.* July 11, 2012.

18. Priestley, Joseph. *Memoirs of Dr. Joseph Priestley to the Year 1795, Written by Himself.* Washington, Barcroft Press, 1964.

19. Priestley, Joseph. *The History and Present State of Electricity, with Original Experiments. 3d Ed.,* London, 1755.

20. Whittaker, E T. *A History of the Theories of Aether and Electricity / 1, the Classical Theories.* London, Nelson, 1951.

21. Forbes, Nancy. *Faraday, Maxwell, and the Electromagnetic Field: How Two Men Revolutionized Physics.* Prometheus Books, 2019.

22. Hirshfeld, Alan. *The Electric Life of Michael Faraday.* New York, Walker, 2006.

23. Faraday, Michael. *Experimental Researches in Electricity, Volume 1*. Kindle Edition, 2012.

24. Campbell, Lewis, and William Garnett. *The Life of James Clerk Maxwell, with a Selection from His Correspondence and Occasional Writings, and a Sketch of His Contributions to Science.* London, Macmillan & Co, 1884.

25. Hertz, Heinrich, and D E Jones. *Electric Waves: Being Researches on the Propagation of Electric Action with Finite Velocity through Space.* London ; New York, Macmillan, 1893.

26. Koenigsberger, Leo. Hermann von Helmholtz. Oxford : Clarendon press. Kindle Edition, 2018.

27. Planck, Max. *The Origin and Development of the Quantum Theory: with "A Scientific Autobiography."* Amazon Kindle Direct Publishing; 3rd edition, 2012.

28. Planck, Max. "On the Theory of the Energy Distribution Law of the Normal Spectrum." *Annalen der Physik*, vol. 4, 1901. English translation from "The Old Quantum Theory," ed. by D. ter Haar, Pergamon Press, 1967, p. 82.

29. Einstein, Albert, and John J Stachel. *The Collected Papers of Albert Einstein.* Princeton, N.J., Princeton University Press, 1987.

30. Blaedel, Niels; *Harmony and Unity: The Life of Niels Bohr.* Plunkett Lake Press, 2017.

31. Thomson, J. J. "Cathode Rays." *Philosophical Magazine, vol. 44, 1897.*

32. Bohr, Niels. *The Theory of Atomic Spectra And Atomic Constitution: Three Essays by Niels Bohr.* Cambridge University Press, 1922.

33. Gamow, George. *Thirty Years That Shook Physics; The Story of Quantum Theory. Illus. by the Author.* Garden City, N.Y., Doubleday, 1966.

34. Heisenberg, Werner. *Physics and beyond. Encounters and Conversations.* Allen And Unwin, London, 1971.

35. De Broglie, Louis. "On the Theory of Quanta." *Annales de Physique,* vol. 10, 1925. English translation by A. F. Kracklauer.

36. Gribbin, John. *In Search of Schrödinger's Cat. London, Black Swan, 2012.*

37. Pagels, Heinz R. *The Cosmic Code: Quantum Physics as the Language of Nature.* Am Oved Publishers, 1991.

38. Nelson, Craig. *The Age of Radiance: The Epic Rise and Dramat-*

*ic Fall of the Atomic Era.* Scribner, 2014.

39. Curie, Marie. *The Discovery of Radium.* Address by Madame M. Curie at Vassar College May 14, 1921. Published by Vassar College.

40. Crawford, Elisabeth. "German Scientists and Hitler's Vendetta against the Nobel Prizes." *Historical Studies in the Physical and Biological Sciences,* vol. 31, no. 1, 2000, pp. 37–53.

41. Rife, Patricia. *Lise Meitner and the Dawn of the Nuclear Age.* Boston, Mass., Birkhäuser, 2007.

42. Clark, Ronald W.. *Einstein: The Life and Times.* New York: Avon Books, 1970

43. Joseloff, Michael. *Chasing Heisenberg: The Race for the Atomic Bomb.* Amazon Publishing, 2018.

# BOOK III

1. Yukawa, Hideki. *Hideki Yukawa "Tabibito" (The Traveller).* Singapore, World Scientific Pub, 1982.

2. Yukawa, Hideki. "On the Interaction of Elementary Particles. I*." *Progress of Theoretical Physics Supplement,* vol. 1, 1955, pp. 1–10.

3. Brown, Laurie M. "Hideki Yukawa and the Meson Theory." *Physics Today,* vol. 39, no. 12, Dec. 1986, pp. 55–62.

4. Feynman, Richard Phillips. *The Strange Theory of Light and Matter.* Princeton University, 1988.

5. Smolin, Lee. *The Trouble with Physics : The Rise of String Theory, the Fall of a Science, and What Comes Next.* London, Penguin, 2008.

6. Cham, Jorge, and Daniel Whiteson. *We Have No Idea: a Guide to the Unknown Universe.* Riverhead Books, 2018.

7. Knox, Dilwyn. "Giordano Bruno," The Stanford Encyclopedia of Philosophy (Summer 2019 Edition), Edward N. Zalta (ed.).

8. Kerszberg, Pierre. "The Cosmological Question in Newton's Science." *Osiris,* vol. 2, 1986, pp. 69–106.

9. Smith, Robert W. "Edwin P. Hubble and the Transformation of Cosmology." *Physics Today,* vol. 43, no. 4, Apr. 1990, pp. 52–58.

10. Hubble, E. "A Relation between Distance and Radial Velocity among Extra-Galactic Nebulae." *Proceedings of the National Academy of Sciences*, vol. 15, no. 3, 15 Mar. 1929, pp. 168–173.

11. Hawking, Stephen. *A Brief History of Time.* New York: Bantam Books, 2017.

12. Gamow, George. "Expanding Universe and the Origin of Elements." *Physical Review*, vol. 70, no. 7-8, Oct. 1946, pp. 572-73.

13. Guth, Alan H. *"Inflationary Universe: A Possible Solution to the Horizon and Flatness Problems." Physical Review D, vol. 23, no. 2, 15 Jan. 1981, pp. 347-356.*

14. Ijjas, Anna, Paul J. Steinhardt, and Abraham Loeb. "Cosmic Inflation Theory Faces Challenges." *Scientific American,* 2017.

15. Ahmadi, M., Alves, B.X.R., Baker, C.J. *et al.* "Characterization of the 1S–2S transition in antihydrogen." *Nature* 557, 2018, pp. 71–75.

16. Wheeler, J. Craig. "Astrophysical Explosions: from Solar Flares to Cosmic Gamma-Ray Bursts." *Philosophical Transactions: Mathematical, Physical and Engineering Sciences*, vol. 370, no. 1960, 2012, pp. 774–799.

17. Hajdukovic, Dragan Slavkov. "Antimatter Gravity and the Universe." *Modern Physics Letters A*, vol. 35, no. 08, 18 Nov. 2019.

18. Einstein, Albert, and John J Stachel. *The Collected Papers of Albert Einstein.* Princeton, N.J., Princeton University Press, 1987.

19. O'Raifeartaigh, Cormac, et al. "Einstein's 1917 Static Model of the Universe: A Centennial Review." *The European Physical Journal H*, vol. 42, no. 3, 20 July 2017, pp. 431–474.

20. Friedmann, Alexander. "On the Curvature of Space." *Zeitschrift für Physik, vol. 10, 1922, pp. 377-386.* Translated by Brian Doyle.

21. Turek, Józef. "Georges Lemaitre's Contribution to the Formation of the Dynamic View of the Universe." Roczniki Filozoficzne / Annales De Philosophie / Annals of Philosophy, vol. 33, no. 3, 1985, pp. 59–74.

22. Van den Bergh, Sidney. "The Early History of Dark Matter." *Publications of the Astronomical Society of the Pacific*, vol. 111, no. 760, 1999, pp. 657–660.

23. de Swart, J. G., et al. "How Dark Matter Came to Matter." *Nature Astronomy*, vol. 1, no. 3, Mar. 2017.

24. Clowe, Douglas, et al. "A Direct Empirical Proof of the Existence of Dark Matter." *The Astrophysical Journal*, vol. 648, no.

2, 30 Aug. 2006, pp. L109–L113.

25. Kirshner, R. P. "Supernovae, an Accelerating Universe and the Cosmological Constant." *Proceedings of the National Academy of Sciences*, vol. 96, no. 8, 13 Apr. 1999, pp. 4224–4227.

26. Abbott, B. P. et al. "Observation of Gravitational Waves from a Binary Black Hole Merger." *Phys. Rev. Lett.* vol. 116, 11 February 2016.

27. Kalogera, Vassiliki, and Albert Lazzarini. "LIGO and the Opening of a Unique Observational Window on the Universe." *Proceedings of the National Academy of Sciences of the United States of America*, vol. 114, no. 12, 2017, pp. 3017–3025.

28. Aguiar, Odylio Denys. "Past, Present and Future of the Resonant-Mass Gravitational Wave Detectors." *Research in Astronomy and Astrophysics*, vol. 11, no. 1, 22 Dec. 2010, pp. 1–42.

29. Abbott, B. P., et al. "GW170817: Observation of Gravitational Waves from a Binary Neutron Star Inspiral." *Physical Review Letters*, vol. 119, no. 16, 16 Oct. 2017.

30. Weinstein, Galina. "Einstein's Discovery of Gravitational Waves 1916-1918" (unpublished)

31. Eddington, A. S. "The Propagation of Gravitational Waves." *Proceedings of the Royal Society of London. Series A, Containing Papers of a Mathematical and Physical Character*, vol. 102, no. 716, 1922, pp. 268–282.

# BOOK IV

1. Reichenbach, H. "Contributions of Ernst Mach to Fluid Mechanics." *Annual Review of Fluid Mechanics*, vol. 15, no. 1, Jan. 1983, pp. 1–29.

2. Mach, Ernst, and Philip E B Jourdain. *History and Root of the Principle of the Conservation of Energy*. Cambridge, Cambridge University Press, 2014.

3. Blackmore, John T. *Ernst Mach: His Work, Life, and Influence*. Berkeley, Univ. Of Calif. P, 1972.

4. Mach, Ernst, and Thomas J. McCormack. *The Science of Mechanics: A Critical and Historical Exposition of Its Principles*. Open Court, 1919.

5. Mach, Ernst, and Thomas J Mccormack. *Popular Scientific*

*Lectures.* Chicago, The Open Court Publishing Company, 1898.

6. Heinrich Hertz, et al. *The Principles of Mechanics : Presented in a New Form by Heinrich Hertz; with an Introduction by H. von Helmholtz. Authorised English Translation by D.E. Jones ... and J.T. Walley. [Edited by P. Lenard.]* London, Macmillan & Co., Ltd, 1899.

7. Einstein, Albert, and Paul Arthur Schilpp. *Autobiographical Notes.* Open Court Printing, 1996.

8. Heitler, W. "Erwin Schrödinger. 1887-1961." *Biographical Memoirs of Fellows of the Royal Society*, vol. 7, 1961, pp. 221–228.

9. Barbour, Julian B, and Herbert Pfister. *Mach's Principle: From Newton's Bucket to Quantum Gravity.* Boston, Birkhäuser, 1995.

# BOOK V

1. Berends, Frits. "Hendrik Antoon Lorentz: His Role in Physics and Society." *Journal of Physics: Condensed Matter*, vol. 21, no. 16, 31 Mar. 2009.

2. Fölsing, Albrecht, and Ewald Osers. *Albert Einstein : A Biography.* New York, Penguin Books, 1998.

3. Einstein, Albert, and John J Stachel. *The Collected Papers of Albert Einstein.* Princeton, N.J., Princeton University Press, 1987.

4. Einstein, Albert, and Carl Seelig. *Ideas and Opinions.* New York, Crown Trade Paperbacks, 1995.

5. Hoffmann, Banesh. *Albert Einstein Creator and Rebel.* New York Viking Press, 1974.

6. Sciama, Dennis. "Dennis Sciama." *American Institute of Physics (Oral History Interviews).* Interviewed on 14 Apr. 1978 by Spencer Weart.

7. Ellis, George F. R., and Sir Roger Penrose. "Dennis William Sciama. 18 November 1926 — 19 December 1999." *Biographical Memoirs of Fellows of the Royal Society*, vol. 56, Jan. 2010, pp. 401–422.

8. Sciama, D. W. "On the Origin of Inertia." *Monthly Notices of the Royal Astronomical Society*, vol. 113, no. 1, 1 Feb. 1953, pp. 34–42.

# BOOK VI

1.  Hallerberg, Arthur E. "The Metric System: Past, Present—Future?" *The Arithmetic Teacher*, vol. 20, no. 4, 1973.
2.  Palmer, Chris. "Redefining the Kilogram." *Engineering*, vol. 5, no. 3, June 2019, pp. 361–362.
3.  Klein, Herbert Arthur. *The Science of Measurement: A Historical Survey.* Dover Publications, 1988.
4.  Quinn, Terry, and Jean Kovalevsky. "The Development of Modern Metrology and Its Role Today." *Philosophical Transactions: Mathematical, Physical and Engineering Sciences*, vol. 363, no. 1834, 2005, pp. 2307–2327.